Praise for *Vaccine Epidemic*

In a book bursting with intelligence, Louise Habakus and Mary Holland reframe the entire autism-vaccine debate turning the conversation away from the tired clichés of the past and opening bright new avenues of ideas and thought that delve deep below the surface of pro-this or anti-that. Ever purposeful and in some ways masterful, by reframing the debate, the authors get to the heart of the matter that aims for your head and hits your heart.

—Ed and Teri Arranga, AutismOne.org

As vaccine mandates are increasing every year and healthcare workers live with the threat that they will be fired if they don't get a flu shot, it is essential for Americans to educate themselves about our loss of informed consent rights. We need adequate scientific research and clinical evidence for any public health policy, not an attack on personal belief or medical exemptions to vacci[illegible] by special interest groups, which benefit pharma[illegible] a true, unbiased investigation into vaccine sci[illegible]u are concerned about making an informed d[illegible]garding vaccines then the book *Vaccine Epidem*[illegible]

—Miranda Baile[illegible] Entertainment

Vaccine Epidemic provides readers with an array of valid well researched perspectives on the vaccine safety and personal choice issue. Whether you are a parent who is pondering a vaccine decision, a parent of a child who has suffered a vaccine injury, or an injured adult, this book provides factual first person accounts from doctors, lawyers, parents, scientists, victims, activists, and both civilian and enlisted personnel who have experienced serious adverse effects from vaccines. *Vaccine Epidemic* emphasizes how important it is to seek, preserve and defend our personal rights when it comes to making individual choices about vaccines.

—Claire Dwoskin, philanthropist and children's health activist

Vaccine Epidemic is a timely reminder that the same way Wall Street conspires with the government for its own benefit at the expense of the working and middle classes, the healthcare industry conspires with the government for the sake of its profits.

—Marc Faber, Ph.D., *Barron's* magazine Roundtable member, international investment strategist, and author *Gloom, Boom and Doom Report*

There are unanswered questions about vaccine safety. We need studies on vaccinated populations based on various schedules and doses as well as individual patient susceptibilities that we are continuing to learn about. No one should be threatened by the pursuit of this knowledge. Vaccine policy should be the subject of frank and open debate, with no tolerance for bullying. There are no sides—only people concerned for the well-being of our children.

—Bernadine Healy, M.D., former director, National Institutes of Health (NIH), and former health editor, *U.S. News & World Report*

Not since *Evidence of Harm* have I been so incensed when reading a book after learning, yet again, of more times our government has turned its head to the devastating collateral damage happening every day to our nation's children from vaccines. Yet at the end of this account, instead of resentment, I felt empowered to do something constructive. *Vaccine Epidemic* enables parents to become informed and decide whether "the greater good" is worth the sacrifice of one's child. The common thread that links all of the personal and historical reports is the same: for our children to have any worth in society, parents must have a choice. I highly recommend *Vaccine Epidemic* for all parents to use as the "go to" handbook on this important, yet controversial, issue.

—Jennifer Keefe, Esq, cofounder, Elizabeth Birt Center for Autism Law and Advocacy

Readers will learn how vaccination resembles dogmatic religion rather than fact based science and how Americans are paying the price in declining health from childhood through the "golden" years.

—Kim Stagliano, managing editor, *Age of Autism,* and author, *It's All I Can Handle: I'm No Mother Theresa*

This is a brilliant book! Louise Kuo Habakus and Mary Holland are the Woodward and Bernstein of the autism epidemic. Buy it for the "there is no science" people in your life. The science is there—in spades. Habakus and Holland meticulously dissect every element of this debate. The evidence they have uncovered is astounding and deeply disturbing.

—Katie Wright, writer and board member, National Autism Association

The most ardent supporter of vaccination programs only needs to witness or be touched by one serious vaccine reaction to fully understand and respect the wishes of those who support freedom of choice relative to routine pediatric vaccinations.

—F. Edward Yazbak, MD, board-certified pediatrician

VACCINE EPIDEMIC

How Corporate Greed, Biased Science, and Coercive Government Threaten Our Human Rights, Our Health, and Our Children

Edited by
Louise Kuo Habakus, MA
and
Mary Holland, JD
with
Kim Mack Rosenberg, JD

Skyhorse Publishing

Center for Personal Rights, Inc. is a nonprofit organization founded to advance personal rights, including the human right to vaccination choice. Its cofounders are Louise Kuo Habakus, Mary Holland, and Kim Mack Rosenberg.

Please Note: The essays in this book reflect the views of individual authors and do not necessarily reflect the views of the Center for Personal Rights, Inc., (CPR) or Skyhorse Publishing, Inc., (Skyhorse). The authors bear sole responsibility for their essays. CPR and Skyhorse assume no liability for any actions you may take or omissions you may make as a result of reading this book.

Copyright © 2011, 2012 by Center for Personal Rights, Inc.

All Rights Reserved. No part of this book may be reproduced in any manner without the express written consent of the publisher, except in the case of brief excerpts in critical reviews or articles. All inquiries should be addressed to Skyhorse Publishing, 307 West 36th Street, 11th Floor, New York, NY 10018.

Skyhorse Publishing books may be purchased in bulk at special discounts for sales promotion, corporate gifts, fund-raising, or educational purposes. Special editions can also be created to specifications. For details, contact the Special Sales Department, Skyhorse Publishing, 307 West 36th Street, 11th Floor, New York, NY 10018 or info@skyhorsepublishing.com.

Skyhorse® and Skyhorse Publishing® are registered trademarks of Skyhorse Publishing, Inc. ®, a Delaware corporation.

www.skyhorsepublishing.com

20 19 18 17 16 15 14 13 12

Library of Congress Cataloging-in-Publication Data is available on file.

ISBN: 978-1-62087-212-3

Printed in China

To all who are asking the questions
and all who are demanding the answers.

CONTENTS

On résiste à l'invasion des armées; on ne résiste pas à l'invasion des idées.

"Greater than the tread of mighty armies is an idea whose time has come."

—Victor-Marie Hugo, 1852, *Histoire d'un Crime*
French novelist, statesman, human rights activist

FOREWORD

Louise Kuo Habakus, MA, and Mary Holland, JD

Vaccination evokes strong opinions and emotions. Some are grateful and relieved to get vaccines. Others are angry and resentful about vaccination mandates. Many are caught in the middle—afraid, confused, and looking for the straight scoop. If you want to understand the real issues in the vaccination debate, read this book. Whether you are a parent, student, employee, soldier, or anyone else, you have a stake in this debate. No matter who you are, sooner or later you will wonder,

Should I get a vaccine?
Do I have a choice?
Who decides?

The vaccination debate has two distinct viewpoints, one "pro-mandate" and the other "pro-choice." The camps usually hurl the same insults at one another, characterizing the other side as irresponsible, dangerous, denialist, corrupt, and antiscience. Although this book favors vaccination choice and explains why, we attempt to objectively outline each basic view by setting derogatory labels aside.

THE PRO-MANDATE CAMP

The pro-mandate camp strongly supports U.S. vaccination policy. Two books published in early 2011 advance this position: Paul Offit's *Deadly Choices* and Seth Mnookin's *The Panic Virus*. These books join Arthur Allen's *Vaccine*, Susan

Jacoby's *The Age of American Unreason,* and Michael Specter's *Denialism* in their critique of those who question vaccine safety. Here are their key arguments:

1. **Government officials are best qualified to make vaccination decisions.** Only government can ensure that a sufficiently high percentage of people vaccinate to preserve societal herd immunity.
2. **Vaccines are overwhelmingly safe and effective**, and the benefits vastly outweigh the risks. Adverse events are vanishingly rare.
3. **Science proves the benefit of vaccines beyond a reasonable doubt.** The science is in—vaccines are not responsible for any increase in common childhood health problems, including autism. They're safe and effective.
4. **Vaccine refusers are dangerous and selfish.** People who elect not to vaccinate are parasites. They are selfish, irrational, and threaten others with deadly disease.
5. **Only "false prophets" suggest that vaccines may cause disorders like autism.** Quack healthcare practitioners concoct unfounded treatments for autism and prey on vulnerable parents desperate for help. People should disdain and shun them.
6. **Vaccine exemptions should be abolished.** Dr. Offit and others suggest that states should abolish philosophical and religious exemptions. People abuse them and put others at risk. Because vaccines have been proven safe and effective, people have a social responsibility to vaccinate.

THE PRO-CHOICE CAMP

This book details the pro-vaccination choice perspective in law, religion, science, ethics, philosophy, and personal experience. Here are its key points:

1. **Vaccination choice is a human right.** Because vaccination poses a risk to life, liberty, and security of person, only an individual or guardian may decide how, when, and whether to vaccinate.
2. **Society as a whole benefits from the cumulative impact of free and informed individual healthcare choices.** The theory of herd immunity is not an adequate rationale for state compulsion to vaccinate.
3. **Vaccine safety science is flawed and incomplete.** The Institute of Medicine as well as informed scientists, doctors, and officials have repeatedly acknowledged that fundamental questions about vaccine safety remain unanswered.

4. **The U.S. vaccine program is rife with conflicts of interest.** Vaccines are big business and all of their promoters—pharma, government, medicine, and science—get their cut. The vaccine program does not put children's safety first.
5. **Biomedical interventions are valid.** Considerable science and anecdotal evidence support biomedical interventions for autism spectrum disorders and other chronic conditions. These interventions include diet, vitamin and mineral supplementation, chelation, and gastrointestinal treatment. Individuals are entitled to the practitioners and remedies of their choosing.
6. **Vaccination exemption rights must expand, not contract.** Individuals have the right to free and informed consent for all medical interventions, including vaccination. In practice, not just in theory, individuals must have the right to make their own decisions.

• • •

As polarized as these camps are, we do share common ground. Both sides want a healthy, vibrant society. Both sides want responsible health policy grounded in ethics and science. Both sides see the vaccination issue as hugely important and seek to sway the public. Both sides are doing the work to understand each other's analysis and arguments. Progress will not come from shadowboxing in our respective corners. It is now time for engagement and dialogue.

So, we invite the leaders of the pro-mandate camp—in government, industry, and the professional medical associations—to publicly debate our different views. And, we invite the reader to read on.

Introduction

THE CASE FOR VACCINATION CHOICE

Louise Kuo Habakus, MA, and Mary Holland, JD

VACCINATION CHOICE IS A HUMAN RIGHT

Vaccine Epidemic addresses an issue of serious public concern for parents and all Americans. It builds the case that it is your right—not the government's—to decide whether to vaccinate yourself or your children. Only you, with your healthcare practitioner, can make the appropriate risk-benefit tradeoff, and only you can be fully accountable for your choice, as with any other medical intervention.

All medical interventions, including vaccination, require free and informed consent. To abridge that right is to violate the essential human rights to life, liberty, and bodily integrity. By denying truly free and informed consent to vaccination, U.S. vaccine policy violates fundamental rights. This book brings together for the first time in-depth discussion of the ethical, legal, and scientific dimensions of the vaccination debate from the perspectives of those who insist on the right to decide.

VACCINES ARE EPIDEMIC

Starting on the day of birth, nearly all American children receive vaccines. The Centers for Disease Control and Prevention (CDC) recommends seventy doses of sixteen different vaccines between birth and age eighteen. The federal government recommends vaccines for adults too. It is the states, however, and increasingly private employers, that compel them. All fifty states mandate from thirty to forty-five doses of about a dozen different vaccines for admission to day care and school.

Public health officials proclaim that vaccines are safe and effective, but the truth is far more complicated. Vaccination is a serious medical intervention for healthy individuals against a disease that *might* occur. The decision whether to vaccinate has important, potentially life-altering ramifications. While vaccines protect some, they unquestionably harm and cause death to others. For government to compel a potentially life-threatening product sets vaccination apart from other medical procedures. Public health officials tell us that those who suffer adverse reactions are exceedingly few. The basis for such an optimistic claim is doubtful.

Science on vaccine safety is inadequate. There are no randomized controlled studies comparing the total health outcomes of the vaccinated versus the unvaccinated. Scientists do not know the cumulative impact of the childhood vaccination schedule. Vaccines are approved and licensed individually, yet the CDC recommends that doctors give multiple vaccines at once. Furthermore, the CDC recommends vaccines to populations for which there are no safety data, such as flu shots for pregnant women. Research confirms that fewer than 10 percent of doctors report adverse vaccine events. Even so, the Vaccine Injury Compensation Program has paid over $2 billion in damages to more than 2,500 families since 1988.

Vaccines are *epidemic*—they have spread rapidly and extensively to almost all children and to many adults in the United States and around the world. In the 1800s and early 1900s, vaccines were a medical intervention of last resort during infectious disease outbreaks. In the last fifty years, however, governments have imposed multiple preventive vaccines on all children and some adults for the "greater good." Today, vaccines and their advertisements are everywhere—from the doctor's office to the Internet to the corner drugstore. The primary definition of an epidemic is a contagious disease that spreads rapidly. The Center for Personal Rights (CPR) does not suggest that vaccines are equivalent to disease. But it does suggest that vaccines are a widespread popular development. The book describes a modern *epidemic* of vaccines—recent, prevalent, and unprecedented in human history.

THE NO-CHOICE CHOICE

Governments, schools, employers, doctors, and officials concede that Americans have a choice about vaccination—but, in practice, they make that choice nearly impossible to exercise. There is pressure to begin vaccinating children literally from the moment of birth. Formally, all states honor medical exemptions from vaccination as the law requires. Further, forty-eight states recognize religious exemptions, twenty recognize philosophical exemptions, and some recognize exemptions for documented immunity (blood tests showing sufficient antibodies). Choice is honored in the breach, however. States punish doctors who grant "too many" medical exemptions, schools don't inform parents that exemptions exist, and states impose vaccination on the unwilling through school exclusion and child removal.

Compulsory vaccination for schoolchildren, soldiers, and healthcare workers heightens the pressure to vaccinate all people for whom vaccines are recommended. It also reinforces the stigma for those who do not comply. Government and the medical profession exercise potent financial, legal, and social levers to ensure that people make the "right choice." Should you wish to make the "wrong choice"—no day care, no public school, often no private school, often no access to your doctor. You even risk possible charges of child neglect and the threat of child removal. While in theory the country acknowledges the virtue of vaccination choice, in practice, it indulges the vice of coercive medicine. Renowned human rights scholar Louis Henkin understood: "hypocrisy is the homage that vice pays to virtue." The U.S. vaccine program hypocritically gives lip service to choice while violating the fundamental principle of free and informed consent to medical intervention.

BASIC QUESTIONS ABOUT VACCINATION CHOICE

This book is for people who want to explore the basic question: Do I have a choice? From this simple question cascades a torrent of more difficult ones:

- Is it ethical for the government to compel individuals to accept medical procedures that could cause injury and death without free and informed consent?
- May the government legally withhold essential rights and privileges, such as a public school education, to induce vaccination compliance?
- What is the science for vaccine safety and efficacy?

- Are mandates ethical if people do not know they have choices?
- Is there a legal duty to protect those most at risk for vaccine injury?
- What ethical and legal responsibilities does society bear to compensate those injured?
- Should doctors and vaccine makers be shielded, as they are, from legal liability?

These are some of the challenging questions that this book addresses through the contributions of thirty-one authors—credentialed professionals and mainstream parents—who share their specialized expertise and personal experiences.

PART I: THE CASE FOR VACCINATION CHOICE

In Part I, we outline the essential components of the claim that all people are entitled to vaccination choice. The right is well-founded in human rights, civil rights, religious law, science, history, ethics, and philosophy. Mary Holland, lawyer and CPR cofounder, explains why vaccination choice necessarily implicates the rights to life, liberty, and security of the person. Sookyung Song places vaccination choice in the context of contemporary international human rights norms supporting free and informed consent for all medical interventions. Louise Kuo Habakus, CPR executive director and cofounder, assesses U.S. vaccine policy from a health and human rights perspective, finding that it comes up short.

Jim Turner, lawyer and health advocate, explores the civil rights dimension of the national vaccine program through the lens of the U.S. Constitution and the due process required to impose the death penalty. Holland and Robert Krakow, lawyer and disability rights advocate, address what happens when vaccines cause injury. They outline the remarkable liability protections the pharmaceutical industry and medical profession lobbied for and won in the 1986 National Childhood Vaccine Injury Act. They explain how, as a result of the act, the vaccine-injured and their families face unprecedented hurdles to obtain compensation. William Wagner, a law professor, considers vaccination choice a parental right under the Constitution and religious law.

Carol Stott, an epidemiologist, and Andrew Wakefield, doctor and scientist, address the science—what we know, what we don't know, and what we appear not to want to know. Based on peer-reviewed science, they argue that the case for a link between vaccines and autism and other childhood health problems is robust and cries out for further research. They provide a primer on epidemiology and important guidance about the use of epidemiology to answer questions of causation. The appendix to this chapter contains scientific abstracts of nearly two dozen peer-reviewed studies, which raise grave questions about the wisdom of current vaccination policy.

Historian Robert Johnston writes about the critical role of dissent in a democracy. There is a long tradition of dissent from orthodox healthcare in the United States, including the advocacy movement that sprang up against coercive vaccine programs as early as 1721. Human rights advocate Vera Sharav illuminates what we know from painful historical experience: that any digression from humanitarian medicine with its commitment to "first, do no harm" is fraught with danger. She shows us examples not only from the Nazi past but also from contemporary U.S. medicine of doctors who caused grave harm for financial enrichment under the banner of "the greater good." Sharav makes clear that current U.S. practice fails to honor human rights obligations. Only free and informed consent to vaccination, as to all medical interventions, fulfills medicine's Hippocratic oath.

Allen Tate discusses the utilitarian ethic that underlies all vaccine mandates. When we speak of the greater good, we invoke a higher moral authority claiming that we will benefit more people and save more lives. Nevertheless, there is an uncomfortable underside to the argument. Can we sacrifice one to save ten? Five to save six?

PART II: BREAKING THE SILENCE OF VACCINE INJURY—PERSONAL NARRATIVES

We break the silence of vaccine injury through the personal stories of seven contributing authors. These personal accounts are by turns painful, heartbreaking, courageous, and hopeful. We hear from Russ Bruesewitz, Gay Tate, Judy Converse, and Sonja Hintz, parents who watched in agony as their babies suffered severe vaccine reactions. They have fought mightily, guided by professional and parental dedication, to heal their children, fight on their children's behalf, and help others.

Bruesewitz describes his family's heroic eighteen-year-long battle to try to win justice for their daughter Hannah. Imagine that your child develops a catastrophic seizure disorder a mere two hours after receiving a federally recommended vaccine. Now imagine that this same vaccine is pulled from the market several years later because it was insufficiently safe. Bruesewitz took his case all the way to the United States Supreme Court, only to be told that their claims were preempted and that there is no court in the land that would be allowed to hear his case. He writes for thousands of families who present their cases before the Vaccine Injury Compensation Program, only to be turned away.

Tate and Converse express alarm at what is happening to children. How can there be no outcry against rising infant mortality and skyrocketing childhood diagnoses for attention deficit disorder, autism, learning disabilities, and asthma? How can it be that patients who go to doctors with vaccine injuries—Hintz, Tate, Pingel, Smith, Rovet, Converse—meet with denial and accusations that they are "faking it"? The U.S. government tells its citizens that they have an obligation to protect herd immunity while it denies the existence of vaccine injury. Instead of receiving gratitude, the vaccine-injured are scorned and called liars. What does this say about us?

The Hintzes describe Alex's remarkable recovery from autism. Through biomedical interventions to address his medical conditions, Alex is now a neurotypical teenager. His recovery is cause for celebration and enormous optimism for the hundreds of thousands of young people with autism today. Alex is not alone—thousands of children on the autism spectrum have benefited from effective biomedical interventions. Yet conventional medicine rejects the efficacy of such approaches and challenges those medical practitioners who use them. The profession today is ostracizing and punishing practitioners, like Dr. Wakefield, who have pursued inconvenient truths and treatments, just as the establishment persecuted other revolutionary thinkers in their day.

Expanded vaccination recommendations and mandates increasingly place older children and adults at risk too. Amy Pingel recounts her teenage daughter's devastating neurological damage after receiving the Gardasil vaccine against human papillomavirus. Lisa Marks Smith, a forty-eight-year-old mother, tells how she almost died after a routine flu shot from a local pharmacy. Richard Rovet, retired medic, nurse, and U.S. Air Force captain, who cared for hundreds of victims of the military's experimental anthrax vaccine program, speaks out. His commanding officer, Colonel Felix M. Grieder, responds with a powerful afterword to Captain Rovet's chapter. These deeply personal stories depict the extraordinary burdens vaccines have placed on some.

PART III: THE TOPICS IN DEBATE

Part III explores the exceptionally contentious terrain of the vaccination debate from the perspective of those who adamantly support choice. Michael Belkin, prominent investment analyst and parent of a newborn who died from a vaccine, explains how a corrupt market economy has shaped vaccine policy. Read about the "blowback" that is sure to follow today's vaccine rage. All investment bubbles burst in time, and the vaccine bubble is no exception.

Mark Blaxill and Dan Olmsted write about astonishing conflicts of interest inherent in the "public-private partnership" that made the U.S. government and Merck close business partners in the fast-tracked delivery of the human papillomavirus vaccine to the marketplace.

Ginger Taylor examines government statements on vaccines and autism and the media's failure to investigate them. When the government and media mislead the citizenry, true democracy no longer exists. Even though government and medicine would like to slam the book shut on this controversy, it is not going away.

Boyd Haley, former chair of the Department of Chemistry at the University of Kentucky, argues that thimerosal, the mercury-containing preservative, has played a key role in the autism epidemic.

Toni Bark, a medical doctor, explores the creation of new compulsory influenza vaccine mandates for healthcare workers.

Kim Mack Rosenberg, CPR cofounder argues that child protective services sometimes label the choice not to vaccinate as "medical neglect" and seek child removal. She explains the rights families have.

Dr. Julian Whitaker discusses the links between childhood vaccines and autism, and the utter lack of scientific basis for the HPV vaccine. He underscores the collusion between government and Big Pharma that has made mandated vaccination programs a serious threat to our children's well-being and our country's future. Whitaker offers solutions, including returning tort liability to the pharmaceutical companies and restoring parental rights. Sherri Tenpenny provides a doctor's view of public health and challenges the belief that health must come through a needle. Annemarie Colbin, noted healthcare lecturer and author, assesses vaccines from a holistic health perspective. Children today are sicker than ever before, with chronic disabilities and nutritional disorders, yet few in mainstream medicine dare ask whether there might be a vaccine link. Habakus provides resources for the critical question, "What should parents do?" She summarizes the government's position and eight books by leading physicians who span the spectrum of medical advice and offer recommendations about vaccination and the options that parents have. The chapter highlights the wide and conflicting range of medical views, further underscoring the need for choice.

We close Part III with an examination of the sound and fury behind the story of Dr. Andrew Wakefield. Holland examines the charges against him and concludes that medical regulators and the media made Wakefield a scapegoat. He dared to dissent from medical orthodoxy by suggesting that parents have a choice; the establishment retaliated. Dr. David Lewis explains how the exoneration of John Walker-Smith only partially rights a terrible wrong. In March 2012, the High Court of England and Wales overturned the U.K. General Medical Council's (GMC) findings of serious professional misconduct against Professor John Walker-Smith, one of the coauthors of the controversial 1998 *Lancet* study linking autism with MMR (measles, mumps, rubella) vaccination in eight of twelve children studied. Lewis discusses previously unpublished documents that completely vindicate Wakefield as well, and which journalist Brian Deer had concealed from the GMC. In the closing chapter, Wakefield discusses the suppression of vaccine safety science from the unique vantage point of a scientist who has lost his career, his medical license, and his home country simply because he sought to pursue possible links between vaccines and autism.

TOOLS FOR VACCINATION CHOICE

CPR shares "tools of the trade" for those who support an individual's right to choose or refuse vaccination. It contains scientific abstracts, supporting documentation, Harris Interactive poll data, and CPR's principles and calls for action. We provide a website link to CPR's top slides on vaccination choice and a list of frequently asked questions including responses about the return of deadly diseases, vaccine safety, and vaccination ethics.

* * *

Vaccination, like all medical interventions, involves a risk-benefit assessment that only individuals, in conjunction with their healthcare practitioners, are qualified to make. Government coerces no other one-size-fits-all medical program; its justification for this one is unconvincing. A growing body of scientific literature points towards the contribution of vaccines to rising childhood immune and neurological disorders. Using the precautionary principle as a guide, people must be able to choose.

For those who do choose vaccines, the burden to prove safety must be on the government—not on those injured to prove that vaccines caused damage. Today, the government has a lightweight burden to show safety and efficacy, and those injured have a heavyweight burden to prove harm. These burdens must shift, and the government must fund impartial science to study the unintended, as well as the intended, effects of vaccination.

SURVIVING BETRAYAL

Taken together, the authors make a reasoned case for vaccination choice. Although the book seeks discourse on a thoughtful plane, neither the book nor the debate is about reason alone. They are also about powerful human emotions—parents' love for their children, a soldier's dedication to fellow soldiers, and the righteous anger of people betrayed by those in whom they placed sacred trust. Parents trust their doctors, soldiers trust their superior officers, and citizens trust their governments to ensure medical safety. When people reluctantly reach the conclusion that their trust was betrayed, they have a fierce and courageous anger. Vaccines needlessly harmed the lives of many of the authors and their children and numerous others. By raising legitimate questions about vaccination, CPR seeks to make these voices heard.

TOWARDS A REASONED MAINSTREAM DIALOGUE

Americans are rightfully concerned. There are no safe vaccines but only difficult choices to weigh and make. Americans want and deserve a more accurate, measured, and responsible national dialogue concerning vaccination. To dismiss as "anti-vaccine" those who advocate for choice, safety, transparency, and restraint harms us all. The legitimate battlegrounds in the vaccine debate are between compulsion and choice, transparency and nondisclosure, caution and recklessness, and accountability and impunity. Debasing an important and vast subject into a pro-vaccine versus anti-vaccine shouting match serves no one.

Our current vaccination schedule is in many ways a grand experiment—we do not know whether widespread, mandatory vaccination will ultimately prove beneficial or harmful. Beyond the science, however, is the question of individual rights. Does any country, and in particular a democracy, have the right to impose vaccination on unwilling individuals, especially in the absence of disease epidemics and adequate safety science? Foremost in the minds of the country's founders was the imperative to protect citizens' inalienable rights. Government today apparently does not trust individuals to serve their own best interests.

The mainstream media portray concerns about compulsory vaccination as an anti-vaccine agenda. This is grossly inaccurate. The reality is that a majority of American parents support vaccination choice. According to an online Harris Interactive poll sponsored by CPR in May 2010, 52 percent of American parents believe that parents should have the right to decide which vaccines their children receive without regard to government mandates. In this landmark study, Harris reached a representative sample of 1,114 parents with children up to age seventeen. Through this poll, CPR also learned that over half of American parents are concerned that pharmaceutical firms have undue influence over vaccine mandates. Moreover, a majority believe that the federal government should fund an independent scientific study of fully vaccinated versus unvaccinated individuals to assess long-term health outcomes. With only modest variations by education, socioeconomic status, and other criteria, these concerns span the gamut of American parents.

Those deeply concerned about vaccination choice, parental rights, vaccine safety, and undue corporate influence are the majority. Vaccination choice is a broad-based American movement.

Vaccination choice is not a partisan issue—and it is not even just about vaccines. It is about ethics, human rights, science, freedom, dignity, and democracy. It is about the common ideals that Americans hold dear. It is about who we are and what we stand for. And it is literally about the biological integrity of the next generation and those that succeed it.

Part I

THE CASE FOR VACCINATION CHOICE

HUMAN, CIVIL, AND RELIGIOUS RIGHTS, SCIENCE, HISTORY, ETHICS, AND PHILOSOPHY

Since Jacobson v. Massachusetts *in 1905, the world has embraced human rights in many forms—women's suffrage, the Nuremberg Code, free and informed consent, medical autonomy, the human rights revolution, and public health revolutions in sanitation, hygiene, and antibiotics. A thorough reconsideration of compulsory vaccination mandates is long overdue, based on the language of the* Jacobson *decision, which calls on courts to end vaccination mandates that are oppressive and unreasonable.*

—Mary Holland, JD

The U.S. government's mandatory vaccination policies do not appear to comply with international human rights standards. The government has limited the exercise and enjoyment of specific rights—including children's right to obtain a free, public education—based on vaccination status.

—Louise Kuo Habakus, MA

A country that requires all children to receive a product—no matter how beneficial—knowing that some children will die and others' lives will be destroyed by the use of that product, risks losing all moral authority.

—James Turner, JD

Science lies at the heart of the vaccination controversy, and there can be no substitute for it. The ethical practice of medicine requires full and informed consent. It is not possible to have the informed part of informed consent without the science. That vaccine safety science has been and remains inadequate is not in dispute.

—Carol Stott, PhD, MSc, CSci, CPsychol, and
Andrew Wakefield, MB, BS, FRCS, FRCPath

Chapter One

VACCINATION CHOICE IS A FUNDAMENTAL HUMAN RIGHT

Mary Holland, JD

Consider the meaning of the phrase *"fundamental human right." "Fundamental"* means essential, basic, and inalienable. *"Human"* means what we have simply by virtue of the fact that we are human beings. We are born human; we need not be of a particular age, nationality, gender, or class. A "right" is a claim that we may enforce against governments and other people. Thus we assert that we have inalienable claims we can make within society, simply because we are human beings. The decision of when and whether we vaccinate ourselves and our children is a fundamental human right.

Vaccination choice is a fundamental right because it implicates our most precious rights—to life, liberty, and security of person. Basic laws—such as religious laws, the Universal Declaration of Human Rights, the U.S. Constitution, and international laws that all countries must obey—exist to protect inherent human dignity and the equal and inalienable rights of all members of the human race. Having witnessed the war crimes, genocide, and crimes against humanity committed during World War II, nations of the world explicitly embraced human rights in the United Nations Charter.[1] They recognized that it would be impossible to secure a peaceful, just world without the Universal Declaration of Human Rights that proclaims a "common standard of achievement for all peoples and all nations."[2] In the aftermath of

World War II, the world embraced the human rights principles of the Nuremberg Code, a set of ethical principles that forbids experimentation on human subjects without free and informed consent.[3]

Today's human rights laws reject many institutions of the past—slavery, genocide, piracy, torture, inhuman treatment, and systematic discrimination based on race or gender. Some of these laws—such as the prohibition against slavery—have become international norms that apply regardless of whether a particular country has signed a treaty or law to that effect. Today, nations also reject medical experimentation on human subjects without informed consent. Increasingly, nations are recognizing that all healthcare interventions must be based on free and informed consent.[4]

It is an undisputed, scientific fact that vaccines in their current state of development injure and cause death to certain vulnerable people. U.S. law considers vaccines to be "unavoidably unsafe."[5] Vaccination mandates today seek to protect the majority while sacrificing the unknowable, and unknowing, genetically vulnerable few. This utilitarian practice can only be justified if it is based on free and informed consent.

With a complete understanding of the risks and benefits of vaccines, individuals, parents, or guardians may elect to undergo the risks of vaccination to protect against possible disease. They are free to engage in a risk-benefit calculus with their healthcare practitioners and to accept the consequences of their choices. It is unjustifiable, however, for the state to deprive individuals of accurate information and then to coerce them to accept potentially life-threatening medical interventions. Compulsory state vaccination policies violate the rights to liberty and security of person, and when vaccinations result in death, such policies violate the right to life. Even in circumstances of epidemic disease, compulsory vaccination policies would be suspect without consideration of less invasive alternatives, such as self-quarantine and even coercive quarantine.

In the United States, several vulnerable groups—such as children, military personnel, and immigrants—do not have a choice whether to receive vaccinations. Children cannot attend school and adults cannot keep some jobs without fulfilling vaccination mandates. The Vaccine Information Statements that healthcare workers are required by law to give with each federally recommended vaccination are grossly incomplete and often not given at all.[6] Unlike manufacturers of almost all other products, vaccine manufacturers in the United States are legally free from ordinary tort liability for their "unavoidably unsafe" products.[7] The absence of free choice whether or not to use dangerous products, particularly when the manufacturers have exceedingly little liability for them, violates our fundamental rights.

States' rights to compel vaccination in this country stem from a 1905 U.S. Supreme Court decision, *Jacobson v. Massachusetts.*[8] In that case, the Supreme Court upheld Massachusetts' right to require adults to be vaccinated during a smallpox epidemic. If the adults failed to comply, the state required them to pay a $5 fine. The Supreme Court found Massachusetts' regulation reasonable for protecting the public's health, but it also pointed out that its decision would not justify "regulations so arbitrary and oppressive" that they would be "cruel and inhuman in the last degree." Based on *Jacobson*, every state now offers at least the formal right to medical exemption from vaccination.

Overall, *Jacobson* has been interpreted expansively. *Jacobson*, justifying reasonable use of a state's police power, is now the basis for condoning mandates for up to forty-five doses of about one dozen vaccinations for children—in the absence of epidemics, at the cost of public school attendance, based on inadequate and incomplete science, and with significant evidence of undue corporate influence. One may argue today that state vaccination mandates—including compulsory vaccinations for sexually transmitted diseases, such as hepatitis B, and noncontagious diseases, such as tetanus—are oppressive, unreasonable, and disproportionate to the public's health needs. *Jacobson* does not justify today's compulsory vaccination program.

Since *Jacobson* in 1905, the world has embraced human rights in many forms—women's suffrage, the Nuremberg Code, free and informed consent, medical autonomy, the human rights revolution, and public health revolutions in sanitation, hygiene, and antibiotics. A thorough reconsideration of compulsory vaccination mandates is long overdue—based on the language of the *Jacobson* decision, which calls on courts to end vaccination mandates that are oppressive and unreasonable. Even in the context of a military draft, which the *Jacobson* Supreme Court decision refers to by analogy, U.S. citizens have the right to conscientious objection. We must demand the right to philosophical exemption from vaccination mandates in every state as a first step towards truly free and informed vaccination choice.

On many occasions, Martin Luther King, Jr. quoted an abolitionist minister from the 1850s, Theodore Parker, saying, "The arc of the moral universe is long, but it bends toward justice."[9] Ending compulsory vaccination is its own human rights struggle. It is time to bend the arc of the moral universe toward the justice of vaccination choice.[10]

Chapter Two

THE INTERNATIONAL HUMAN RIGHTS STANDARD

Sookyung Song, JD

International law asserts the fundamental human rights to life, liberty, and security of person. National governments must protect and uphold these fundamental rights according to universal standards. International human rights law plays an increasingly important role in all countries, including the United States. Almost all countries have signed basic international human rights conventions, holding them to common goals and standards, and have strived to live up to them.

International law has long recognized that prohibitions against abusive human experimentation are necessary. On numerous occasions, from experiments by Nazi doctors to the U.S. Tuskegee Syphilis Study to the recent Pfizer trovafloxacin experiments in Nigeria,[1] the modern world has observed the dangerous tendency to exact individual sacrifice in medical experiments for the alleged benefit of humankind. The interests of individual participants are not necessarily in line with those of the scientists and broader society, which may attach more value than the human subjects do to scientific progress and the so-called common good. Balancing these conflicting interests, international law has for decades incorporated the critical norm to protect human subjects from experimentation without free and informed consent.

The international community first set down fundamental human rights after World War II. Having witnessed unspeakable human rights violations and having suffered the worst war in human memory, world leaders codified fundamental human rights in the Universal Declaration of Human Rights under Eleanor Roosevelt's leadership. Later, building on this declaration, countries adopted the

International Covenant on Civil and Political Rights and the International Covenant on Economic, Social and Cultural Rights. Together, the international community refers to these documents as the International Bill of Human Rights.

The modern international human right to bodily integrity affords the individual a right to make informed choices about vaccination and all medical interventions. The underlying idea is that those who undergo the risk of experimentation must make the final decision about their own participation after they are informed of the purpose, risks, and benefits of the experiment. While this right to choice in international law sprang from the Nuremberg Code, the international right to informed consent now encompasses the right to free and informed consent for *all* medical decision making.

This chapter provides a brief overview of the development of the human rights to bodily integrity and free and informed consent and shows why these rights apply to vaccination.

POST–WORLD WAR II HUMAN RIGHTS NORMS

Modern bioethics began in 1947 at the criminal trial of Nazi doctors. These practitioners had carried out medical experiments on human subjects in which they intentionally injured and killed individuals for the purpose of scientific discovery. The judges in these trials established ten principles, later called the Nuremberg Code, on ethical standards for medical research.[2] The foremost principle in the Nuremberg Code is that "the voluntary consent of the human subject is absolutely essential." The Nuremberg Code elaborated that the subject's consent must be competent, informed, and free from coercion or inducement.

One year later, in 1948, the United Nations National Assembly adopted the Universal Declaration of Human Rights (UDHR)[3] out of a global effort to represent the expression of rights to which all human beings are entitled. Although the after effects of the Nuremberg trials were still felt, the UDHR did not expressly address human experimentation but simply prohibited any "cruel, inhuman or degrading treatment,"[4] with the belief that this language was broad enough to cover experimentation on human subjects.

The International Covenant on Civil and Political Rights (ICCPR)5 subsequently laid out specifics on the rights proclaimed in the UDHR and directly picked up the issue of human experimentation. In Article 7, the ICCPR clearly dictates that "no one shall be subjected to torture or to cruel, inhuman or degrading treatment or punishment. In particular, no one shall be subjected without his free consent to medical or scientific experimentation." Article 7 articulates that medical experimentation without free consent may at times amount to torture or inhumane treatment. Unlike the UDHR, however, the ICCPR drafters did not stop there. In addition to the prohibition against certain kinds of treatment, Article 7 affirmatively creates an individual right to informed consent for medical research. The structure of this provision, with the second sentence complementing the first, reflects an effort to address medical conduct that may not rise to the level of torture but nonetheless infringes on the right to bodily integrity. The drafting history of Article 7 illuminates how seriously the drafters considered the consent principle:

> Early in the drafting process, government representatives discussed whether this sentence should be included at all. Those in opposition to the addition of the second sentence believed that it was not necessary, since it aimed to prohibit that which was already covered by the first sentence . . . What is notable about this aspect of the development of Article 7 is that those who objected to the second sentence opposed it not because they felt it was irrelevant but because they believed it was already covered by the prohibition on "torture or cruel, inhuman or degrading treatment or punishment." The drafters finally agreed that the matter was so important as to require a special provision, even at the risk of repetition.[6]

THE COUNCIL OF EUROPE'S 1997 CONVENTION ON HUMAN RIGHTS AND BIOMEDICINE

In part because of the pace of medical innovation, the international community has returned to these core principles on several occasions to further elaborate them. The European community was at the forefront of this movement. In 1997, the Council of Europe adopted the Convention on Human Rights and Biomedicine (the Oviedo Convention).[7] While this document is binding only in European countries that signed it, the Oviedo Convention represents the most up-to-date development of biomedical law. The Oviedo Convention reflects the ethical requirements in a society where the research community has become more aggressive in pursuing scientific knowledge, the boundary between science and medicine has become blurred, medicine has become more complex and difficult for patients to comprehend, and medical research is no longer confined to national or jurisdictional boundaries.

In response to these social changes, the Oviedo Convention provides greater protection of individual rights by guaranteeing free and informed consent to *all* medical interventions. Unlike the UDHR and ICCPR, this recent convention is not limited to human experimentation—it abolishes the distinction between research and therapy and addresses the need for individuals to have adequate information about all medical interventions. The Oviedo Convention requires consent for any medical intervention, including prevention, diagnosis, treatment, rehabilitation, and research.[8] It even recognizes that a medical intervention may be psychological as well as physical.[9] This comprehensive approach is necessary today because medical treatment and research are rapidly converging and no longer have static, distinct meanings. To maintain adequate protection, individuals must have the right to free and informed consent regardless of the historical distinction between therapy and research.

THE 2005 UNESCO DECLARATION

In 2005, the global community followed suit and adopted many of the expansive principles of the Oviedo Convention. The United Nations Education, Scientific, and Cultural Organization (UNESCO), a specialized United Nations agency, adopted the Universal Declaration on Bioethics and Human Rights in the UNESCO Declaration.[10] The UNESCO Declaration made it clear that the interests of individuals cannot give way to the "sole interest of science or society."[11] Article 6 states that individuals must give "prior, free and informed consent" to "any preventive, diagnostic and therapeutic medical intervention" and "scientific research." Moreover, requirements for informed consent for participation in scientific research on human subjects include even higher procedural safeguards than requirements for consent to treatment.

Taken together, the message is clear: International bioethics norms are moving away from distinctions between research and treatment and now require free and informed consent to all medical interventions.

THE CONTEMPORARY INTERNATIONAL STANDARD IS FREE AND INFORMED CONSENT FOR ALL MEDICAL INTERVENTION

While neither the Oviedo Convention nor the UNESCO Declaration has binding legal effect in the United States, they both carry unique persuasive power in that they highlight the global trend in an area where the traditional legal system has not kept pace with rapid social and scientific change.

One hundred ninety-three countries are members of UNESCO. While the UNESCO Declaration does not establish enforceable rights, it is persuasive about what the global standard for informed consent should be. Drafters of the UNESCO Declaration strategically chose to write a declaration rather than a treaty for the same reasons that UDHR drafters chose to write a declaration more than fifty years earlier. In both instances, legislators perceived an urgent need to reform the existing law but viewed the treaty process as too slow and arduous—it would require much more time for negotiation and consultation, and less than the majority of states would be willing to sign it. However, the UNESCO Declaration's declaratory form does not undercut its legitimacy or potential legacy. The UDHR, though nonbinding, not only has become a foundational document for many subsequent human rights treaties and for customary international laws that all countries must obey but also has influenced many national constitutions and laws.

In the same vein, we should see the declaratory form of the UNESCO Declaration as an intermediate step for further legal development. We should understand the UNESCO Declaration's principles, which are largely built on the Oviedo Convention, as precursors to international human rights norms of the near future.

There is a growing global trend toward higher standards for informed consent for all medical interventions, and growing impatience with medical paternalism. This trend logically applies to vaccination, an invasive medical procedure that carries both known and unknown risks and benefits. The case is even stronger for vaccination than other medical interventions because historical justifications to waive the right to free and informed consent do not necessarily hold true. The medical community plays only a marginal role in determining the effectiveness and safety of vaccines. Instead, the vaccine industry, the very party that has a financial stake in the outcome, primarily drives safety verification, creating a huge conflict of interest. Depriving society's most vulnerable citizens—infants and young children—and their guardians of informed consent is particularly egregious under the developing international law norms providing free and informed consent for all medical interventions.

Thus medical practitioners must first inform individuals receiving vaccines of the purpose and nature of the intervention, as well as the risks and benefits associated with the vaccination, and then must obtain consent. Emerging international law norms and a faithful reading of the Nuremberg Code require no less.

Chapter Three

A HUMAN RIGHTS ASSESSMENT

Louise Kuo Habakus, MA

Mandatory vaccination represents a limitation on human rights. Any limitation of fundamental rights, such as the rights to liberty and security of person, is serious, regardless of the supposed public good involved. The distinguished health and human rights professor Jonathan M. Mann, MD, MPH, assumed that a government must justify any restriction on human rights in the name of public health. He wrote,

> Unfortunately, public health decisions to restrict human rights have frequently been made in an uncritical, unsystematic, and unscientific manner. Therefore, the prevailing assumption that public health . . . is an unalloyed public good that does not require consideration of human rights norms must be challenged. For the present, it may be useful to adopt the maxim that health policies and programs should be considered discriminatory and burdensome on human rights until proven otherwise.[1]

The discipline of human rights offers a powerful set of widely accepted norms and principles with which to assess public health policies. Human rights scholars have developed assessment tools to evaluate public health policies based on the core international human rights documents. Mann further states,

> The specific rights that form the corpus of human rights law are listed in several key documents. Foremost is the Universal Declaration of Human Rights (UDHR), which, along with the United Nations Charter (UN Charter), the International Covenant on Civil and Political Rights (ICCPR)—and its Optional Protocols—and the International Covenant on Economic, Social and Cultural Rights (ICESCR), constitute what is often called the "International Bill of Human Rights." The UDHR was drawn up to give more specific definition to the rights and freedoms referred to in the UN Charter. The ICCPR and the ICESCR further elaborate the content set out in the UDHR, as well as set out the conditions in which states can permissibly restrict rights.[2]

When we apply a state-of-the-art human rights framework to U.S. mandatory vaccination policies, will the policies pass muster?

THE PUBLIC HEALTH–HUMAN RIGHTS IMPACT ASSESSMENT INSTRUMENT

Scholars at the Harvard School of Public Health developed a Public Health–Human Rights Impact Assessment Instrument[3] to evaluate possible human rights violations that occur when governments take actions in the name of public health that limit individual rights. They argue that such actions must be taken as a last resort and must only occur when they meet the following specific, stringent human rights conditions:

1. The goal of limiting rights may not be contrary to purposes and principles of the United Nations Charter.
2. The limitation must be justified by protection of a legitimate goal such as national security, public safety, and protection of public health or public order.
3. Limitations may be allowed only in a democratic society that presumes a participatory decision process and capacity for redress.
4. A right may be restricted only if the limitation is provided for by law.
5. The limitation of rights must be strictly necessary in order to achieve the public good, which must be carefully assessed on a case-by-case basis.
6. The limitation of individual rights must be proportional to the public interest and its objective (the so-called proportionality test).
7. The limitation must be the least intrusive and least restrictive measure available, which will accomplish the public health goal.

8. The limitation of rights must not be applied in a discriminatory manner.

What follows is a discussion of each of these eight core human rights principles as they apply to mandatory vaccination policies in the United States.

1. The goal of limiting rights may not be contrary to purposes and principles of the United Nations Charter.

The UN Charter calls for the maintenance of peace and international security and respect for human rights. The preamble states,

> We the peoples of the United Nations determined: to save succeeding generations from the scourge of war, which twice in our lifetime has brought untold sorrow to mankind, and to reaffirm faith in fundamental human rights, in the dignity and worth of the human person . . .[4]

The Charter recognizes the central role of human rights protection. Every human being possesses dignity and worth. Since mandatory vaccination represents a restriction of rights, the UN Charter would require a careful and thorough evaluation of any public health policy's appropriateness.

2. The limitation must be justified by protection of a legitimate goal, such as national security, public safety, and protection of public health or public order.

The 1905 U.S. Supreme Court decision, *Jacobson v. Massachusetts*, is the United States' legal foundation for vaccination mandates and determines that mandates are necessary for public health or safety. Although *Jacobson* upheld a vaccination mandate for smallpox, it warned against potential governmental abuse of public health regulations. The decision noted that the smallpox mandate was justified because there was an imminent health hazard that "imperiled the entire population."[5] The Supreme Court warned, though, that such mandates must not be arbitrary, oppressive, unreasonable, or "go far beyond what was reasonably required for the safety of the public."[6] While the protection of society from infectious disease is a legitimate goal, there are important questions about whether the current measures adopted by federal and state governments are excessive.

3. Limitations may be allowed only in a democratic society, which presumes a participatory decision process and capacity for redress.

On the surface, vaccination mandates appear to meet this criterion. The United States is a democratic country, and states impose vaccination mandates through the legislative process. The reality of the decision-making process, however, is less democratic than it appears. Vaccine manufacturers have many opportunities to influence legislation through lobbying and private financing of political campaigns. Recent changes to campaign finance laws have made large corporations even more influential in electoral politics. Many advisers on governmental advisory committees have deep professional and financial stakes in vaccination policy. Truly public participation in such committees is extremely limited. Voting members frequently have conflicts of interest,[7] including stocks, patents, grants, and honoraria from industry. Public health officials often benefit from a lucrative "revolving door" that may convert their public service into a private sector sinecure. Congressional interventions, including the 1986 National Childhood Vaccine Injury Act, have severely restricted the capacity for redress from a vaccine-induced injury. The net result is that public health decisions regarding vaccination policy may not be as truly democratic as they appear.

4. A right may be restricted only if the limitation is provided by law.

State governments rely on the Supreme Court precedent of *Jacobson v. Massachusetts* to uphold the constitutionality of vaccination mandates. There are reasons, though, that a decision mandating the vaccination of adults in 1905 does not apply to childhood vaccination mandates today. *Jacobson* upheld Cambridge's regulation, imposing a smallpox vaccination mandate during an epidemic. The town assessed a relatively small fine to adults who decided not to comply, and the regulation contained an exemption for children. Some argue that states' reliance on the *Jacobson* decision today is misplaced.[8] Much has changed since 1905—there are no infectious disease epidemics in the United States, and state governments impose vaccination mandates *only* on children.

5. The limitation of rights must be strictly necessary in order to achieve the public good, which must be carefully assessed on a case-by-case basis.

We must challenge the assumption that vaccination mandates are strictly necessary for the public good. These mandates have not been studied empirically. Infectious disease mortality had a precipitous and sustained decline during the entire twentieth century. Before the advent of federal vaccination policy, health officials attributed these gains to the significant investments the country made in modern sanitation and hygiene infrastructure.

The last century brought forth dramatic developments that revolutionized public health in households and communities.[9] These included hand washing, flush toilets, good plumbing, covered cesspools, and sewage treatment. The country implemented new medical procedures, such as the sterilization of surgical equipment. Refrigeration brought about important food safety changes. Modern roads and railways helped to deliver fresh produce to the cities. Housing improvements and decreased crowding led to a reduction in deaths from airborne diseases. Moreover, as specific, effective antibiotics were developed and perfected, the number of fatalities due to infectious diseases plummeted. All of these changes occurred before vaccine use became widespread.

Pediatrics, the journal of the American Academy of Pediatrics, admitted in 2000 that "vaccination does not account for the impressive declines in mortality seen in the first half of the century."[10] Infectious diseases were on their way out before vaccination became commonplace. Some diseases for which vaccines were never developed or routinely administered—such as typhoid fever, scarlet fever, and the bubonic plague—have almost completely disappeared. Public health officials simply do not know what would have happened in the absence of widespread vaccination. They cannot assert with certainty that vaccines are responsible for disease eradication or that vaccines are required to maintain herd immunity.

The second portion of the condition specifies that the limitation of rights must be assessed on a case-by-case basis. This means assessing the public health restriction as well as the individual facing the restriction. There are three important reasons why a case-by-case approach is a critical, unmet condition for vaccination mandates in the United States today.

First, not all mandates that exist today are "strictly necessary." Many state vaccination mandates are for diseases from which American children do not typically die. Some diseases, such as chickenpox and rotavirus, have never caused a significant number of deaths in children. Other diseases, such as pertussis, have become less virulent and more easily treated with modern medical technology and antibiotics.

Furthermore, young children are required to receive vaccines for diseases that do not routinely afflict them. Such mandates are not "strictly necessary." State mandates exist for vaccination against hepatitis B and human papillomavirus, diseases that spread primarily through intravenous drug use and sexual contact. Similarly, the vaccine for rubella, a mild disease in children, is administered to toddlers primarily to protect the unborn fetuses of pregnant women. Immunity following vaccination is not lifelong, and the administration of the rubella vaccine at an early age has the potential to render older individuals susceptible at a time when the disease may pose a greater health risk.[11]

Second, an individualized assessment is appropriate because some states have more burdensome mandates than other states do, imposing substantially greater vaccine-related risks on the children who reside in those states. New Jersey, for example, mandates the most vaccinations for day care and school admission. In 2008, it became the only state in the country to mandate children to receive annual flu shots. Twenty states offer philosophical exemptions from mandatory immunization—allowing parents the right to opt out of required vaccinations for their children because of moral or philosophical beliefs—and forty-eight states allow religious exemptions, but two states, West Virginia and Mississippi, only offer medical exemptions.[12] The variation in state requirements highlights the need for case-by-case inquiry in order to protect fundamental human rights.

Last, without screening mechanisms to identify children most at risk for a vaccine-related injury, mandates may do more harm than good. A case-by-case approach is critical to be able to screen for those whom vaccines are mostly likely to harm. Medical professionals acknowledge that some people—including infants, those with severe allergies to vaccine ingredients, and those with immunodeficiencies—are unable to receive certain or all vaccines. Doctors do not know how children at risk will respond to a specific antigen or to the cumulative impact of multiple, simultaneous vaccines. Since adequate vaccine safety research has not been done, it is impossible to identify with greater precision those who should use alternative vaccination schedules or avoid vaccines altogether. Case-by-case analysis would allow for consideration of individual vulnerabilities.

6. The limitation of individual rights must be proportional to the public interest and its objective (the so-called proportionality test).

Proportionality is the quality of being in proportion, balance, or appropriate in size or magnitude relative to other items. In the United States today, no widespread infectious epidemic of any single disease exists, and yet, state governments impose vaccination mandates for a dozen diseases or more. The burden of diseases is not proportional to the number of vaccination mandates. To date, public health officials have not demanded randomized, controlled trials to evaluate the long-term health outcomes of vaccinated versus unvaccinated individuals. Additionally, the government has not commissioned research to prove that outbreaks will occur if childhood vaccination mandates stop. In the absence of science, the government cannot prove the proportionality of childhood vaccination mandates.

7. The limitation must be the least intrusive and least restrictive measure available that will accomplish the public health goal.

In the absence of a public health emergency and compelling science, vaccination mandates for young children are unduly intrusive and restrictive. Most families cannot choose whether or not to comply because they cannot do without day care and public education. While the government does not literally force vaccination, its coercive means are restrictive and intrusive.

Moreover, there does not appear to be an ongoing assessment of the public's health needs. Since the implementation of vaccination mandates for school admission, state governments only add vaccines to the schedule; mandates are almost never removed.

8. The limitation of rights must not be applied in a discriminatory manner.

Current compulsory childhood vaccination policies discriminate against children. If the theory of herd immunity is correct, then vaccination and revaccination of the adult population must be necessary to achieve the high, designated threshold levels deemed protective. There would need to be repeated boosters for adults, since vaccine-induced immunity fails in some percentage of the population and wanes over time regardless. Applying vaccination mandates only to children is discriminatory and contradicts the theory of herd immunity, which underlies the necessity of childhood vaccination mandates.

Mandatory vaccination policies do not appear to comply with international human rights standards. The government has limited the exercise and enjoyment of specific rights—including children's right to obtain a free, public education—based on vaccination status. According to this human rights assessment framework, we must consider reforming the vaccination program because specific and stringent conditions have not been fulfilled. Existing mandates are discriminatory and disproportionate to the public health goal.

A WORD ABOUT HERD IMMUNITY

The validity of herd immunity undergirds all compulsory vaccination policies. The theory of herd immunity posits that there is resistance to the spread of an infectious disease when a sufficiently high percentage of people in a community are immune to that disease. Herd immunity is achieved when the vaccinated portion of the population acts as a protective cordon that prevents a resurgence of the disease and, as a result, protects vulnerable individuals who are unable to receive vaccines (or whose vaccinations failed). Public health officials invoke the legitimacy of herd immunity to justify mandatory vaccinations. By definition, a high number of people must receive vaccines to attain herd immunity.

Herd immunity is a questionable theory. Researchers may point to epidemiological studies and anecdotes that appear to support it, but its validity remains in doubt. Public health historian James Colgrove writes[13] that target immunization levels of 70 to 80 percent, the "magic numbers," are accepted on faith but not proven. He quotes a county health commissioner who conceded, "We must perforce accept . . . [assumptions about herd immunity] until some future controlled study gives us a definite answer."[14]

What is most salient and troubling about the theory of herd immunity is that the original basis for its validity has nothing to do with vaccines. People observed a protective effect in the community when a sufficiently high number of individuals contracted the wild form of a disease and secured lifelong immunity. This enduring, cumulative protection from the natural acquisition and resolution of disease is what public health officials called "herd immunity." In the 1930s and 1940s, health officials began using "techniques of mass persuasion"[15] and "aggressive salesmanship"[16] to increase the public's compliance in receiving vaccines. They reviewed the scant literature and limited empirical data and enlisted epidemiological theory to help endorse vaccinations and achieve target rates.[17] Since vaccine-induced immunity appeared to afford protection similar to natural immunity, officials presumed that vaccines could also create herd immunity. Vaccine-induced immunity, however, is not the same as natural immunity. A person who contracts measles in early childhood will never get the disease again; a person with vaccine-induced immunity will need to receive boosters or assess his or her antibody levels periodically. Vaccine-induced immunity is qualitatively different. Vaccines do not always work, and even when they do, their protection weakens over time.

There are many examples that throw the validity of herd immunity into question. Outbreaks of diseases such as measles, mumps, and chickenpox, among others,[18] routinely occur in fully vaccinated communities. Pertussis, or whooping cough, is another case in point. Federal vaccination guidelines recommend five doses of the pertussis vaccine, in the combination of diphtheria, pertussis, and tetanus, for children by age six and a sixth dose of each at age eleven. Despite repeated boosters and high compliance rates,[19] public health officials are now discussing the need for additional boosters because of sporadic pertussis outbreaks[20] among the vaccinated and unvaccinated. The disease is most dangerous in early infancy, and babies typically contract it from their parents—not schoolchildren.[21] While vaccine promoters see another booster as the best answer, perhaps a superior approach would be to reconsider the effectiveness of vaccines and the validity of herd immunity.

• • •

In conclusion, vaccination mandates are public health measures intended to fulfill a legitimate goal that restricts and intrudes on fundamental human rights. There is significant value in applying a human rights framework to vaccination mandates. The necessity for childhood vaccination mandates today cannot be justified on human rights principles. Therefore, until public health policies can pass muster under a serious human rights examination, all citizens should have vaccination choice.

Chapter Four

DUE PROCESS AND THE AMERICAN CONSTITUTION

James Turner, JD

Over the past two decades the United States government has paid over $2 billion to families of children who died or have been permanently harmed by vaccinations.[1] A government that requires individuals—particularly children—to be vaccinated, knowing that some will die and others will be permanently disabled as a result, risks losing all moral authority.

VACCINES ARE LESS THAN PERFECT

If our national, state, and local governments—including schools, hospitals, prisons, and other government-funded enterprises—are to maintain their moral authority, we must adopt important new procedures and answer troubling questions regarding vaccines. I first began looking carefully at this public policy issue in 1970. What I found deeply troubled me.

John Gardner, President Lyndon Johnson's secretary of Health, Education, and Welfare (HEW) from 1965 to 1968, called me in the summer of 1970. He said that after leaving office, he had learned that unfortunate, indeed dangerous activities involving vaccines had been going on while he was secretary. He asked if I would be willing to look into them.

At the time, I was working with Ralph Nader, and we had just published my book, *The Chemical Feast.* We had accomplished some visible successes: the U.S. Food and Drug Administration (FDA) had removed cyclamate sweeteners from the market until they could be proven safe, the food industry had voluntarily removed monosodium glutamate (MSG) from baby food, and the president had ordered a review of all food chemicals said to be generally recognized as safe by the FDA.

Secretary Gardner wanted to know if the same kinds of lapses we had discovered for food were also occurring in vaccine regulation, as he had been told. In 1970, the National Institutes of Health (NIH) Division of Biologics Standards, not the FDA, regulated vaccines. This agency, charged with ensuring vaccines' availability rather than their safety and efficacy, acted more as an NIH research group than a national drug regulator.

Secretary Gardner introduced me to J. Anthony Morris, PhD, a microbiologist who, at the time, was a senior vaccine regulatory researcher and a control officer for influenza vaccines at the NIH. He, in turn, introduced me to the officials in charge of the other vaccines, who were some of the most dedicated and conscientious government officials I have ever met. Each spelled out story after story of problems. The diphtheria-pertussis-tetanus (DPT) vaccine control officer called the DPT shot then in use the "dirtiest material put into humans." The cholera vaccine reviewer said, "A glass of proper mineral water would work better than the shot," which masked symptoms and might even spread the disease. The quality control officer worried about the amount of mercury in vaccines—in 1971! Dr. Morris gathered years of research, casting doubt on the safety and efficacy of flu vaccines.

Others outside the agency added their stories. The scientist who won the Nobel Prize for discovering "slow viruses" worried that these agents—related to problems such as mad cow disease—might be contaminating certain vaccines, such as the yellow fever vaccine. The key scientific advocate for the creation of the Centers for Disease Control and Prevention (CDC) said he had been fired in the early 1960s for refusing to make the annual announcement urging Americans to get the flu shot. The attorney general found that NIH vaccine regulators were illegally exempting vaccines from drug regulation.

We communicated these facts to the Government Research Subcommittee of the Government Operations Committee of the United States Senate chaired by Abraham Ribicoff of Connecticut, also a former HEW secretary. He held hearings where the HEW secretary at the time, Elliot Richardson, announced that because of these concerns, vaccine regulation would be transferred from the NIH to the FDA. The former NIH regulatory body became the FDA's Center for Biologics Evaluation and Research (CBER).[2]

Dr. Morris began his professional career with the elite vaccine group at Walter Reed Army Medical Center in Washington DC during World War II. A large number of the world's leading vaccine researchers and advocates began their careers in this group, which, under the leadership of the renowned Dr. Joseph E. Smadel,[3] became the Walter Reed Army Institute of Research in 1953.

In 1956, Dr. Smadel left Walter Reed to join the NIH as an associate director. He took this job to help clean up the vaccine regulation that had allowed the Salk vaccine to paralyze nearly two hundred people and kill ten. NIH scientists who had discovered paralysis as a possible side effect of the Salk vaccine before market launch were ignored. Scientists with connections to Salk and the vaccine's manufacturer overruled their less-conflicted colleagues who wanted to stop the program until they had eliminated its potential for harm. Once people died, it only took thirty days to find the problem.

Shortly after arriving at the NIH, Dr. Smadel arranged for his colleague Dr. Morris to move from Walter Reed to join him. Drs. Morris and Smadel were particularly concerned about data that raised questions about the safety and efficacy of flu vaccines. They knew, for example, of armed forces data that suggested a connection between an increase in Guillain-Barré syndrome (GBS) and flu vaccine inoculation.

In 1963, Dr. Smadel became chief of the Laboratory of Virology and Rickettsiology at the NIH in order to more closely study vaccine safety and efficacy. He took this step because he had discovered that polio vaccines grown in live monkey's kidney cell culture contained simian virus 40 (SV-40), a cancer-causing agent. At the NIH, he did bench science alongside Dr. Morris and other vaccine researchers. While a demotion, this position allowed him greater access to vaccine regulatory activity.

Unfortunately, Dr. Smadel died later that same year, leaving Dr. Morris to carry on a review of vaccine safety alone. For the next few years, Dr. Morris worked mostly on respiratory vaccines and viruses, primarily the many varieties of flu. This was the work he was doing when John Gardner introduced us in 1970. At that time, his superiors were trying to fire him for continuously raising embarrassing questions about vaccine safety and efficacy. As the vaccine regulation battle unfolded, Dr. Morris was not fired, but several of his superiors were. Once vaccine regulation moved to the FDA, the battle continued. Dr. Morris turned down an opportunity to become part of the senior elite that ran the agency, in favor of bench research and continuing his critique of the ongoing, faulty vaccine process. His critique cost him his job. He was fired in 1980, though he was able to preserve his retirement rights.

Dr. Morris's final battle inside the government occurred during the buildup, collapse, and aftermath of the ill-conceived 1976 effort to inoculate all Americans against swine flu. Congress balked at funding the program until the outbreak in Philadelphia of what became known as Legionnaires' disease. The CDC refused to rule out swine flu as the cause of this event until the day after Congress approved funds for the program to vaccinate all Americans against swine flu. That day, the CDC announced that Legionnaires' disease was not swine flu. The government terminated the swine flu program when the vaccine caused injury and death to some. The vaccine only reached about 25 percent of the two hundred million people intended to receive it.

MONEY FOR VICTIMS ENTERS THE DEBATE

The 1976 swine flu debacle set off another round of anger and lawsuits, which ended in 1986 when Congress passed the National Childhood Vaccine Injury Act. Dr. Morris vigorously fought against this legislation. He believed that the government's use of cash payments to escape its legal responsibility to provide safe and effective vaccines would end any possibility of ensuring no harm from vaccines, which was his goal. Time has proven him right.

Barbara Loe Fisher, the tireless founder of the National Vaccine Information Center, testified before the California State Senate Committee on Health and Human Services in 2002:

> I worked with Congress in the early 1980s on that [vaccine compensation] law and have watched it be turned into a cruel joke as 2 out of 3 vaccine injured children are denied federal compensation for their often catastrophic vaccine injuries because HHS [Department of Health and Human Services] and the Department of Justice officials fight every claim, viewing every award to a vaccine injured child as admission that vaccines can and do cause harm.[4]

Barbara Loe Fisher's story about the development and neutering of the 1986 National Childhood Vaccine Injury Act ("1986 Act") is a truly distressing object lesson in do-good legislation. The vaccine industry promoted the legislation ensuring compensation because it provided protection to them against mounting lawsuits. At the insistence of the parents of dead and injured children, the compensation law design was broadened to include vaccine safety provisions as well as financial assistance to families of vaccine-injured children. But time has shown that those added provisions lacked teeth.

The 1986 Act required doctors to give parents information on vaccination benefits and risks. It required doctors to record the vaccine manufacturer's name and vaccine lot number in the child's permanent medical record and to document and report hospitalizations, injuries, and deaths following vaccination to a centralized federal Vaccine Adverse Events Reporting System. Today this set of procedures works minimally, if at all.

The 1986 Act also called on the Institute of Medicine at the National Academy of Sciences to review the medical literature for evidence that vaccines can cause immune and brain dysfunction. It seemed designed to help reduce, if not eliminate, unnecessary vaccine-induced injuries and deaths, but in fact, the bill's momentum for passage came from its shift of liability from manufacturers to taxpayers for vaccine-induced disabilities and deaths.

The 1986 Act intended that parents of vaccine-injured children would receive federal compensation on an expedited, no-fault, fair basis as an alternative to lawsuits. The act promised vaccine manufacturers that they were no longer liable for the vast majority of vaccine-induced injuries and deaths. The act included a table of Compensable Events that listed clinical symptoms of adverse reactions to vaccines. A child having symptoms on this table would receive compensation unless the government could present compelling evidence that proved the vaccine had not caused the symptoms. The Department of Health and Human Services and the Department of Justice opposed the bill to the end. Yet, through rule-making authority, they were ultimately given the power to change almost everything once the law was passed.

In 1999, Ms. Fisher summarized a decade's worth of experience for parents trying to use the compensation system:

> Today, the bitter truth is that, although more than $1 billion has been paid out to some 1,000 families whose loved ones have been harmed by vaccines, three out of four vaccine victims are turned away. Although parents pay a surcharge on each vaccine their child gets and the money from that surcharge is put into the vaccine injury trust fund, there is more than $1 billion languishing in the trust fund because HHS and Justice pay expert witnesses and lawyers to fight every vaccine injury claim. And to make it easier for compensation to be denied to vaccine injured children, under rule making authority these federal agencies gutted the Table of Compensable Events in 1995 and

> arbitrarily rewrote the definition of encephalopathy (brain dysfunction) that had been used by medicine [for] decades.[5]

Today the government surrenders its moral authority by requiring children to receive vaccines it knows will cause death or permanent damage to some. Vaccines may have value for those who can use them safely. However, as the government continues to defend vaccine-caused deaths and disabilities and coerce families to put their children at risk, confidence in vaccines will decline, resistance to their use will grow, and what value they might have will be lost. In short, the way we are conducting mandatory vaccination now is not only causing the death and disability in children, it is undermining the very programs for which advocates claim great benefits. At a minimum, routine vaccination mandates must end. We must create a sound system for identifying all vaccine-induced injuries, and for giving proper care to all children, both before and after vaccination and vaccine-induced injuries.

IS MANDATORY VACCINATION CONSTITUTIONAL? AN APPROACH TO MINIMIZE VACCINE-INDUCED INJURIES AND PUBLIC OUTRAGE

The state of Arkansas requires that vaccination "shall not apply if the parents or legal guardians of that child object thereto on the grounds that immunization conflicts with the religious or philosophical beliefs of the parent or guardian."[6] To obtain an exemption, parents must sign a notarized statement requesting a religious, medical, or philosophical exemption from the Arkansas Department of Health (DOH). The parents must then complete an educational component prepared by the DOH, sign a statement of informed consent with a refusal to vaccinate, and provide a signed statement of understanding that the unvaccinated child may be removed from school during a disease outbreak. For the pertussis vaccine, the rule says that if a child's sibling or half sibling had an adverse reaction, the child is exempt. The Arkansas law contains a "philosophical" exemption, which defeats mandatory vaccination. Currently philosophical exemptions exist in twenty states and should be the law in every state.[7]

The campaign that led to the Arkansas philosophical exception culminated in the law authorizing exemption signed on April 1, 2003. It began when a federal judge threw out the previous religious exemption law because it required a "statement . . . from a pastor or church official that the parents or guardians are members or adherents of a recognized church or religious denomination whose tenets are opposed to immunization." The court found that an individual could have a religious belief precluding vaccination without belonging to a recognized church. The families of 123 children who had religious exemptions appealed the law, and while the appeal was pending, the state legislature passed the "philosophical" exemption law.

One constitutional argument presented to the appeals court specifically urged the court to find mandatory vaccination unconstitutional. The case called into question the constitutionality of current mandatory childhood immunization programs required of children to attend school. All five-year-olds must have received approximately sixteen viral or bacterial agents, even though health authorities acknowledge that a random and significant number of children will die or be disabled each year by the required shots. This acknowledgement of vaccine-induced injury is underscored by the federal government's payout of over $2 billion to families of dead and disabled children since 1986.

The Fifth Amendment of the U.S. Constitution says, "No person shall . . . be deprived of life, liberty or property without due process of law." The Fourteenth Amendment extends this protection to persons in the states, saying, "No State shall make or enforce any law which shall . . . deprive any person of life, liberty, or property, without due process of law." Mandatory vaccination programs, as conducted, fail to provide due process to individuals.

The case law justifying mandatory vaccination programs contains limitations on the state power to compel vaccination. It refers to the need for an emergency (e.g., a threatened or actual epidemic), it gives vaccine-endangered individuals the choice to opt out, and it recognizes conscientious objection. While current programs formally recognize these elements, in practice, their procedures fail to meet the due process requirements necessary to protect life.

THE DEATH PENALTY ANALOGY

The stark contrast between the process required to obtain compensation after a vaccine-induced tragedy—including documentation, hearings, counsel, and appeals—and the process before compulsory immunization underscores the procedural paucity of mandatory vaccination programs (see chapter 5: The Right to Legal Redress). Capital punishment's procedural history offers insight into the constitutional problems of mandatory immunization. A 2002 New York district court decision found federal capital punishment unconstitutional because of lack of due process. At the same time that the Arkansas vaccination case was pending before the appeals court, two federal district courts—one in New York and one in Connecticut—declared capital punishment unconstitutional. They based their decisions on the fact that over two hundred individuals nationwide had been released from death rows across the country after their convictions were overturned. The courts found that these innocent individuals had not received due process.

Appellate courts overturned these cases saying that, even though the juries that convicted them had made clear factual mistakes, the existence of a jury trial met the requirements of due process. Still, the capital punishment victims received more due process than any child receiving mandatory vaccines, who receive no due process at all. The argument that the U.S. Constitution requires due process for vaccine recipients, including informed consent and the right to opt out, is straightforward. Because a small, predictable, but individually somewhat unforeseeable number of children will die or be permanently disabled by mandatory vaccines, an increasing number of parents are seeking ways to avoid vaccination.

The Arkansas federal court declared the Arkansas vaccine exemption unconstitutional because it applied only to members of established churches. The court held that individuals could have a religious belief that justified a vaccine exemption even if they belonged to no established mainstream church. The language of the court could be read to suggest that unconstitutionality might be best corrected by expanding the exemption to a broader number of religions. Left unaddressed was the broader question of when, if at all, the state may mandate behavior that subjects a minority of citizens to the certitude of death or permanent disability.

The vaccination exemption argument continues. Mandatory vaccine regulation today is comparable to that of capital punishment in 1972, when the U.S. Supreme Court decided *Furman v. Georgia*.[8] That is, state procedures used to implement the policy of moral and medical exemptions are so lacking in effectiveness that they lead to the unconstitutional taking of life. *Furman* led to a four-year-long moratorium on capital punishment until states adopted procedures for implementing the death penalty that satisfied the Supreme Court's reading of due process.

Current mandatory immunization programs have major procedural failings. These failings include a lack of clarity about exemption standards and a bias against informing families of their exemption rights. It also leaves life-and-death decisions about medical exemptions and the assessment of the risk-benefit ratio solely in the hands of government bureaucrats. Bureaucrats decide the fate of children without the substantive input of the families most likely to be affected. In this procedural framework, as with recipients of capital punishment sentences before *Furman*, children who become disabled or die from vaccinations have not received the due process the Fifth and Fourteenth Amendments require.

In this situation, the proper course is to suspend the mandatory nature of the immunization program until and unless states adopt procedures to protect against the unnecessary death and disability of children required to receive vaccines. A rational procedure for effective immunization should include a right to opt out of vaccination.

PROCEDURES TO MEET DUE PROCESS REQUIREMENTS

First and foremost, every state must adopt the right to philosophical exemption.[9] There ought to be several procedural steps that states must take before mandating vaccination. The government must prove that those who are vaccinated will be harmed by those who are not vaccinated. There must be an open public procedure to evaluate how likely an epidemic threat is and how effective the respective vaccine is against that threat compared to how dangerous the vaccine is to individuals. In such a proceeding, the government must prove that the benefits for the subject vaccine in the prevailing conditions outweigh the harms. Interested families who may be harmed by state mandates must have input into the decision-making process. States must establish procedures to balance fairly the benefits and risks and to create appropriate opt-out rights. Without these and related procedural reforms, the mandatory vaccine program cannot withstand constitutional scrutiny.

Under the Due Process Clause of the Fourteenth Amendment, no state may "deprive a person of life, liberty or property without due process of law." Courts have decided that the Fourteenth Amendment protection encompasses a constitutionally protected right to refuse unwanted medical interventions, including the "principle that a competent person has a constitutionally protected liberty interest in refusing unwanted medical treatment."[10] In the case of a minor, courts have consistently held that the parents have the capacity to act in the child's best interest, which includes making decisions regarding appropriate healthcare. A child is not "the mere creature of the State."[11]

An Arkansas case concerned the right to refuse unwanted medical treatment, namely, state-mandated medical vaccinations, which courts have determined cause death and serious bodily injury.[12] The issue of who is or is not susceptible to vaccine-induced injury is in constant flux. The number of mandated immunizations per individual child has increased from four in the 1950s to approximately thirty to forty-five for most children today. HHS has reduced the number of compensable injuries since the 1986 Act took effect. As a result, there are more exposures to possible harm and fewer officially recognized injuries. The government has provided no effective procedural framework with public participation to adjust the compensation scheme to cover today's expanded vaccination mandates.

By contrast, when families suspect a vaccine-induced death or injury and seek compensation, the 1986 Act's detailed and meticulous procedures come into play. The Act pays for an attorney to represent the family of the dead or injured child. The attorney must submit a petition for compensation to the Vaccine Injury Compensation Program of the Court of Federal Claims. If the Department of Justice contests the petition, there is a hearing process and an administrative judge, or special master, decides whether to provide compensation. The family may also remove its claim from this program to civil court.[13] The government's decisions on which vaccinations to compel include no analogous procedural safeguards despite the knowledge that a predictable number of children will die or become disabled.

LEGAL PRECEDENTS FAVOR PROCEDURAL SAFEGUARDS

Jacobson v. Massachusetts is the seminal U.S. Supreme Court case addressing the state's authority to mandate vaccination.[14] Recent Arkansas federal district court decisions on exemptions cited *Jacobson* with very little discussion, as the basis for Arkansas's authority to mandate vaccinations.[15] It is unfortunate that the Arkansas courts failed to thoroughly review *Jacobson*, because the Supreme Court emphasized that it was upholding the Massachusetts statute under the "necessities of the case," namely, that an epidemic was prevalent and increasing. In fact, Massachusetts did not force vaccination on any adults—it merely required them to pay a relatively small fine for noncompliance if they elected not to receive vaccination.

The Supreme Court "observed that the legislature of Massachusetts required the inhabitants of a city or town to be vaccinated *only when*, in the opinion of the

Board of Health, that was *necessary for the public health or the public safety.*"[16] Recognizing the public health emergency, the Supreme Court was careful nonetheless to warn that even during an epidemic, the police power of the state "might be exercised in particular circumstances and in reference to particular persons in such an arbitrary, unreasonable manner, or might go so far beyond what was reasonably required for the safety of the public, as to authorize or compel the courts to interfere for the protection of such persons." The Supreme Court apparently did not contemplate that compulsory vaccination would be imposed outside of a clear public health emergency.[17]

Similarly, shortly after the *Jacobson* decision was made, Arkansas cases upheld a board-of-health regulation enforcing a smallpox vaccination mandate, only after determining that there was a public health emergency.[18] Since there was no smallpox in the schools at issue or in the surrounding community, the courts looked to the statewide threat to find a basis for upholding the regulation as reasonable and necessary. One court noted that it was "commonly known at the time smallpox was prevalent in the State and that unless preventive measures were adopted a smallpox epidemic might result."[19] It was only under these expanded parameters that the court was able to find that the rule was not "unreasonable or unnecessary."[20] Other cases have similarly held that state-mandated vaccination of children requires exigent circumstances, such as an epidemic or outbreak of a contagious disease, in order to be reasonable and necessary.[21] Courts in the early twentieth century invalidated vaccination mandates when there was no imminent danger from a disease.[22]

In spite of legal authority requiring emergencies for mandatory vaccination and in spite of the fact that some children will die and some will be permanently disabled, states have made mandatory immunization a routine requirement for attending school. They have done so bureaucratically, without implementing any procedure for members of the community to participate in the decision for mandatory immunization. As a result of the vaccination program, lives are harmed and lost without even the pretense of due process.

VACCINATION MANDATE STAKES ARE HIGH

In 2002, District Judge Jed S. Rakoff found the Federal Death Penalty Act unconstitutional because it created too great a risk that innocent people would be executed.[23] This opinion and order reaffirmed and incorporated Judge Rakoff's original decision of April 25, 2002, in the same case that new science had rendered capital punishment unconstitutional because it established as irrefutable that innocent people were on death row.[24]

Judge Rakoff's capital punishment opinion underscored the consternation of judicial authorities that innocent people might be executed. He wrote,

> It is therefore fully foreseeable that in enforcing the death penalty a meaningful number of innocent people will be executed who otherwise would eventually be able to prove their innocence. It follows that implementation of the Federal Death Penalty Act not only deprives innocent people of a significant opportunity to prove their innocence, and thereby violates procedural due process, but also creates an undue risk of executing innocent people, and thereby violates substantive due process.[25]

The potential execution of an individual for a crime he or she did not commit, even after a fair trial and posttrial scrutiny, created a sense of failure that led Judge Rakoff to declare the Federal Death Penalty Act unconstitutional:

> If, instead, we sanction execution, with full recognition that the probable result will be the state-sponsored death of a meaningful number of innocent people, have we not thereby deprived these people of the process that is their due? Unless we accept—as seemingly a majority of the Supreme Court in *Herrera* was unwilling to accept—that considerations of deterrence and retribution can constitutionally justify the knowing execution of innocent persons, the answer must be that the federal death penalty statute is unconstitutional.[26]

Obvious distinctions exist between the state's action in seeking the death penalty and the state's action in ordering mandatory immunization. However, the distinctions seem to make the death and disability that follow mandatory vaccination more, rather than less, shocking to the conscience. In the case of mandatory immunization, innocent children die or are disabled with no opportunity to be represented in fighting for their lives and without proving that the state interest in advancing public health requires such an outcome. Such deprivation of life in the absence of due process violates substantive due process and shocks the conscience. The fact that the federal appeals court overturned Judge Rakoff because the death row victims received jury trials underscores the due process failure for those who are required to receive vaccines.[27] The dead and disabled children had no hearing at all.

Justice Sandra Day O'Connor said of capital punishment:

> I cannot disagree with the fundamental legal principle that executing the innocent is inconsistent with the Constitution. Regardless of the verbal formula employed—"contrary to contemporary standards of decency," "shocking to the conscience," or offensive to a "principle of justice so rooted in the traditions and conscience of our people as to be ranked as fundamental"—the execution of a legally and factually innocent person would be a constitutionally intolerable event.[28]

The death of an innocent child in the service of mandatory immunization is shocking. It is contrary to contemporary standards of decency and it violates a principle of justice deeply rooted in tradition—that affected parties must be allowed to participate in decisions that affect their lives, liberty, and property. Parents and the public must participate in the decisions regarding vaccination risk-benefit tradeoffs. They must decide which, if any, vaccines to mandate, at what age the vaccines should be administered, how many doses should be administered, and reasons for exemptions. Just like the execution of an innocent inmate, the death of an innocent child from a mandatory immunization program violates the U.S. Constitution.

CONCLUSION

Based on my life experience and my awareness of vaccination regulatory failure for the past forty years, this I know to be true: In a moral society, there can be no mandated vaccinations. In a moral society, citizens must be allowed to choose which vaccines they and their children receive and when. Without vaccination choice, society places both public and individual health at risk. Vaccination choice is a fundamental human right.

Today, Dr. J. Anthony Morris, with whom I began my vaccination policy sojourn, is ninety-two years old and living in the same house he has occupied since he began working for the U.S. government during World War II. Tony continues his sharp critique of vaccination policy that uses slovenly science to create unnecessary risks for the lives and well-being of the people, especially children, it intends to help. But a long life is the best revenge. When I asked Tony what message he would give to the American people, he said, "Tell them I do not take the flu shot."

Chapter Five

THE RIGHT TO LEGAL REDRESS

Mary Holland, JD, and Robert Krakow, JD

Vaccines, like all prescription medicines, carry risks—the law considers them to be "unavoidably unsafe."[1] Because the government and medical community want to ensure high vaccination rates, they do not publicize this legal fact. To the contrary, they tell the public that "vaccines are safe and effective."[2] The public is lulled into believing that vaccines are almost perfectly risk free. That, however, is public relations.

Congress passed the 1986 National Childhood Vaccine Injury Act ("the Act") in part to compensate families for "vaccine-related injury or death."[3] The Act includes a Vaccine Injury Table listing brain damage, paralytic disorders, anaphylaxis, seizures, and death, which are the basis for compensation if they occur within specific time periods after vaccination.[4] The Act establishes a National Vaccine Injury Compensation Program (VICP) to compensate injured children for specified "on-table" injuries and for "off-table" injuries where the petitioners can prove causation.

In the name of protecting children's health, the Act changed the legal landscape fundamentally. Instead of keeping doctors and the vaccine industry directly liable for adverse reactions to vaccines, the Act created a taxpayer-financed compensation program for injuries. Unprecedented at that time, the Act was, in effect, a corporate bailout for the pharmaceutical industry, because it forced the public—rather than the industry—to pay for damage from "unavoidably unsafe" products. Thus the Act deprived children of two of the most significant legal protections they had to ensure safety and remedial compensation: informed consent and the right to sue manufacturers directly.

The U.S. Department of Justice (DOJ) represents the Department of Health and Human Services (HHS) in claims of vaccine-induced injury or death before the U.S. Court of Federal Claims. Neither the vaccine industry nor doctors are defendants in the VICP. So far, HHS has compensated about 2,500 claims of vaccine-induced injury and has committed over $2 billion in compensation and legal fees.[5]

The idea behind the compensation program is simple—if children are injured in the "war on disease," society has an ethical duty to care for them, just as society has a duty to care for soldiers injured while serving in the military. Congress gave bipartisan support to the Act and intended for compensation to be quick, easy, certain, and generous. In principle, there was congressional and public consensus that caring for vaccine-injured children was the right thing to do.[6]

The reality has not lived up to the principle. The message that "vaccines are safe and effective" does not reflect the reality that vaccines injure and cause death. Due to this tension, perhaps, the government goes to great lengths to compensate as few cases as possible, maintaining the fallacy that vaccine-induced injury occurs in a vanishingly rare number of cases.

If a parent or guardian asserts that vaccines caused his or her child's injury,

> they [the DOJ] will deny your reality. They'll deny your word. They'll say you're lying. They'll say you made it up. They'll say you're mistaken. They'll say you're very well-educated so you know how to game the system. And then you'll come up against the full weight of their authority—expert witnesses with unlimited funds who will say that your child's injury is genetic, genetic, genetic. You'll find obstruction in the Department of Justice. You'll find resistance. And you'll find scorn.[7]

The VICP has been a dismal failure. Almost four out of five claimants lose in what was meant to be a petitioner-friendly administrative forum.[8] The tenor of VICP proceedings is exceptionally hostile and adversarial—the exact opposite of what Congress intended. Petitioners must litigate almost every case—almost no injuries are "on-table" administrative claims anymore. Though Congress intended cases to resolve within a year, cases now take many years to litigate.

There is a long list of things that are wrong with the VICP. First and foremost, while it is referred to as a court, it simply is not a court. There is no judge, no jury, no right to require the adversary to provide information, and no formal rules of evidence and civil procedure. It is an administrative tribunal meant to handle simple injury cases based on certain presumptions. In a completely inappropriate way, the VICP is being used today as a forum for exceedingly complex health litigation, such as the Omnibus Autism Proceeding.[9] The VICP is woefully inadequate for handling such litigation.

The following list outlines some of the major problems with the VICP:

1. THE LACK OF JUDICIAL INDEPENDENCE

In lieu of judges, the U.S. Court of Federal Claims appoints "Special Masters" to decide questions of fact and law and oversee VICP proceedings. The U.S. Court of Federal Claims and the U.S. Court of Appeals for the Federal Circuit only review legal decisions for abuse of discretion and factual mistakes. Special Masters do not have the kind of judicial independence they need to decide these extremely controversial and serious cases. They are appointed to four-year terms, they have no specialized training in medicine, they primarily come from governmental law jobs, and they sometimes appear to defer to the DOJ and HHS.[10] Decisions in cases like the Omnibus Autism Proceeding potentially affect national vaccination policy. Special Masters, lacking the life tenure of judges or even long-term appointments, are not suited to make decisions that might affect federal policy.

2. THE LACK OF EQUALITY BETWEEN THE GOVERNMENT AND PETITIONERS

There are gross financial inequalities between the DOJ and petitioners in these cases. The DOJ has a virtually unlimited budget and can unilaterally retain and pay for expert witnesses. The VICP pays the fees of the petitioners' experts and lawyers, but there's a catch: The VICP often compensates petitioners' lawyers years after a case has started. Moreover, the DOJ and the U.S. Court of Federal Claims have the opportunity to review and cut petitioners' legal fees. There is no reciprocal right for petitioners' lawyers to review and approve or disapprove the DOJ's fees. In other words, the VICP, the DOJ, and HHS have strong financial levers to exert control over petitioners' lawyers and experts. The financial playing field steeply tilts in the government's favor.

3. THE LACK OF ADEQUATE ACCESS TO EXISTING SCIENCE

The most important information about vaccine safety is the taxpayer-funded Vaccine Safety Datalink, the government repository of epidemiological data on the vaccine program.[11] In the Omnibus Autism Proceeding, which aggregated over five thousand claims, petitioners were not granted adequate access to the Vaccine Safety Datalink information about early thimerosal studies while the DOJ presumably had access.[12] Without this data, petitioners' lawyers were at a severe disadvantage in making their argument for a thimerosal-autism link. In a normal court, a judge would have compelled access to such critical information as part of the civil discovery process, in which the parties are forced to share relevant information about causation. Because the VICP was not set up as a court to deal with complex questions of causation—it was set up to review recovery for presumptive injuries—it failed to address the heart of the vaccine-autism question.

4. THE ABSENCE OF ESSENTIAL SCIENCE

In order to prevail in the VICP on a claim of vaccine-induced injury, a petitioner must present a plausible scientific theory, a logical sequence of cause and effect, and a connection in timing between the vaccination and the injury.[13] The burden of proving a vaccine-induced injury is on the petitioner. There are many universally acknowledged gaps in vaccine safety science, which leave petitioners without the ability to definitively prove their cases and, therefore, without compensation, even when there is no credible explanation for what happened to the child except vaccine-induced injury.

5. THE LACK OF TRANSPARENCY AND THE PERCEPTION OF ARBITRARINESS

To protect families' privacy, medical records in the VICP are generally sealed and unavailable. The DOJ may settle cases rather than litigate them and may enter in to stipulations to keep information confidential, as happens in most kinds of civil litigation. The discrepancy between settled cases and dismissed cases leads to an appearance of arbitrariness, with the inability to compare and contrast VICP and DOJ decisions on full, factual records.

For instance, the DOJ appropriately agreed to compensate one child diagnosed with an autism spectrum disorder in the Omnibus Autism Proceeding, based on an individualized expert report.[14] The child's family appears to have received a significant amount of compensation that will cover the enormous cost of caring for the child, who was severely injured.[15] The expert report is sealed. This outcome for the family diametrically contrasts with the case of another child in the Omnibus Autism Proceeding, for which the government produced another individualized expert report from the same expert.[16] The two expert reports appear to have been completely different in the two cases. In the compensated case, the expert's report led to compensation for encephalopathy leading to regressive autism. In the uncompensated case, the expert's report led to a denial of compensation for encephalopathy leading to regressive autism. The lack of transparency in the decision-making process creates an appearance of capriciousness and injustice. It is inexplicable to families with claims and to the public why HHS conceded one case of vaccine-induced autism while aggressively opposing five thousand markedly similar cases of vaccine-induced autism.

Questions concerning arbitrariness in the VICP took on even greater urgency in the wake of the publication, in the peer-reviewed *Pace Environmental Law Review Journal*, of "Unanswered Questions from the Vaccine Injury Compensation Program: A Review of Compensated Cases of Vaccine-Induced Brain Injury."[17] In "Unanswered Questions," the authors reported and analyzed the findings of an empirical investigation examining claims that the VICP had compensated for vaccine-induced encephalopathy and seizure disorder. This study found eighty-three cases of acknowledged vaccine-induced brain damage that included autism. In twenty-one published cases of the Court of Federal Claims, the Court stated that the petitioners had autism or described autism unambiguously. In sixty-two additional cases, the authors identified settlement agreements where HHS compensated children with vaccine-induced brain damage as well as autism or an autism spectrum disorder. Using publicly available information, the investigation showed that the VICP has been compensating cases of vaccine-induced brain damage associated with autism for more than twenty years. This investigation raises the question of whether officials at HHS and the Department of Justice may have been aware of this association but failed to publicly disclose it.

This finding of autism in compensated cases of vaccine injury is significant, as U.S. government spokespersons have been asserting the absence of a vaccine-autism link for more than a decade. Thousands of families allege that their children sustained post-vaccinal injury that resulted in the diagnosis of autism. This paper calls into question the decisions of the Court of Federal Claims in the Omnibus Autism Proceeding in 2009 and 2010 and the statement of Health and Human Services on its website that "HHS has never concluded in any case that autism was caused by vaccination."[18]

The authors of "Unanswered Questions" have called on Congress to thoroughly investigate the VICP, including a medical investigation of compensated claims of vaccine injury. The authors ask Congress to get answers to these critically important unanswered questions.

6. THE LACK OF DISCOVERY

Although discovery—a civil court's ability to require the parties to produce documents, answer questions, and make witnesses available for questioning under oath—is theoretically possible in the VICP, the court is not obligated to make it available. The ability to subpoena documents or witnesses absent a court order is rarely granted. Unlike in normal courts, where civil litigants win and lose on discovery, discovery and subpoena powers are all but absent in the VICP.

7. THE LACK OF ADEQUATE PROCEDURAL SAFEGUARDS

The VICP does not have normal rules of evidence or civil procedure. While Congress created this informality for the petitioners' benefit, the lack of rules now severely hampers petitioners. For example, in the Omnibus Autism Proceeding, the DOJ was able to introduce expert reports without producing the underlying data[19] and information that almost certainly would have been deemed irrelevant and excluded in a normal court proceeding.[20] The VICP's lack of formal rules for civil procedure and evidence harms petitioners. Despite the lack of evidentiary rules, the VICP has begun to impose standards of admissibility applied in normal courts,[21] such as the *Daubert* rule, which governs the admissibility of scientific evidence. Thus, the full burden of producing proof subject to evidentiary rules is being imposed on petitioners who have neither the resources nor the legal mechanisms through discovery to meet these formidable challenges.

8. THE GOVERNMENT'S LACK OF BURDEN TO PROVE CAUSATION

The Court of Federal Claims assumes that once the FDA and the CDC have approved vaccines, they are safe. Therefore, the government is not required to prove the safety of a vaccine—and the petitioner must demonstrate the harmfulness of a vaccine, even though the U.S. Court of Appeals for the Federal Circuit acknowledges this scientific area is "bereft of complete and direct proof of how vaccines affect the human body."[22] Only when the petitioner has made a very strong case that suggests compensation is necessary does the government have to provide an alternate theory of causation. This burden of proof does not correspond with Congress's intention to afford parents the presumption of recovery. Congress never intended that parents would litigate every case; on the contrary, they expected that the Vaccine Injury Table would afford families' presumptions of recovery that would streamline the entire legal process. Congress expected that certain injuries from specific vaccines within certain time frames would lead to almost automatic compensation.

9. THE LACK OF A JURY OF PEERS

The VICP permits no juries. Most families who have had the misfortune of bringing a claim to the VICP believe that more citizen participation would ensure more fairness. While courts have upheld the constitutionality of the VICP to date, many petitioners who have gone before it consider it to be fundamentally unfair. As one parent with a claim on behalf of her child in the Omnibus Autism Proceeding said, "The deck is stacked against families in Vaccine Court. Government attorneys defend a government program, using government-funded science, before government judges."[23]

10. THE INAPPROPRIATELY SHORT STATUTE OF LIMITATIONS

Petitioners have only three years from the time a vaccine-induced injury occurs to file a claim. The Vaccine Injury Table was set up in a time when it was thought that all vaccine-induced injuries happened immediately. This concept of vaccine-induced injury does not reflect many of the disabilities petitioners allege today—autism, seizure disorders, learning disabilities, arthritis—which are conditions that may appear years after vaccination and that have gradual and indistinct onsets. Furthermore, many doctors are unfamiliar with the emerging scientific evidence linking certain disabilities to vaccines.[24] While many lawmakers have proposed a longer statute of limitations, the three-year window remains—and excludes many families who think their children's injuries were due to vaccines. Even if a claimant is not in a position to know that a vaccine may have caused an injury, the three-year limitation will strictly apply.

* * *

In creating the 1986 National Childhood Vaccine Injury Act, Congress struck a balance between diametrically opposed interests. The Act provided a program to compensate victims of vaccine injury that would avoid the perils, delay, and costs of litigation. The compensation program would ensure quick and generous compensation to children injured by vaccines. Justice would be delivered in a manner, so it was thought, superior to the traditional tort system. Vaccine manufacturers would have protection from costly litigation that carried heavy financial risk. In this way vaccine manufacturers would be encouraged to continue to develop and produce vaccines, and the vaccine program, seen as the foundation of public health in the United States, would be stabilized.

A cornerstone of the 1986 Act was preservation of the right of vaccine-injured persons to retain, at their option, access to the traditional legal mechanisms to obtain justice against vaccine manufacturers. Children could obtain compensation through a new streamlined system. If it did not work, they could still sue the company that made the vaccine.

Advocates for injured children fought hard to preserve the traditional legal rights of injured children. They properly considered compensating children—those injured "in the line of duty" of the vaccine program—as the Act's primary purpose. As argued in the amicus curiae (friend of the court) brief filed on behalf of parents' advocacy groups in *Bruesewitz* v. *Wyeth*, the Act created a compensation system as an *alternative*, not a substitute, to the tort system.

Those who fought to preserve the legal rights of children never contemplated that a court could determine that the plain language of the Vaccine Act eliminated a child's right to sue a vaccine manufacturer for producing and distributing a defectively designed vaccine.

Unfortunately, a majority of the members of the Supreme Court of the United States saw it differently. On February 22, 2011, in a 6-2 decision by Justice Scalia, the Supreme Court made a sweeping policy decision in *Bruesewitz* v. *Wyeth LLC*.[25] The Court determined that giving liability protection to vaccine manufacturers for all design defect claims is the paramount purpose of the Act. The Court eliminated an injured child's right to seek compensation from vaccine manufacturers in all but extremely narrow circumstances.

Fortunately, the judicial process allows dissenting justices to articulate the reasons for their disagreement with the opinion of the majority. Justice Sonia Sotomayor's dissent in *Bruesewitz*, in which Justice Ruth Bader Ginsburg joined, stands as a resoundingly articulate rebuttal to the majority's elimination of an injured child's right to obtain justice in a Court of Law. It is redeeming that in Justice Sotomayor's opinion Hannah Bruesewitz and all children found a powerful voice.

Justice Sotomayor described the law that is intended to hold vaccine manufacturers to account:

> Vaccine manufacturers have long been subject to a legal duty, rooted in basic principles of products liability law, to improve the designs of their vaccines in light of advances in science and technology. Until today, that duty was enforceable through a traditional state-law tort action for defective design. In holding that [the Vaccine Act] pre-empts all design defect claims for injuries stemming from vaccines covered under the Act, the Court imposes its own bare policy preference over the considered judgment of Congress. In doing so, the Court excises 13 words from the statutory text, misconstrues the Act's legislative history, and disturbs the careful balance Congress struck between compensating vaccine-injured children and stabilizing the childhood vaccine market. Its decision leaves a regulatory vacuum in which no one ensures that vaccine manufacturers adequately take account of scientific and technological advancements when designing or distributing their products. Because nothing in the text, structure, or legislative history of the Vaccine Act remotely suggests that Congress intended such a result, I respectfully dissent.[26]

The dissent emphasized that the majority had removed the legal mechanism promoting accountability by vaccine manufacturers. The dissent also affirmed the principle for which advocates had fought in 1986—that the "plain text and structure of the Vaccine Act" provides liability protection to vaccine manufacturers "only if it demonstrates that the side effect stemming from the particular vaccine's design is 'unavoidable.'" Thus, if a vaccine manufacturer produced a vaccine that was properly designed but an "unavoidable" part of the design resulted in injury, it could not be held liable.

In the legal world, this made perfect sense—if society deems vaccines so important to public health, then a vaccine manufacturer should not be punished for a defect that it could not avoid in properly manufacturing the vaccine. But the majority went much further in holding, in effect, that *all* vaccines by their very nature are "unavoidably unsafe" if manufactured to government specifications. Designs defects of particular vaccines should not be legally examined, according to the majority. The majority determined that all vaccines are "unavoidably unsafe" despite a clear finding by Congress that "whether a side effect is 'unavoidable' for the purposes of [the Vaccine Act] involves a specific inquiry in each case as to whether the vaccine in the present state of human skill and knowledge cannot be made safe.'"[27]

Thus, the dissent relied on the rational approach contemplated by Congress that required a court to determine "whether a feasible alternative design existed that would have eliminated the adverse side effects of the vaccine without compromising its cost and utility." Justice Sotomayor cited with approval the argument of the renowned Kenneth W. Starr, arguing in an amicus curiae brief siding with Hannah Bruesewitz, "[i]f a particular plaintiff could show that her injury at issue was avoidable ... through the use of a feasible alternative design for a specific vaccine, then she would satisfy the plain language of the statute, because she would have demonstrated that the side effects were not unavoidable."[28]

The dissent demonstrated that Congress unequivocally intended to allow a court to determine whether the vaccine administered to Hannah Bruesewitz was an avoidable danger for which the vaccine manufacturer should have been held accountable. Congress emphasized that "there should be no misunderstanding that the Act undertook to decide as a matter of law whether vaccines were unavoidably unsafe or not," and that "this question is left to the courts to determine in accordance with applicable law."[29]

The dissent criticized the majority's view that the Vaccine Act provides a means for achieving improved vaccine designs while providing compensation for injured individuals. Justice Sotomayor pointed out that the Vaccine Act places no legal duty on vaccine manufacturers to improve the design of vaccines based on scientific advances. This legal function has "traditionally been left to the States through the imposition of damages for design defects." The States' function is especially important in operating to regulate safety in the vaccine industry because "the normal competitive forces that spur innovation and improvements to existing product lines in other markets thus operate with less force in the vaccine market."[30]

The dissent reasoned that there was no statutory mandate "to think that Congress intended in the vaccine context to eliminate the traditional incentive and deterrence functions served by state tort liability in favor of a federal regulatory scheme providing only carrots and no sticks."[31]

Justice Sotomayor's dissent concluded:

> The majority's decision today disturbs that careful balance based on a bare policy preference that it is better "to leave complex epidemiological judgments about vaccine design to the FDA and the National Vaccine Program rather than juries."[32] To be sure, reasonable minds can disagree about the wisdom of having juries weigh the relative costs and benefits of a particular vaccine design. But whatever the merits of the majority's policy preference, the decision to bar all design defect claims against vaccine manufacturers is one that Congress must make, not this Court. By construing [the Vaccine Act] to preempt all design defect claims against vaccine manufacturers for covered vaccines, the majority's decision leaves a regulatory vacuum in which no one—neither the FDA nor any other federal agency, nor state and federal juries—ensures that vaccine manufacturers adequately take account of scientific and technological advancements. This concern is especially acute with respect to vaccines that have already been released and marketed to the public. Manufacturers, given the lack of robust competition in the vaccine market, will often have little or no incentive to improve the designs of vaccines that are already generating significant profit margins. Nothing in the text, structure, or legislative history remotely suggests that Congress intended that result.[33]

Thus, with powerful logic and allegiance to the intent of Congress, Justice Sotomayor spoke for our children. Her words no doubt will endure until such time that their wisdom will carry the day for all injured children. Congress should radically reform the VICP or abolish it. The Act has failed to achieve its primary goals—to compensate for vaccine-induced injuries and to make vaccines safer. The VICP has achieved only its third goal—to insulate industry and medical professionals from liability for vaccine-induced injuries. In that area, the Act has succeeded, at the grave expense of vaccine-injured children.

Chapter Six

GOD, GOVERNMENT, AND PARENTAL RIGHTS

William Wagner, JD

In 1653, an Englishman built a church and on it placed the following inscription:

> In the year when all things sacred were throughout the nation either demolished or profaned, a man founded this place, whose singular praise it is to have done the best things in the worst of times . . .

Like the English church builder, parents find themselves placed on this earth at a time when all things sacred—including the unalienable truth that parents are best equipped to make decisions in the best interest of their children—face destruction or irreverent mocking.

The sacred and legal underpinnings of American parents' unalienable right to direct the upbringing of their children are embedded in deeply rooted divine, natural, and common law traditions. These timeless truths articulate an inviolable objective standard or reference point: Parents, not the state, have responsibility for and authority over decisions concerning the raising of their children—including vaccination choices.

DIVINE, NATURAL, AND COMMON LAW TRADITIONS

Long before the United States was a nation, the ancient *Holy Bible* articulated the authority and responsibility of parents to direct the upbringing of their children.[1] Thus, as a matter of religious conscience, parents made, and today continue to make, medical and other decisions in the best interest of their children. Indeed they view such decisions as a sacred parental responsibility of the highest order. For example, under divine law, God designates parents, not the state, as his agent to direct and control the upbringing of children. Since ancient times, parents exercised this responsibility as a matter of religious conscience. In doing so, they have, either quietly or boldly, declared the ancient sacred tenet: "As for me and my house, we will serve the Lord."[2]

Many of the natural law writings that influenced America's founders reflect those parental rights revealed first in divine law.[3] Moreover, the English philosopher John Locke, in his *Second Treatise of Civil Government*, recognized an important distinction between a parent's right to govern the upbringing of his or her child, a state's political power to govern for the security of society, and a dictator's despotic power to take for self-enrichment. Locke warned of the threat to liberty from the state usurping parental authority.[4] Looking back, his caution seems no less than prophetic.

English jurist Sir William Blackstone's writing convincingly shows that our nation's common law traditions also reflect the principles from natural and divine laws that the authority and responsibility to direct the upbringing of children rest with parents.[5] This unalienable liberty, and the idea that government exists to preserve it, became part of the American tradition.

AMERICAN TRADITIONS: THE DECLARATION OF INDEPENDENCE, THE U.S. CONSTITUTION, AND THE U.S. SUPREME COURT

America's founders enshrined the liberty recognized in sacred divine, natural, and common law traditions in the Declaration of Independence and then in the U.S. Constitution. Drawing upon these traditions, our founders, in 1776, boldly stated,

> We hold these truths to be self-evident, that all . . . are endowed by their Creator with certain unalienable rights, that among these are life, liberty, and the pursuit of happiness.

> That to secure these rights, Governments are instituted . . . deriving their just powers from the consent of the governed . . .

We see the Declaration's promise in the Constitution's structure and language. We, the people, delegated power to the government to secure our freedom, while expressly limiting the government's ability to deprive individual liberty.

Thus it is not surprising that the Supreme Court traditionally afforded parents a fundamental constitutional right to direct decisions concerning the upbringing of their children—both as the right to the free exercise of religious conscience under the First Amendment and as the implied right to liberty protected by the Fourteenth Amendment. For example, in a 1925 Supreme Court decision, *Pierce v. Society of Sisters*,[6] the Supreme Court stated, "The child is not the mere creature of the State; those who nurture him and direct his destiny have the right, coupled with the high duty, to recognize and prepare him for additional obligations." Almost fifty years later, in *Wisconsin v. Yoder*,[7] the Supreme Court upheld that "the primary role of the parents in the upbringing of their children is now established beyond debate as an enduring American tradition."

In formulating laws related to parents and their children, lawmakers often looked to this objective standard as a benchmark, which was reflected in statutory law. Where lawmakers failed, Supreme Court majorities regularly upheld the fundamental liberty, invalidating state action that violated constitutional parental rights and freedom of religious conscience.

We might pause for a moment to note the incongruity of parents looking to a Supreme Court opinion for the authority to raise their children—an authority they already possess, naturally and divinely. For many American parents, it is a self-evident truth that God endows them with the unalienable right to protect and direct the upbringing of their children.[8]

The government, however, increasingly substitutes itself for God, as the source of our liberty. The paradigm shifts under this evolving legal philosophy. To evaluate state action that has an impact on parental decisions, the state replaces self-evident, unalienable standards with its own morally relative, utilitarian assessments. Thus the freedom of conscience and the sanctity behind parents directing the upbringing of their children no longer serve as moral benchmarks against which to measure whether government vaccination laws are right or wrong, good or bad, just or unjust. Instead, parents are told that questions regarding vaccination laws are public policy matters for the government to decide. Moreover, parents should not bother asking to participate in the debate if our view of the world is informed by religious principles since we are told we must only adopt public policy informed by secular dogma—without regard to any sacred, conscientious, or moral considerations.

Beware. When the government eliminates a self-evident moral element from the law, it removes any moral reference point with which to measure whether laws are right or wrong, good or bad, just or unjust.

In this situation, the government inevitably institutes its own brand of utilitarian-based policies at the expense of individual liberty and the freedom of conscience.[9] With nothing to limit those in power from imposing rules that they deemed to be in the best interest of the public, bad things happened—forced sterilization of healthy women, scientific experimentation on African Americans without their consent, and the institution of slavery are just a few examples—when the American government refused to recognize self-evident, unalienable liberty as a limit on its utilitarian policy making. Likewise, all around the globe, the freedom of conscience and parents' unalienable right to raise their children no longer serve as a reliable limit on governmental action. Thus parents increasingly encounter government action—merely for expressing sincerely held religious tenets or otherwise exercising freedom of conscience.

Freedom of conscience is a fragile thing. I have held in my hands the ashes of faithful parents and children who died because of their consciences. As a diplomat in Africa, I also worshiped in a church where hundreds of men, women, and children were slaughtered as they sought sanctuary. In addition, I can tell you about an African member of my team who was brutally tortured for standing up for freedom of conscience and good governance under the rule of law.

After these experiences, I vowed never to remain silent. Therefore I tell you that despite the deeply rooted traditions recognizing parental rights, a profound threat to your sacred liberty to protect your children exists. Indeed, this threat is one of the most perilous ever to work its way through our nation's political and judicial institutions.

Looming ominously is a growing body of evidence that lower courts and even a majority of the current Supreme Court may no longer recognize parental rights as a fundamental liberty protected by the Constitution. In this regard, contemporary definitions of a person's fundamental rights and liberties frequently vary, depending on a judge's interpretation of the Constitution.

A CHALLENGE

Replacing the sacred parental right to responsibly determine a child's medical treatment with dictatorial government mandates inevitably erodes a country's essential foundations. Those who came before us built our constitutional, democratic republic upon the foundations of faith, family, and freedom. That foundation is under attack by those who seek to transform our pluralistic nation (where everyone may freely participate in public policy development) to a secular nation (where everyone, except those who are religious, may participate). Although structural institutions of a free government may stand for a while, the principles for which they stand can eventually cease to exist.

Proponents of state-mandated vaccinations for children attack a sacred standard. The self-evident, inviolable standard that parents are best equipped to make decisions in the best interest of their children should not be subject to the morally relative, utilitarian whims of a shifting governmental regime. The government is supposed to protect parents' freedoms, not seize them.

Like the English church builder in 1623, this book has the potential to build faith, family, and freedom—to do the best things, even in the worst of times—so that parents may carry out their sacred duty. May each one of us rise to the occasion.

Chapter Seven

AN URGENT CALL FOR MORE RESEARCH

Carol Stott, PhD, MSc (Epidemiology), CSci, CPsychol, and Andrew Wakefield, MB, BS, FRCS, FRCPath

Note: The authors would like to thank Louise Kuo Habakus for her contributions to this chapter.

THE NEED FOR SCIENCE

Science lies at the heart of the vaccine controversy, and there can be no substitute for it. The ethical practice of medicine requires full and informed consent. It is not possible to have the *informed* part of informed consent without the science. That vaccine safety science has been and remains inadequate is not in dispute.

The National Academy of Sciences chartered the Institute of Medicine (IOM) in 1970 to serve as an adviser to the federal government on issues affecting public health and to act independently on issues of medical care, research, and education[1] through the analysis and evaluation of evidence-based information. The IOM has played an important role in guiding vaccine policy over the past several decades. Excerpts from a succession of IOM reports over a fifteen-year period illustrate the continued lack of progress in addressing the paucity of research on vaccine safety.

Institute of Medicine Reports on Vaccine Safety

In 1991:

> In the course of its review, the committee encountered many *gaps and limitations in knowledge* bearing directly and indirectly on the safety of vaccines. These include *inadequate understanding* of the biologic mechanisms underlying adverse events following natural infection or immunization, *insufficient or inconsistent information* from case reports and case series, *inadequate size or length of follow-up* of many population-based epidemiologic studies, and *limited capacity of existing surveillance systems* of vaccine injury to provide persuasive evidence of causation. The committee found *few experimental studies* published in relation to the number of epidemiologic [*sic*] studies published. Clearly, *if research capacity and accomplishment in these areas are not improved, future reviews of vaccine safety will be similarly handicapped.*[2] (emphasis added)

In 1994:

> Committee members were struck by the *lack of evidence*. . . . For about two-thirds of the relations evaluated, the committees found either that there was *no evidence* bearing on the question of causality or that the available evidence was *insufficient or inadequate* to make a determination about causality.[3]
>
> Clinical trials for evaluating vaccine efficacy generally *do not include sufficient sample sizes* to permit adequate evaluations of the risk of adverse reactions. Participants expressed dismay that post-marketing studies of vaccines are generally *not randomized controlled trials.* The role of clinical trials for assessing adverse reactions is *limited.*[4]
>
> Discussion arose about the issue of the *lack of unvaccinated controls* in the studies under consideration. Some participants felt that *a true control group* in a study of vaccines and adverse events would consist of never-vaccinated children (whereas now, control groups often consist of children who had not been vaccinated *recently* [*sic*]). *It would then be possible* to look at the frequencies of adverse events and disease in these children in comparison with the frequencies in those who had been vaccinated. This *might be particularly relevant* in studying adverse events with long latencies from the time of vaccination.[5] (emphasis added)

In 1997:

> A number of factors make it *difficult to detect adverse events* associated with the administration of a vaccine: (1) the *need to study* multiple exposures and mul-

tiple outcomes, (2) the lack of unique vaccine-associated syndromes, making it *difficult to establish causality*, (3) the *need for large sample sizes and lack of large computerized immunization databases* with individual level data including vaccine lot number, (4) brief exposure periods for each individual, (5) high vaccination coverage makes unvaccinated individuals highly selected.[6]

More research could be done on potential long-term adverse events from vaccines as well as the potential of vaccines to induce or worsen immune disorders. *Research also could usefully address* such questions as whether age is a factor in the adverse events experienced following vaccination and whether some groups of individuals are more prone to such adverse effects than others.[7] (emphasis added)

In 2001:

Little is known about ethylmercury (the active component in thimerosal) compared to methylmercury.[8]

There are *no data* that elucidate how much, if any, mercury exposure from all sources contributes to the prevalence of autism, ADHD, or speech or language delay.[9]

As noted in previous IOM reports (IOM, 1994 a, b, 2001), a positive ecological correlation constitutes only *weak evidence* of causality, and *additional research would be needed* to establish a causal association.[10]

The available case reports are *uninformative* with respect to causality. There are *no published epidemiological studies* examining the potential association between thimerosal-containing vaccines and neurodevelopmental disorders. The *unpublished and limited* epidemiological studies provide *weak and inconclusive* evidence. . . .[11]

The committee has found *inadequate* evidence to accept or reject a causal relationship between thimerosal-containing vaccines and neurodevelopmental disorders. Although the available evidence is *indirect and incomplete*, and the relationship is not established, it is biologically plausible. Because thimerosal was used in millions of vaccine doses over several decades, it is *important that additional research be done* to understand the nature of the risk, if any, from this exposure to thimerosal.[12] (emphasis added)

In 2002:

> The committee concludes that the epidemiological and clinical *evidence is inadequate* to accept or reject a causal relationship between multiple immunizations and an increased risk of allergic disease, particularly asthma.[13]
>
> The committee was *unable to address* the concern of some that repeated exposure of a susceptible or fragile child to multiple vaccines over the developmental period may also produce atypical or nonspecific immune or nervous system injury that could lead to severe disability or death. Such adverse health outcomes may not be "classical" diseases but variants of diseases . . . there are *no epidemiological studies that address this,* either in terms of exposure or outcome. That is, there is *no study that compares an unvaccinated control group* with children exposed to the complete immunization schedule, *nor are there any studies* that looked at health outcomes other than those classically defined, such as infections, allergy or diabetes. Thus, *the committee recognizes with some discomfort* that this report *addresses only part of the overall set of concerns* of some who are most wary about the safety of childhood vaccines.[14]
>
> Research on the developing human immune system, especially in relation to vaccines, is *limited.*[15] (emphasis added)

As the immunization schedule expanded, the IOM acknowledged a concomitant "dramatic increase in the complexity of immunization safety issues."[16] Although the IOM identified this complexity and these inadequacies and recommended more research over a fifteen-year period, critical questions about vaccine safety remain unanswered.

In addition, funding for vaccine safety has been and remains insufficient, despite urgent, unanswered questions. In 1995, Dr. Robert Chen, chief of Vaccine Safety and Development at the CDC, said,

> The only line item for vaccine safety research is, I think, on the order of a little less than $2 million per year. That basically covers operation of VAERS (Vaccine Adverse Events Reporting System) period, and nothing else.[17]

In 2008, almost fifteen years later, Dr. Louis Cooper, vaccine inventor and a former president of the American Academy of Pediatrics, wrote that the total vaccine safety science research budget was $20 million or 0.5 percent of the $4 billion total vaccine budget for purchase, promotion, and delivery of vaccines.[18] Despite insufficient scientific knowledge and funding for safety research, the compulsory vaccination program continues to expand.

Epidemiology and the Vaccine Safety Discourse

Frequently, those who have expressed concerns about the validity of studies claiming "no association" between vaccines and autism have claimed that, because the majority of these studies are "epidemiological," they cannot be used to establish or refute causality. This is misleading and needs clarification.

Epidemiology is the study of the distribution and determinants of disease in the human population; it is the basic science and fundamental practice of public health.[19] As such, it is a discipline that has a much wider scope and range of methods than is generally acknowledged, certainly with respect to vaccine safety discourse.

Epidemiological studies may be descriptive or analytical, and analytical studies include not only case-control and cohort designs (see the section Epidemiological Primer below) but also randomized clinical trials and animal experimentation. All these types of study fall within the scope of "epidemiology." This is more than a semantic point. Referring to epidemiology as though it were an inappropriate method to evaluate the possible association between vaccines and autism has, at best, meant that many arguments against the validity of the literature have been confused. At worst, it has meant a failure to articulate what has and has not been demonstrated by the data so far. Ultimately, it has resulted in a failure to identify the real gaps in our knowledge.

Epidemiological data can be used (1) to describe the characteristics and distribution of a disease, (2) to describe its possible association with various factors at the population level, (3) to generate hypotheses about the causal nature of observed associations, (4) to test the validity of these hypotheses in observational or experimental studies, (5) to describe the clinical and biological characteristics of an identified disease, (6) to evaluate proposed treatment regimens, and (7) to directly test the effect of exposures in experimental designs.

The studies that are frequently referred to as indicating "no association between vaccines and autism" have, for the most part, been population-based, analytic, observational studies; almost all of them have reported "no association" and almost all of them are flawed. The reason they are flawed, however, is not that they are epidemiological. It is not that epidemiology is a "blunt tool, incapable of testing causal hypotheses," and it is not that there are so few people affected that "epidemiology couldn't be expected to pick them up." To the contrary, the studies are flawed either because they have been badly designed or because they have not been designed with the right hypothesis in mind in the first place. Their aims, design, analytic procedures, and conclusions, singly or together, have been consistently inappropriate and in some instances, plain wrong. If these studies had been designed, conducted, analyzed, and interpreted properly, they would almost certainly have furthered our understanding of the issue and helped to confirm or refute the role of vaccines in causality.

Epidemiology Primer

It will be helpful at this stage to go back to the books—a basic epidemiological text will suffice.[20] Epidemiological studies may be descriptive or analytical. Descriptive epidemiology aims to describe the general characteristics of disease distribution in relation to person, place, and time. Descriptive studies are valuable. They provide information to healthcare providers and those responsible for resource allocation, and they may also be used to generate hypotheses about disease causality, but their design precludes them from being used to test hypotheses.

Descriptive epidemiological studies include the following:

- Ecological (or correlational) studies, in which characteristics of entire populations are used to describe a disease in relation to a factor or exposure; no data are evaluated at the individual-patient level.
- Case reports or case series, which describe the experience of a single patient or a group of patients.
- Cross-sectional surveys, in which diseases and exposures are assessed simultaneously in a well-defined population. These may be considered analytical in one circumstance: when the factor or exposure is static, unchangeable, and consistent over time (e.g., birth weight).

Analytical epidemiology involves using comparative studies to test hypotheses about associations between an exposure and a disease. Analytical studies can be observational or experimental. Observational analytic studies include:

- Prospective cohort studies, in which an exposed group and an unexposed group are followed over time to determine how many people in each group develop the disease of interest.
- Retrospective cohort studies, in which records from exposed and unexposed individuals are examined for evidence of the disorder of interest.

- Case-control studies, in which a group of cases (the "diseased" group) is compared with a group of controls (the "nondiseased" group) in terms of the exposure of interest.

Experimental epidemiological studies include:

- Clinical trials, in which human subjects are allocated to separate groups with different treatment regimens.
- Experimental animal models.

Experimental studies actively manipulate exposures in the studies' designs. Manipulation of potentially harmful exposures has obvious ethical implications and is most often carried out in experimental animal models. It has been argued that these experimental studies must not be done because the manipulation of a vaccine's exposure is unethical—the vaccination would, by necessity, be withheld from one of the groups. However, it's also possible to argue that such a design is unethical because one group would be exposed to an insufficiently tested vaccine—though 80 percent to 95 percent of the population in developed countries is currently exposed to such vaccines. Experimental studies of vaccine safety on humans have been dismissed as not feasible for many reasons.

Given the urgent need to address questions of causation using higher standards of scientific inquiry, more scrutiny and resources to this important question are required. Other designs such as experimental studies using animal models or analytic studies of human populations, in which vaccinated and unvaccinated groups are examined, are appropriate and ethical, but they need to be done well.

2004 IOM Immunization Safety Review Committee Report on Thimerosal and Autism

In view of the challenges of epidemiological research generally and the paucity and poor design construction of the epidemiological research examining the relationship between vaccines and developmental injury specifically, it came as a surprise to many when the eighth and final report of the IOM's Immunization Safety Review Committee was released in May 2004. In drawing its conclusions on the vaccine-autism connection, the IOM chose to rely upon the research at a level that was not supported by the science. Despite severe limitations of the research that the IOM itself had identified, the IOM announced,[21]

> The committee concludes that the body of epidemiological evidence favors rejection of a causal relationship between the MMR vaccine and autism. The committee also concludes that the body of epidemiological evidence favors rejection of a causal relationship between thimerosal-containing vaccines and autism. The committee further finds that potential biological mechanisms for vaccine-induced autism that have been generated to date are theoretical only. (Note: when the IOM uses the term "epidemiological" above, it means "population-based.")

This 2004 IOM study is the basis upon which the government made its decisions to *not* recommend

1. a policy review of the current vaccination schedule,
2. a policy review for administration of the MMR vaccine or of thimerosal-containing vaccines, and
3. further funding for vaccine-related autism research.

A Fear of What the Research Might Show?

Evidence-based medicine[22] (EBM) is the highest standard of medicine. It uses peer-reviewed published scientific research to maximize accurate prediction of outcomes in medical treatment. EBM categorizes different types of clinical evidence and grades them according to the integrity of their results, as defined by minimization of bias and error. While a cornerstone of EBM is the randomized, controlled trial (RCT),[23] the RCT model doesn't lend itself easily to addressing the vaccine-autism issue for reasons given earlier. The challenges do not mean, however, that we have reached the limits of scientific inquiry. To the contrary, there is critical evidence-based research that can and must be done.

In the case of vaccine safety, the evidence base should come from adequate assessment of total health outcomes using designs that are appropriate to address the issue of causality, with clearly and appropriately defined case versus control groups (in case-control designs) or exposed versus unexposed groups (in cohort designs). The problems with the majority of studies aiming to address the issue of causality include the failure to define appropriate case groups and the failure to include potential cases as controls. For cohort studies, poorly defined exposure events, with little consideration of the timing of exposures or the cumulative effect of several exposures over time, have contributed to design flaws. Other problems include poor understanding of the actual hypotheses and retrospective analysis of data gathered in studies not originally designed to address the issue.

The absence of an adequately designed vaccinated versus unvaccinated study, which many have argued to be the ideal for assessing causality, led to a bill in Congress which sought to remedy this situation. Congresswoman Carolyn Maloney (D-NY) introduced H.R. 2832, the Comprehensive Comparative Study of Vaccinated and Unvaccinated Populations Act of 2007. It never became law. She reintroduced the bill in 2009.[24] To date, the forces of political influence have not prevailed to support the passage of this proposed legislation.

As parent advocate and author Mark Blaxill wrote,

> So the obvious research project of comparing the total health outcomes in vaccinated vs. unvaccinated individuals has been rejected not merely as too expensive, [now] it simply must not be done. In the Orwellian logic of the CDC, such studies in humans would be "prospectively unethical" and "retrospectively impossible." Let's be frank here. This is an epistemological obscenity: It's not just that we don't know some very basic things about the safety of the sacred program, we also cannot know and should not seek to know. This stance should offend even the most skeptical scientists.[25]

This is where misunderstandings about what epidemiological studies can and can't do become crucial. The accepted position, promoted by the CDC, is that a study of vaccinated versus unvaccinated individuals can't happen because it would need to be an RCT. Of course, it wouldn't. What is needed is a well-designed prospective cohort study; there are no negative ethical implications of this type of study whatsoever. This does not preclude the importance of clinical studies, which are also of crucial importance, but a well-designed population study that identifies accurately defined exposure groups and compares their clinical and biomedical presentations would be extremely informative. No exposure manipulation is required, so there are no ethical issues to address in that regard. So why are these studies not being carried out? The work of Gayle DeLong, PhD, also has significant resonance in the current debate. In a recent report published in *Accountability in Research,* DeLong (2012)[26] outlines the competing interests that may impede objective study of vaccine side effects, listing a number of financial and bureaucratic reasons why vaccine manufacturers, health officials, and editors of medical journals may have financial and bureaucratic reasons for failing to acknowledge risks associated with vaccine exposure. DeLong points to how this may also extend to advocacy groups having legislative and financial reasons to sponsor research to identify vaccine risk. DeLong identifies a number of potential conflicts of interest and suggests how "bureaucratic restructuring and enforced transparency"[27] could help support a move towards increased trust in vaccine safety research from all sides.

The late Bernadine Healy, MD, a former chief of the National Institutes of Health and member of the IOM, offers some insight. In May 2008, Dr. Healy said to CBS reporter Sharyl Attkisson:

> There is a completely expressed concern that they don't want to pursue a hypothesis because that hypothesis could be damaging to the public health community at large by scaring people. First of all, I think the public's smarter than that. The public values vaccines. But more importantly, I don't think you should ever turn your back on any scientific hypothesis because you're afraid of what it might show.
>
> I think that the public health officials have been too quick to dismiss the [vaccine-autism] hypothesis as irrational.[28]

In response to the claim that there is no vaccine-autism link, Healy responded, "You can't say that." As a member of the IOM, she would know. The IOM's 2001 review of the MMR vaccine and autism stated, "While the committee did not exclude the possibility that MMR vaccine could contribute to ASD 'in a small number of children,' it felt comfortable concluding that the evidence 'favors rejection of a causal relationship at the population level between MMR vaccine and ASD.'"[29] Further, it did not recommend a policy review of the licensure of the MMR vaccine or of the current schedule and recommendation for administration of the MMR vaccine.[30] Again, in 2004, the IOM concluded, "The evidence favors rejection of a causal relationship between thimerosal-containing vaccines and autism."[31]

Parents are not convinced. A recent report estimated that one in four parents believe vaccines cause autism[32] and another estimated that 89 percent choose vaccine safety as their top priority for medical research in children's health.[33] Remarkably, government today remains unable to answer these basic questions about vaccine safety:

- Do vaccines cause autism?
- How do the total health outcomes of those who are vaccinated compare with those who are unvaccinated?
- Is the CDC's recommended vaccination schedule safe?

In the absence of evidence-based research, U.S. vaccine policy rests on a theory, not on empirical evidence.

Most extraordinarily, in the absence of adequate science, doctors are blaming parents. The American Academy of Pediatrics (AAP), the professional association of U.S. pediatricians, issued a press release in the fall of 2010 stating, "Deadly diseases are making a comeback," and parents who choose not to vaccinate their children are "at the heart of all these outbreaks."[34] The AAP is promoting the message of drug industry representatives and vaccine inventors, that "the antivaccine movement threatens America's children."

Vaccines Can Cause Harm

The government and the IOM clearly state that vaccines can cause harm. The IOM states, "No vaccine is perfectly safe or effective, and vaccines may lead to serious adverse effects in some instances."[35] Its committee concluded that the evidence either favors or establishes causation between vaccines and adverse events (see table).[36] The government's Vaccine Injury Table codifies this information.

Vaccines	Adverse Events
DPT vaccine	Encephalopathy, shock, and anaphylaxis
Rubella vaccine	Chronic and acute arthritis
DT vaccine	Guillain-Barré syndrome and brachial neuritis
Measles vaccine	Anaphylaxis and death from a vaccine-strain viral infection
Oral polio vaccine	Guillain-Barré syndrome, polio, and death from a vaccine-strain viral infection
Hib vaccine	Susceptibility to Hib disease
MMR vaccine	Thrombocytopenia, anaphylaxis
Hepatitis B vaccine	Anaphylaxis

The Vaccine Injury Table[37] lists the injuries and conditions that are presumed to be caused by certain vaccines along with the time periods in which the first symptoms must occur after vaccination, and these criteria must be met in order to receive compensation. It further indicates that an acute complication or sequela of any of the listed injuries, including death, is compensable. The government acknowledges that other vaccine-induced injuries may occur, but the onus is then on the vaccinee to prove that the vaccine caused the condition.

Science That Supports Previously Unrecognized Vaccine Risk and Injury

There is emerging scientific evidence that vaccination may be causing more widespread harm than is generally recognized by the medical community. Recent peer-reviewed studies, published within the past three years, suggest the possibility that vaccines are causing a greater degree of neurodevelopmental and immune disorders than is generally appreciated (see appendix to this chapter). Peer review is the process by which manuscripts submitted for publication or abstracts submitted for presentation at scientific meetings are judged for scientific and technical merit by other scientists in the same field.[38] As such, the peer-review process confers acceptability from peers and is assumed to be the arbiter of fairness and objectivity. Despite its peer-reviewed status, this information tends to be ignored by those promoting vaccines.

1. **Primate studies evaluating the impact of a vaccination schedule based on the complete 1994 to 1999 U.S. vaccination schedule**

As Kennedy et al.[39] discuss, studies that evaluate the safety of vaccines in the human population should use the species most closely related to humans. While the great apes (chimpanzees, orangutans, gorillas, and gibbons) represent our closest "relatives," they are now endangered species. The Old World monkeys (macaques, baboons, mandrills, and mangabeys) are now frequently used in preclinical primate research. By studying primate outcomes, researchers are able to infer human exposure risks.

Recently, Laura Hewitson and colleagues reported on a pilot study in which male infant rhesus macaque monkeys who received the complete 1994 to 1999 U.S. vaccination schedule were compared to an unexposed control group on a number of measures. A series of analyses were undertaken as part of the study and these have generated two reports to date.

In the first report, "Delayed acquisition of neonatal reflexes in newborn primates receiving a thimerosal-containing hepatitis B vaccine: Influence of gestational age and birth weight,"[40] the authors report on data collected after a single birth dose of the hepatitis B vaccine, containing the mercury-based preservative thimerosal. Thirteen vaccinated animals received a weight-adjusted amount of thimerosal containing 2 micrograms of ethylmercury (the dose human infants receive contains 12.5 micrograms). Seven unexposed animals received either a saline placebo injection or no shot at all. Compared to the controlled animals, the exposed animals exhibited significant delays in the development of major survival reflexes—root, snout, and suck. The exposed animals took more than twice as long as the control group to acquire these reflexes, which are typically used to measure infants' brain development and are vital to primate infants' survival in the wild.

The study did not attempt to separate out the effects of the vaccine as a whole, of thimerosal specifically, or of a combination of the two. However, the researchers established that primate infants of lower birth weight and gestational age were at a greater risk. These findings, when combined with the findings of the hepatitis B study by Gallagher and Goodman (see below), raise serious questions about the wisdom of neonatal vaccination and demand an urgent call for more research.

The second paper, "Influence of pediatric vaccines on amygdala growth and opioid ligand binding in rhesus macaque infants: A pilot study,"[41] reports that, after controlling in the statistical analysis for the volume of the left amygdala, and after correcting for multiple comparisons in the longitudinal design, specific opioid-antagonist [11C] diprenorphine (DPN) binding in animals recently exposed to the MMR vaccine remained relatively constant following exposure; however, a statistically significant decrease in [11C] DPN binding in unexposed animals occurred at that time. There was also evidence of a greater total brain volume in the exposed animals following earlier exposure to various other vaccines.

The results, which require confirmation in a larger design, suggest that maturational changes in amygdala volume and the binding capacity of [11C] DPN in the amygdala were significantly altered in infant macaques receiving the vaccination schedule, and therefore, exposed animals did not follow the same neurotypical trajectory as controlled (unvaccinated) animals whose amygdala volumes decreased over time.

2. Hepatitis B vaccine and developmental disability

In a 2008 study, "Hepatitis B triple series vaccine and developmental disability in U.S. children aged 1-9 years,"[42] Gallagher and Goodman found evidence to suggest that boys in the United States who were vaccinated with the thimerosal-containing triple series hepatitis B vaccine were more susceptible to developmental disabilities than unvaccinated boys. (No increased risk was identified in girls.) The fully vaccinated boys' odds of requiring special education services were more than twice as high as the unvaccinated boys' odds. After adjusting the findings for confounding variables, the vaccinated boys' odds of receiving special education services were 8.63 times greater than the unvaccinated boys' odds.

The authors analyzed 1,824 children between ages one and nine who participated in the National Health and Nutrition Examination Survey (NHANES) in 1999 to 2000. They asked if the children received the triple series hepatitis B vaccine and if the children received special education or early intervention services. While the birth dosage was not specifically queried, the CDC's birth dose recommendation was in place during the entire study period.

The same authors have since published their evaluation of the association between the hepatitis B vaccination (HBV) of male neonates and parental reports of autism diagnosis among boys aged three to seventeen who were born before 1999.[43] They found that "HBV in U.S. male neonates born before 1999 was positively associated with almost 3-fold greater odds for autism."

3. DPT vaccine and asthma

In an epidemiological cohort study sponsored by the Canadian Institutes of Health, "Delay in diphtheria, pertussis, tetanus vaccination is associated with a reduced risk of childhood asthma,"[44] McDonald et al. applied a retrospective longitudinal design using vaccinated versus unvaccinated children to assess the impact of different time schedules for vaccinations and their associations with asthma. A cohort of nearly 14,000 children born in Manitoba, Canada, in 1995, for whom complete vaccination and healthcare records were available, was identified and evaluated. Among the 11,531 children who received at least four doses of DPT, the risk of asthma at age seven was reduced by 50 percent in children whose first DPT immunizations were delayed by more than two months. The risk was further reduced if the first three doses of DPT were delayed.

According to the authors, the study generates some interesting hypotheses for the biological mechanisms behind early childhood vaccinations and the development of asthma. They caution that it is premature to make recommendations until these findings can be confirmed with the newer DTaP combination vaccine, which contains the acellular pertussis vaccine. DPT vaccines containing the whole-cell pertussis vaccine are no longer used in many countries because the DTaP vaccine has a lower incidence of adverse events. However, the incidence of asthma in Canada, as elsewhere, continues to rise, suggesting the introduction of the DTaP has no mitigating effect.[45]

The study points out that vaccination schedules vary by country. Scientists may glean potential insights from varying disease prevalence in different countries. For example, Japan recommends three doses of the DTaP vaccine six to nine months after birth, but the first dose can be given no earlier than three months after birth. Between 1975 and 1988, Japan did not vaccinate children under two years of age. In 1982, Japan's asthma-prevalence rate in children was about 3.2 percent and by 2002 it was 6.5 percent. While Japan's asthma rate doubled over twenty years, the rate is significantly lower than the rates in North America. The authors also noted that children who received the pertussis vaccine after age two had higher rates of pertussis than children who received the vaccine when they were younger.

4. Meta-Analyses

A number of reviews also suggest that vaccines possibly fill a role in explaining the increased numbers of people diagnosed with autism.

In a recently published paper, "Sorting out the spinning of autism: heavy metals and the question of incidence,"[46] DeSoto and Hitlan reviewed empirical research and carried out a meta-analysis to address autism incidence and its possible association with an exposure to environmental toxins. They conclude that the weight of relevant research has strongly shifted, and "convergent evidence" now supports a link between exposure to environmental toxins and a serious increase in the population-frequency of autism, which is occurring in numerous countries. Their analysis concludes that:

- Independent labs have found that autism rates in school districts are predicted by the distance from and exposure to toxic emissions within states.[47]
- Exposure to toxins during pregnancy or early infancy is a factor used to predict later ASD symptoms.[48]
- Detectable levels of mercury in blood, present in 8 percent of American women, are demonstrated to cause specific damage to developing brain cells.[49]
- Levels of mercury and other heavy metals in the blood of a human fetus may be up to 70 percent higher than the mother's levels.[50]

A PubMed search of "autism AND heavy metals OR autism AND mercury" yielded 163 articles; 58 were research articles with empirical data relevant to the hypothesis of a link between autism and one or more toxic, heavy metals. Of these articles, fifteen were offered as evidence against a link between exposure to heavy metals and autism, and forty-three were offered as evidence supporting a link between exposure to heavy metals and autism.

The authors review the acknowledged errors in studies that were previously used to refute this hypothesis; these errors included lacking an objective confirmation of diagnosis (i.e., relying upon clinical judgment), treating a continuous variable as dichotomous, defining nonzero numbers as zero, and employing methods that resulted in having 95 percent of samples returned by the lab because they were too low to detect. In their reanalysis, DeSoto and Hitlan found that the data "support the contention that those with autism had higher levels of heavy metals." They urge professionals to refer to the original studies in order to clearly understand the methods and results before weighing the evidence.

In a paper titled, "Timing of increased autistic disorder cumulative incidence," McDonald and Paul investigate the timing of the increased incidence of autism.[51] The authors, who work for the U.S. Environmental Protection Agency, suggest that pinpointing the "changepoint" year would help to narrow the focus in the search for possibly causal environmental factors, within and among countries. They examined the cumulative incidence (sum of all cases for a specific birth cohort up to a given age) of autistic disorder (AD) based on published data from different locations around the world. Three studies met the criteria: one each from Japan, Denmark, and the United States (California). The changepoint years for each country were consistent and "surprising." The data strongly suggest a major causal environmental change that affected children born in those locations in the late 1980s. The authors point out that the changepoint data are also consistent with data on AD incidence in children born after 1987 in Minnesota, in children born in the mid- to late-1980s in Sweden, and in children born between 1987 and 1992 across the United States. The findings are clearly indicative of a major and relatively circumscribed causal environmental change affecting children born in the late 1980s.[52]

Vaccinations represent both novel and increasing exposures that occur while children are still in utero (i.e., they are given to pregnant women) and then are directly administered to children starting on the day of birth. In debating the role of vaccines in these autism trends, the following events are notable:

- Denmark introduced the MMR vaccine in 1987.
- Japan introduced the MMR vaccine in 1989.

- MMR vaccination recommendations in the United States changed in the 1980s: those who received the vaccine at twelve months or younger were revaccinated at fifteen months; a second vaccination, to be administered by the age of four, was introduced in 1991; and the current recommendation to vaccinate at twelve months was introduced.
- In 1990, vaccine manufacturer Merck quadrupled the amount of the mumps virus included in the U.S. MMR vaccine; it was increased from 5,000 to 20,000 units. The viruses contained in the MMR vaccine interfere with each other and may increase the risk of adverse reactions. The U.S. birth cohort of 1989 was the first to receive this new vaccine.
- Among many changes and additions to the vaccination schedule, the hepatitis B vaccine, containing the mercury-based preservative thimerosal, became part of the recommended schedule in 1991.

Due to the significant economic and societal costs of autism, McDonald and Paul stress the importance of determining whether a preventable exposure to an environmental factor may be associated with the increase in AD incidence. Such environmental factors would need to disrupt early human neural development. The authors conclude that future studies should focus on novel or increasing exposures, occurring from in utero to at least three years of age in the birth cohorts of 1988 and 1989. Vaccines surely cannot be ruled out. This study raises important questions and emphasizes the need for more science, including vaccine science.

These studies do not prove a causal association between vaccines and autism, any more than those cited by the IOM and CDC prove that there is no causal association between them. What they do indicate is that the true status of vaccine safety science has been systematically misrepresented by those whose priority is not vaccine safety. They also suggest the possibility that vaccines fill a substantial, causal role in the autism epidemic. One might rationally think that these hypotheses would drive substantial funding into vaccine-autism research. Yet the opposite appears to be the case (see above).

Statistical Manipulation in Whose Interest?

The lengths to which epidemic denialists are prepared to go to disguise the facts are astonishing. Due to concerns about a possible autism cluster in Brick Township, New Jersey, in 1997, the CDC investigated the cases and possible envi-

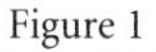

Figure 1

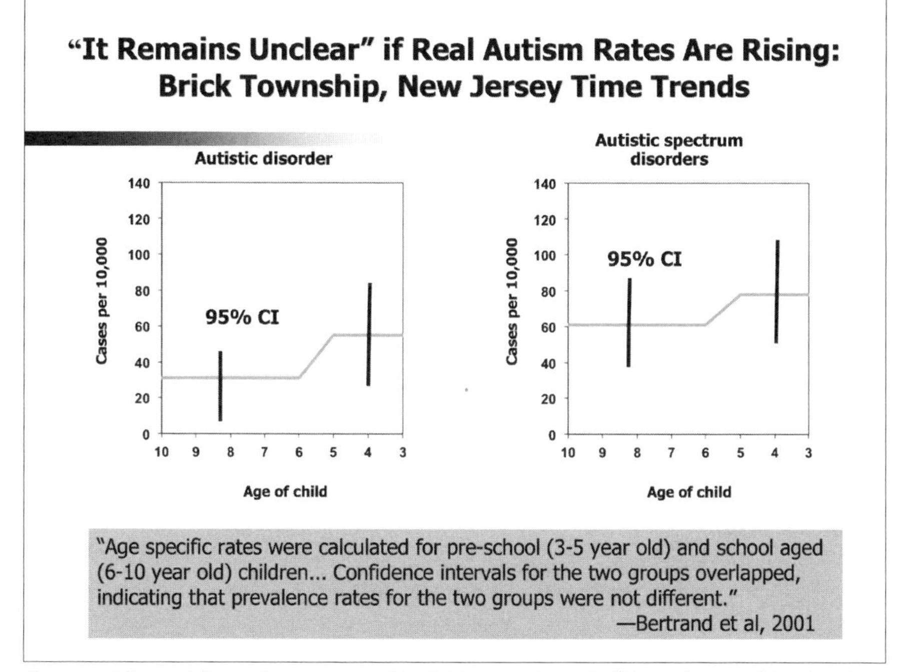

Source: Adapted from data reported in J. Bertrand et al., "Prevalence of Autism in a United States Population: The Brick Township, New Jersey investigation," Pediatrics (2001) 108:1155-61; and from data provided by the authors to Sallie Bernard of SafeMinds.

ronmental causes, and determined that there was "no cluster" and no increasing trend of autism rates in Brick.[53]

After finding all of the autism cases in Brick, the CDC scientists separated them into two groups. Age-specific rates of autism and autistic spectrum disorder were calculated for preschool children (aged three to five) and elementary school children (aged six to ten). When the rates for each group were compared, the confidence intervals overlapped, indicating that the numbers were not significantly different between age groups, and that there was no evidence of an increasing rate (see figure). In addition, no environmental factors had been identified that could account for the apparent autism cluster in Brick. Yet, the CDC failed to identify any concerns and instead chose to conclude its investigation.

Perplexed by the reported findings, parent researchers Mark Blaxill and Sallie Bernard obtained and reanalyzed the original data that the CDC collected regarding cases of full-syndrome autism. Instead of dividing the cases into two unequal and arbitrary groups, they plotted the data for autism prevalence by

Figure 2

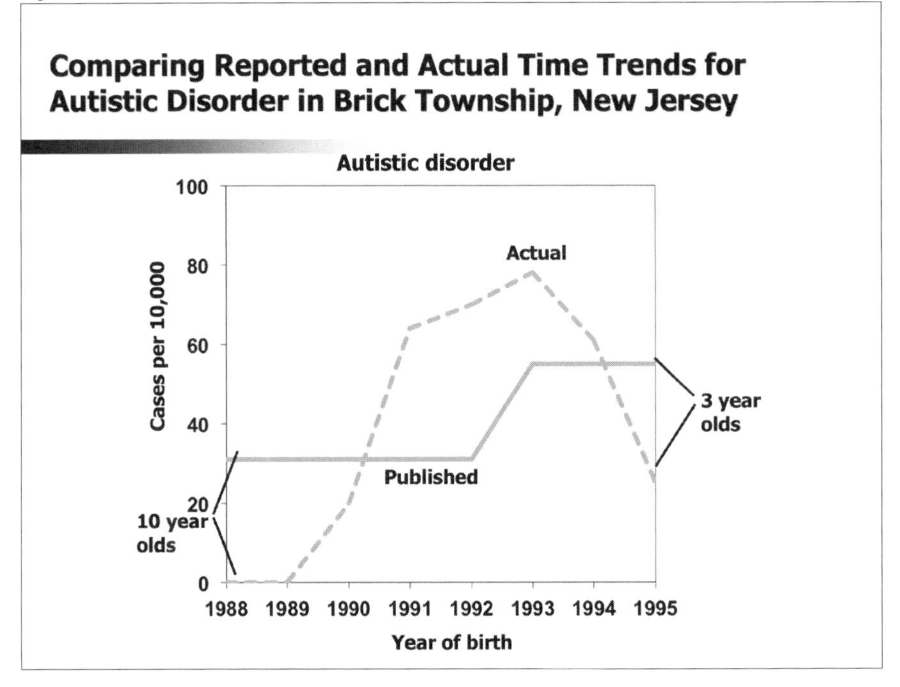

Source: Adapted from data reported in J. Bertrand et al., "Prevalence of Autism in a United States Population: The Brick Township, New Jersey Investigation," Pediatrics (2001) 108:1155-61; and from data provided by the authors to Sallie Bernard of SafeMinds.

year of birth. The findings were dramatically different (see figure 2). The CDC's average rate for preschool children was determined by a range that included zero cases in the oldest children to almost one in one hundred cases in children born only three years later! The CDC's average rate for school-age children was determined by a range in which the highest number approached one in one hundred, and low numbers were for three-year-olds, who were not yet diagnosed and therefore should not have been included in the analysis.

The simple fact is that Brick Township's autism rates went from zero to nearly one in one hundred in a few short years, with a highly significant increasing trend. The CDC's actions can only be interpreted as a deliberate effort to obfuscate this fact. Had they presented an honest analysis of the data, the source of the autism epidemic might have been identified and dealt with by now.

The mainstream press routinely draws conclusions about vaccine safety in the absence of sufficient evidence. For example, we are told that:

1. The removal of thimerosal from pediatric vaccines did not result in a reduction in the population frequency of autism. Therefore thimerosal-containing vaccines do not cause autism.
2. Ten infants died in a pertussis outbreak in California.[54] A CNN article quotes a health official who says that pertussis is vaccine-preventable and criticizes parents for choosing not to vaccinate their children. The health official concludes that everyone who comes into contact with diseases needs to be vaccinated to protect newborns.

In both of these cases, there is simply not enough information to support the conclusions.

Regarding the first example, at the same time that thimerosal was removed from most pediatric vaccines, the number of vaccines given to pregnant mothers and children—some containing thimerosal—dramatically increased. Thus the cumulative burden of these vaccines' other exposures, such as the aluminum salts used as vaccine adjuvants, increased as thimerosal exposures declined. We cannot separate out the impacts of these other, potentially "confounding" factors.

Regarding the second example, it is not appropriate to tie disease outbreaks to reduced vaccination rates based on ecological data (an earlier section of this chapter, "Epidemiology Primer," explains why ecological studies can't be used to address causality). There is no publicly-available evidence that these children did, in fact, have pertussis. The willingness of public health authorities, on one hand, to freely bandy such numbers about and, on the other hand, to dismiss possible adverse reactions to vaccines as "coincidence," is lamentable. Within a large population that may be experiencing a change in vaccination compliance, we cannot assume that the deaths were actually related to pertussis; and even if they were, we cannot assume that the cases were all caused by unvaccinated and infected individuals. Ecological data cannot be used to indicate causality because it is not evaluated at the level of individual exposures and outcomes and is, therefore, subject to misinterpretation (this is known as the ecological fallacy). This is an error of inference, in which an association observed at the population level is assumed to exist among individuals. A hypothetical example of this fallacy would be the assertion that drunk driving is safe because fewer road accidents occur in areas where alcohol consumption is high.

CONCLUSION

This chapter highlights the well-recognized gaps in vaccine safety that the Institute of Medicine has acknowledged for over fifteen years. But it also points to the growing body of peer-reviewed science that suggests previously unrecognized vaccine risks and injuries. The Hewitson primate studies, the Goodman and Gallagher hepatitis B studies, the DeSoto meta-analysis on mercury and autism, and the EPA "changepoint" year in autism prevalence study all support the great need for more research.

Center for Personal Rights (CPR) strongly supports urgent funding for evidence-based vaccine safety science. These studies should be funded by the government but conducted by independent researchers who have no stake in a particular outcome. The studies cannot be conducted by agencies that promote vaccination and have a mission to ensure public confidence in vaccines.

The government's failure to fund the basic scientific studies of these medical products while simultaneously recommending or mandating them raises profound ethical questions. CPR believes this underscores the imperative for vaccination choice.

NEW PEER-REVIEWED STUDIES SINCE 2010

Since the first edition went to press in December 2010, new research raising grave concerns about vaccine safety continues to accumulate. Direct evidence of a potential association between vaccine exposure and ASD prevalence is reported in the *Journal of Inorganic Biochemistry* by Tomljenovic and Shaw (2011).[55] The authors present data indicating that (a) children from countries with the highest ASD prevalence appear to have the highest exposure to aluminum adjuvants in vaccines;[56] (b) an increase in exposure to aluminum adjuvants correlates significantly with the increase in ASD prevalence in the United States ($r=0.92$; $p < 0.0001$); and (c) there is a significant correlation between the amount of aluminum administered to preschool children via vaccination and current ASD prevalence in seven Western countries—particularly at three to four months of age ($r= 0.89–0.94$; $p = 0.002$). On the basis of these data, the authors call for a more rigorous evaluation of aluminum adjuvant safety for all ages at the population level.

Additional evidence of potential harm from vaccine exposure comes from a report on the association of infant mortality rates (IMR) and the number of routine vaccine doses administered.[57] Miller and Goldman point to the IMR as an important indicator of socio-economic well-being and public health conditions. They call for a closer inspection of their observations, whereby linear regression analyses of unweighted mean IMRs generated a statistically significant correlation between increasing number of vaccine doses and increasing mortality rates ($r=0.99$, $p=0.0009$).

The substantial contribution of environmental factors to the observed rise in ASD diagnoses is offered by Stanford University authors in the *Archives of General Psychiatry.*[58] In providing rigorous quantitative estimates of the genetic heritability and shared environmental effects on autism in twins, the authors conclude that while ASD has a moderate genetic heritability, the contribution made by shared environmental components is substantial and much higher than previously estimated. The particular significance of this finding lies not in the possibility of previous underestimates, but in the implications of a change in heritability concordance between identical and nonidentical twin pairs over time. As Lathe (2006)[59] points out, a real decline in heritability (versus an apparent decline due to previous overestimation), is indicative of an increasing environmental contribution. That this has taken place during a period of increasing exposure to environmental toxins is of significant interest.

Schieve et al. (2012)[60] found an association between parental nativity and autism spectrum disorder. The study compared ASD prevalence among U.S.-born non-Hispanic white versus Hispanic children and identified a substantially lower prevalence for Hispanic children with both parents foreign-born compared to Hispanic children with both parents U.S.-born. The possible differences between these groups with regard to early vaccine exposure—with children of foreign-born parents being likely to have lower levels of exposure—is worthy of further investigation.

In evaluating a possible association between vaccine exposure and prevalence of autism and speech/language impairment (SLI) in each U.S. state between 2001 and 2007, DeLong (2011)[61] used regression analysis and controlled for family income and ethnicity. She observed a positive and statistically significant relationship whereby the higher the proportion of children receiving the recommended vaccination schedule, the higher the prevalence of an autism or SLI diagnosis. Specifically, a 1 percent increase in the proportion exposed was associated with an additional 1.7 percent, or 680 children, receiving a diagnosis of autism or SLI. The authors also report that differences in parental behavior or access to care did not affect the results, nor were vaccine exposures significantly associated with any other disability or with the number of pediatricians practicing in the state. DeLong concludes that the data indicate a possible role for vaccines in the observed rise in prevalence of developmental disorders, despite the removal of thimerosal from recommended schedule, and calls for more research.

Together, these data provide further support for Hertz-Picciotto's and Delwiche's (2009)[62] observation that the rise in autism cumulative incidence (from 6.2 per 10,000 for 1990 births to 42.5 per 10,000 for 2001 births) cannot be accounted for by potential artifacts, such as younger age at diagnosis, differential migration, changes in diagnostic criteria, or inclusion of milder cases.

These observations serve to underline the likely environmental underpinnings of the Centers for Disease Control and Prevention (CDC) Morbidity and Mortality Weekly Report (MMWR) of 1 in 88 children (and 1 in 54 boys) with an autism spectrum disorder (CDC, 2012).[63] According to the Autism and Developmental Disabilities Monitoring Network (ADDM), autism incidence in New Jersey is an astonishing 1 in 49 children (and 1 in 29 boys). New Jersey, the eleventh most populous state in the country, is among the most polluted.[64] New Jersey also mandates the most doses of vaccines as a condition of daycare and school admission.[65]

Chapter Eight

A LONG AMERICAN TRADITION

Robert Johnston, PhD

In this chapter, I hope to offer some important highlights of American history that connect to what is happening right now. Above all, I want to argue that the current movement for vaccine safety and liberty is deeply connected to powerful movements in America's past that have been fueled with intense skepticism about coercive vaccination mandates. Vaccine safety advocates have contributed forcefully and meaningfully to our country's many vigorous debates about personal liberty, medical expertise, and the political roles of ordinary citizens in a healthy democracy.[1]

Citizens who question the established vaccination laws are frequently labeled dangerous and deluded. This, unfortunately, is a long American tradition. Another American tradition, however, is an outpouring of activism from those who have questioned the wisdom of mandating vaccinations.

The story begins as early as 1721, when working class and lower middle class Bostonians protested against the scientific and religious elite's introduction of smallpox inoculation to their city.[2]

Controversies over vaccination went on to animate much of the nineteenth century. Some of the most vulnerable communities in the country took the lead in protesting vaccination mandates. Free blacks, who were profoundly connected to the lack of bodily autonomy that they or their relatives had experienced as slaves, consistently resisted vaccination. Indeed, one of our most important

American heroes, Frederick Douglass, declared himself an implacable opponent of compulsory vaccination.[3]

Other opponents included immigrants who had recently arrived in America from tyrannical countries in Europe and Asia. When German, Polish, or Chinese people came to the United States, they expected personal liberty. When the vaccinators came to their doorsteps, they sometimes quietly resisted, and sometimes noisily rebelled. Compulsory vaccination inspired a major riot in Milwaukee, for example, as immigrants poured out into the streets to fight off coercive and paternalistic public health authorities. They stopped their rebellion only when they had won major concessions from medical officials.[4]

However, it was not just the poor and most oppressed who resisted compulsory vaccination. Increasingly in the late nineteenth and early twentieth centuries, middle-class citizens also made claims for full medical freedom. Some were concerned about the dangers of vaccination, others were insistent on the wisdom of parents in medical matters, others believed in natural healing, and all refused to believe that, in the land of liberty, the government should have the power to compel a controversial and potentially dangerous medical intervention.[5]

All this should sound quite familiar as these themes continue to play out more than a century later.

The result of all this activism of ordinary citizens is that by the 1930s, only nine states (plus the District of Columbia) had laws on the books that mandated vaccination. Moreover, several states that at one time had mandated vaccination, such as California, Washington, and Arizona, actually repealed their laws as the result of popular opposition.

We can now see that such citizen activists were not only remarkably successful but they were also remarkably insightful. For example, they had quite legitimate concerns about the safety of smallpox vaccination—the main vaccination of that era. A century ago, public health officials would not give these safety concerns the time of day. However, as the result of Dick Cheney's post–September 11 attempt to launch a massive antibioterrorism smallpox vaccination campaign, even mainstream newspapers, public health officials, and millions of medical workers came to recognize the significant dangers of the quite crude smallpox vaccine.

There were other important ways that vaccine skeptics of a century ago were correct. Today, more parents and citizens are advocating for personal rights. Again, we should take inspiration from activists of long ago—in this case from my favorite vaccine skeptic, Lora C. Little. After losing her son Kenneth to what she believed to be the side effects of vaccination, Little became a tireless—and quite colorful—worker against the vaccine establishment.

Little's most heroic moment came in 1913, when she got a referendum on the ballot in Oregon. This referendum was designed to overturn a law the legislature had just passed, mandating compulsory eugenic sterilization for, among other things, "habitual criminals, moral degenerates and sexual perverts." Little was able to get the majority of Oregonians to support her in her crusade against the evils of eugenics—the idea of breeding a better "race" that was disturbingly popular in early twentieth-century America and that ultimately helped inspire the Nazis.[6]

This proud history of activism continued throughout the twentieth century, although often in a more subdued way. Many organizations have come and gone, including Little's American Medical Liberty League, based in Chicago during the 1910s and 1920s.

After a period of relative slumbering, popular skepticism about vaccinations reignited in the 1970s and 1980s, inspired by, among others, Chicago pediatrician and self-professed "medical heretic" Robert Mendelsohn, *Mothering* magazine, and the 1982 formation of Dissatisfied Parents Together (DPT). This is a history about which we should be equally proud.[7]

In the end, activists, citizens, and parents should recognize that they are part of a grand American tradition. That tradition transcends left and right, embracing all who care about personal liberty. That tradition shows that Americans have long refused to forget their concerns about democracy, even when it comes to scientific matters, such as medicine, that "the experts" wish to control. And that tradition shows that dissidents are often the wise ones who use their activism to force our scientific and political leaders to recognize their blind spots as well as their need for humility when it comes to doing the will of the people.

Chapter Nine

MEDICAL ETHICS AND CONTEMPORARY MEDICINE

Vera Hassner Sharav, MLS

The cornerstone of medical ethics is the physician's ancient Hippocratic oath: "I will prescribe regimens for the good of my patients according to my ability and my judgment and never do harm to anyone." Often stated as "first, do no harm," this personal and professional commitment to adhere to the precautionary principle of medicine, which puts the patient's interest first, is the basis for our trust in medical doctors.

History has demonstrated that any movement to transform medical ethics from the humanitarian Hippocratic tradition to collective, population-based, utilitarian ethics—which justify compromising the interest of the individual for the so-called "greater good of society"—is the first step down a slippery slope whereby civilized medicine has all too often been derailed, by engaging in unconscionable practices that we, or any community, should never be willing to accept. As Leo Alexander, MD, chief U.S. medical consultant at the Nuremberg Trials who helped formulate the Nuremberg Code, warned in 1949, "From small beginnings the values of an entire society may be subverted and led to the horrors of (active) euthanasia, gruesome and unscientific medical experimentation, and ultimately, death in government clinics, and later—in concentration camps."[1]

Dr. Alexander pointed out that well before the Nazis came to power in Germany, a cultural change under the Weimar government "had already paved the way for the adoption of a utilitarian, Hegelian point of view, with sterilization and euthanasia of persons with chronic mental illnesses being discussed by Bavarian psychiatrists as early as 1931 and extermination of the physically or socially unfit openly accepted and discussed in official German medical journals by 1936."[2] Years before genocide was an official policy, prominent German doctors became willing collaborators in the subversion of medicine from a healing profession to a mechanism of murder, by rationalizing that their actions were for the "good of German society" and to improve "the health of the German nation."

As early as 1920, two prominent German scholars, Karl Binding (a lawyer) and Alfred Hoche (a psychiatrist), co-wrote an influential book, *Permission for the Destruction of Life Unworthy of Life.* The authors reinforced the ideology of social Darwinism and eugenics—which called for population control to prevent the "degeneration" of white racial stock. The book provided the rationale for medical killing that the Nazis later used to justify their "systematic secret execution" of the disabled. The authors dismissed the Hippocratic oath as a vestige of "ancient times" and, following the eugenics hierarchy of human categories, they deemed persons with terminal illnesses and mental disabilities as "unworthy" of life and therefore unworthy of scarce medical and fiscal resources. They discarded medicine's moral commitment to "do no harm," to preserve life, and to prescribe treatments for the good of the individual patient. Instead, they encouraged physicians, who must weigh benefits and risks, to protect "higher values" over the needs of the individual patient. The health of the state, they insisted, is of "a higher civil morality."[3]

INFANTICIDE: THE FIRST LEGAL KILLINGS

The first victims of utilitarian medicine—whose lives were deemed "unworthy of life"—were disabled German infants and children under the age of three, who were murdered by doctors in the service of the state. The murder of these infants required the willing cooperation of doctors and midwives who reported every birth of a child with disabilities to the authorities.[4] A directive issued by Hitler in late October 1939, and designated as a state secret, was to begin the T-4 euthanasia program;5 the three-year age limit for medical murder was soon broadened to include children (ages three to seventeen) whose behavior was deemed abnormal or antisocial, and eventually, to include mentally ill adults. The killing took place at hospitals that were converted to killing centers, where victims were gassed when they entered the "showers." An estimated 275,000 to 350,000 mentally disabled children and adults were murdered[6] with Zylcon B, the gas later used to systematically exterminate six million Jews, the principal target of Nazi hate propaganda.

As Leo Kass, MD, a prominent American ethicist, stated, "It is moral myopia to think that all values must yield to the goals of better health and desirable traits. A cost of such yielding can be the reduction of human beings to the status of just another man-made thing."[7] American medicine has its own dark history of morally indefensible public health policies, ostensibly adopted "for the greater good."

STERILIZED IN THE NAME OF PUBLIC HEALTH

Compulsory surgical sterilization, now recognized as a crime against humanity,[8] was a cornerstone of American eugenics—a sociopolitical racist movement with scientific pretention that was dubbed the "science of improvement." It included state-mandated sterilization of persons deemed to be carriers of (presumed) defective genetic traits, whose reproduction was viewed by the elite establishment as posing "a menace for society." Eugenicists asserted that human beings should be selectively bred, just like plants and animals: The goal was to "weed out defective persons from society just as a farmer would clear a field."[9] Eugenics appealed to American ethos and optimism about scientific and technological progress and its ready-made "scientific" solutions. Eugenics provided theoretical, pseudoscientific underpinnings that medicalized social problems—by offering biological interventions for the prevention of naturally occurring diseases—as well as complex human and societal problems. Eugenicists promoted compulsory sterilization as a prophylactic medical measure—a form of genetic engineering. Their stated purpose was to "protect the public health" by preventing the "unfit" from reproducing. Eugenicists applied veterinarian methods for improving the genetics of animal populations to human beings; they were seeking to "improve the human race . . . through better breeding."[10]

The United States was the first country to carry out eugenics-driven compulsory sterilization programs under state statutes. Indiana enacted the first law in 1907 and the last forced sterilization was performed in Oregon in 1981. Adherents of compulsory sterilization policies were the elite of American society—including prominent academic leaders, "progressive" healthcare professionals, public health officials, and major industrialists. The Model Eugenical Sterilization Law published in 1914 served as the template for most state statutes and included an indemnification clause to protect physicians who performed surgical sterilizations from legal action—the clause is similar to the legal immunity currently provided for physicians who perform government-recommended and state-mandated vaccinations. The first individuals to undergo mandatory sterilization were mentally ill patients in state institutions. By the 1920s, sterilization laws were applied to people with mental and physical disabilities, as well as immigrants, orphans, Native Americans, Mexican Americans, and African Americans—all of whom were deemed to be "feebleminded" and "unfit."[11]

In 1926, the U.S. Supreme Court upheld the legality of Virginia's compulsory eugenic sterilization policy in an eight-to-one decision, *Buck v. Bell*,[12] accepting the dubious argument that sterilization on behalf of the collective health of the citizenry was constitutional. The Supreme Court explicitly based this decision on the *Jacobson* vaccination decision. The majority wrote, "The principle that sustains compulsory vaccination is broad enough to cover cutting the Fallopian tubes. Three generations of imbeciles are enough."[13] This infamous decision greatly accelerated the pace of forced sterilizations and led additional states to enact sterilization laws. It provided an important precedent that prompted state courts to uphold a state's right to control a person's right to procreate when adequate hereditary evidence showed that their procreating threatened the welfare of society. The decision has never been overruled.

By 1944, thirty states had sterilization laws, and more than forty thousand sterilizations had been performed.[14] What's more, state-sponsored compulsory sterilizations in the United States continued long after their suspension in Germany following the Nazi trials and long after eugenics theories and policies had been discredited after the revelation of the Nazi sterilization policies that had devolved into genocide.

A review by two Yale University physicians[15] comparing the eugenic sterilization programs in the United States and Germany from 1930 to 1945 found that before World War II, physicians in both countries participated in state-authorized eugenic sterilization programs:

> A comparison of U.S. and German histories reveals similarities that argue against easy dismissal of a Nazi analogy. On the basis of a review of editorials in *The New England Journal of Medicine* and *Journal of the American Medical Association* from 1930 to 1945, it is difficult to accept the suggestion that the alliance between the medical profession and the eugenics movement in the United States was short-

> lived. Comparison of the histories of the eugenic sterilization campaigns in the United States and Nazi Germany reveals important similarities of motivation, intent, and strategy.[16]

Indeed, in 1933, the Nazis enacted a compulsory sterilization law modeled after the American Model Eugenical Sterilization Law. The Nazis sterilized four hundred thousand people. In the United States, thirty-three states had compulsory sterilization statutes, and an estimated sixty thousand poor and working-class Americans were involuntarily surgically sterilized between the 1920s and the mid-1970s.[17] Both governments justified compulsory sterilization in the name of public health.

"Other people had beautiful babies . . . We felt like we had been robbed in that particular category." These are the words of Kenneth Newman who was forcibly sterilized at Fairview Training Center in Oregon, as was his wife, when they were teenagers. They and thousands like them were declared "unfit to reproduce" by medical experts and public health officials of the day. Only seven states have apologized for forced sterilizations. Only five states have apologized for slavery.[18]

America's litany of abusive—even lethal—experimental medical travesties was conducted by prominent academic physicians, who were often sponsored by public health agencies. The following list provides only a few examples.

- 1932–1972: The notorious Tuskegee syphilis experiment, sponsored by the U.S. Public Health Service, was conducted, in which physicians denied life-saving medicine to African American men.
- 1946–1948: The U.S. Public Health Service and the National Institutes of Health sponsored a vaccination experiment that intentionally infected nearly seven hundred institutionalized Guatemalans with sexually transmitted diseases. The principal researcher was the same doctor who conducted the Tuskegee experiment. The Guatemalan experiment was concealed for sixty-four years. The public only learned about it when the U.S. government apologized on October 1, 2010, calling it "clearly unethical."[19]
- 1941: The "refrigeration experiments" conducted at Longview State Hospital of the University of Cincinnati[20] and at McLean Hospital of Harvard University[21]—like those conducted at Dachau by Nazi doctors—exposed mentally disabled patients to freezing temperatures of thirty degrees Fahrenheit for 80 to 120 hours.

- 1942: U.S. Army and Navy doctors infected four hundred prisoners in Chicago with malaria in experiments designed to obtain "a profile of the disease and develop a treatment for it." The Nazi doctors on trial at Nuremberg cited this experiment in their own defense.[22]
- 1956–1972: Doctors deliberately infected "mentally retarded" children at the Willowbrook State School in New York with hepatitis in an experiment aimed at tracking the development of this viral infection, in pursuit of developing a hepatitis vaccine. Parents were coerced into agreeing to the experiment because participation in the study was a condition for admission to the institution. The researchers defended the study, claiming hepatitis was rampant at the institution and the children would have been infected regardless.[23]

These, and hundreds of other examples, demonstrate the reality of medical travesties that occur when doctors stray from Hippocratic medical ethics to serve "the greater good."

INTERNATIONAL CODES OF MEDICAL ETHICS

The revelations at the Nuremberg Doctors' Trial, which took place from 1946 to 1947, that documented medical atrocities of unprecedented scope and magnitude committed by Nazi doctors, shocked the world. How could highly educated physicians—whose professional oath required them to promote health, alleviate suffering, and protect life—become active instruments of torture and death? In his opening statement, Telford Taylor, the chief prosecutor at Nuremberg, pointed out that this was "no mere murder trial" because the twenty-three defendants were physicians who had sworn to "do no harm" and to abide by the Hippocratic oath. He told the judges that the people of the world needed to know "with conspicuous clarity" the ideas and motives that moved these doctors. He asked, "How could physicians actively and enthusiastically treat other human beings as less than beasts?"[24]

In their verdict, American judges issued the Nuremberg Code (1947) whose first precept is "the voluntary, informed consent to medical research is absolutely essential." The Nuremberg Code, which is deeply rooted in Hippocratic ethics, is the most authoritative document for the protection of human subjects. The Nuremberg Code serves as the cornerstone for all subsequent medical ethics codes, including the Belmont Report, which is the foundational U.S. document for human subject research codified in U.S. federal regulations.[25]

By valuing the liberty and welfare of research subjects above the promise of medical progress, the Nuremberg judges sought to place the interests of individual humans above the interests of society in medical progress. The Doctors' Trial demonstrated in harrowing detail the need to hold physicians to universal standards they might otherwise violate. Legal codes of medical ethics are necessary to guarantee every human being inalienable human rights—especially the right to "life, liberty and security of the person," guaranteed by the Universal Declaration of Human Rights written in 1948. They are also necessary to guarantee the opportunity to give or refuse informed consent to medical intervention.

Children do not have the capacity to exercise their human rights and must rely on surrogates to make medical judgments on their behalves. Because surrogates have not always acted in their best interest, however, children require added legal protections.

In the 1970s, after the hepatitis experiments at Willowbrook were brought to the public's attention, the issue was hotly debated by prominent medical ethicists, such as Paul Ramsey of Princeton University, who argued for the children's best interest. Children should not be exposed to drugs "insufficiently tested on human beings" unless they may "further the patient's own recovery."[26] Father Richard McCormick of the Kennedy Institute of Bioethics at Georgetown University argued from the utilitarian perspective, claiming that parents may consent to their children's participation in non-therapeutic experiments because, if they could, children would (or ought to) consent as long as there were "no discernible risks, no notable pain, no notable inconvenience . . . almost no cost to the child."[27]

In 2001, the Maryland Court of Appeals affirmed, in a strongly worded landmark decision, the rights of children to be protected from any risks of harm caused by non-therapeutic research.[28] The case, *Grimes v. Kennedy Krieger Institute* (KKI), involved the exposure of Baltimorean toddlers to a contaminated environment in a government-sponsored lead abatement experiment, conducted between 1993 and 1995. The experiment sought to determine a cost-effective lead abatement level. It came to light when two families sued KKI, claiming that their children had suffered brain damage from exposure to lead poison. The researchers, who knew that lead paint posed a risk of mental retardation in children, recruited poor families with young children to live in a lead-infested environment. The young children were expected to ingest lead-laced paint dust. During the experiment, the children's blood lead levels were measured over time and those levels rose significantly, but parents, who consented to the research without being fully informed about the risks of exposure, were not informed that the increased levels posed an increased risk of harm. The Environmental Protection Agency sponsored the experiment and Johns Hopkins University's ethics review board approved it. The Maryland court rejected the argument that the research met an important public health need, which unequivocally affirmed the preeminence of an individual child's best interest:

> We have long stressed that the "best interests of the child" is the overriding concern of this Court in matters relating to children . . . Whatever the interests of a parent, and whatever the interests of the general public in fostering research that might, according to a researcher's hypothesis, be for the good of all children, this Court's concern for the particular child and particular case, over-arches all other interests. It is, simply, and we hope, succinctly put, not in the best interest of any healthy child to be intentionally put in a non-therapeutic situation where his or her health may be impaired, in order to test methods that may ultimately benefit all children.

The panel of judges likened this unconscionable lead abatement experiment to Nazi medical atrocities, admonishing the researchers for conducting the experiment and rebuking Johns Hopkins's review board for approving an experiment in which children served as mere measuring tools—or "canaries in the mines."

Although federal regulations were finally adopted in 1983 that restricted the use of children in research that offers a potential benefit to them,[29] the regulations lacked effective checks and balances, and abuses continued. From 1986 to 2000, thousands of children in foster care—wards of the state in at least seven states—were used as human subjects in HIV/AIDS drug and vaccine trials[30] sponsored by corporations and the government. The children, mostly poor minorities ranging from infants to late teens, suffered severe discomfort and some even died. These high-risk phase I trials were conducted without parental consent or the approval of an independent advocate and in violation of federal regulations.

In 1997, bowing to corporate and special-interest pressures, Congress eroded its interpretation of existing federal regulations restricting the use of children as experimental subjects by enacting the FDA Modernization Act (FDAMA). The FDAMA encourages the use of children to test patented drugs and even provides manufacturers huge financial incentives, including six additional months of marketing exclusivity, for testing drugs in children. Children, who had no voice in the adoption of this legislation and are legally precluded to exercise the right to refuse, were forced to bear the burden of drug testing so that drug companies could qualify for a highly lucrative six-month patent extension. Those financial incentives, and the lifting of restrictions on the use of vulnerable children in experiments, also had a major impact on other biomedical research players—such as academic and governmental research institutions, professional associations, and scientific journals—who benefited from the opportunity the FDAMA provided for expanding pediatric drug trials.[31] Thus all research stakeholders have profited financially from a policy that renders children as commodities for commercial exploitation.

In 1997, the FDA's chief of the gastrointestinal drug division acknowledged that "at least" three children in clinical trials had died after being given the heartburn drug Propulsid. The FDA informed Janssen Pharmaceuticals that Propulsid was "not approvable" for use in children, but babies continued to be recruited to test the drug—even as the drug's deadly side effects were known. Janssen Pharmaceuticals used one hundred children and babies in Propulsid drug trials at one hospital alone. The babies had been born with gastroesophageal reflux, which is not a life-threatening condition: It is characterized by "recurrent spitting and vomiting, which occurs in about four out of ten healthy children, and doesn't always require treatment. Propping up the baby after feeding and correctly positioning the baby for sleep usually eliminates the condition. Experts agree that most babies outgrow the condition by their first birthday."[32]

In 1999, Propulsid's sales reached $950 million while the number of deaths reached eighty adults and children. That same year, nine-month-old Gage Stevens was recruited for a controlled, FDA-approved clinical trial and was given Propulsid and Tagamet for four months. He died of cardiac arrhythmia—just as six-month-old Chase Brown had the year before. By the time it was taken off the market, nineteen children who had been given Propulsid had died. After his death, Gage Stevens' parents said, "Little did we know that Gage was basically a guinea pig, and they never told us that [Propulsid] causes dangerous side effects, or [that] there had been deaths."[33] A public policy that puts the well-being of a child at risk—for the good of others or, more likely, for commercial reasons—violates fundamental moral principles and devalues the child as a human being.

MANDATORY VACCINATION POLICIES

Mandatory vaccination policies are a violation of medical ethics and human rights. Mandatory vaccination policies violate fundamental moral principles, and they have a profound and deleterious effect on medical practice and the relationship between patient and physician. When a physician is administering a state-mandated vaccine, that physician is serving as an agent of the state. However, doctors who administer medical interventions without ascertaining that they are for "the good of [the] patient" and "will not do harm" violate the physicians' responsibility under the Hippocratic oath. Doctors who adapt their practice to conform to population-based utilitarian ethics are betraying patients' trust by putting the interests of society above the interest of the individual patient who may be harmed as a result.

Mandatory vaccination policy is predicated on two problematic rationales. The first rationale in support of mandatory vaccination is that vaccines are "safe." Unlike other medical interventions, vaccines are given to healthy children and adults, and they are known to cause severe adverse effects—some of which are endemic to the disease against which they are intended to protect. Indeed, legally, vaccines are classified as "unavoidably unsafe,"[34] an indication that they pose an inherent risk. Vaccines have not been subjected to scientifically rigorous, placebo-controlled safety trials before they have been approved for use in children. Instead, they are tested by exposing healthy human beings to one of two vaccines—each of which poses risks of harm—so the true magnitude of risk is masked. Most of the data, including government-gathered data, are undisclosed and inaccessible for independent scientific scrutiny. This raises serious concerns about the integrity of publicly disseminated claims about vaccine safety.

The second rationale given for mandatory vaccination is the necessity to "protect the herd," that is, for "the greater good of society." The Nazi doctors at Nuremberg used the very same rationale—and it was rejected.

Mass vaccination of healthy children whose lives are put at risk "to protect the herd" is tantamount to a massive, involuntary human experiment in violation of the "absolutely essential" human right to "voluntary, informed consent" under the Nuremberg Code. Mandatory vaccination policies for children have provoked an extremely contentious battle. How can anyone justify a public policy that forces vulnerable infants and young children to bear the burden of risk and suffering—including neurological damage, seizures, permanent disability, and death—for "herd immunity"?

By age eighteen, children in the United States receive seventy doses of sixteen vaccines. Infants, whose immune systems are undeveloped, are exposed to more than thirty-three doses of over one dozen vaccines by fifteen months of age, including hepatitis B, diphtheria, tetanus, pertussis, polio, rubella, measles, mumps, chickenpox, pneumococcal infections, *haemophilus influenzae* type b, and influenza. Moreover, infants are being vaccinated not only against childhood diseases, but also against diseases to which they are unlikely to be exposed for the express purpose to protect adults. For example, why are newborn infants on the day of birth being vaccinated against hepatitis B, a disease contracted by having unprotected sex or sharing needles? Furthermore, why are two-month-old infants vaccinated against tetanus, which is extremely rare, occurs primarily in older populations, is effectively treatable after exposure, and is uncontagious?

To gain even a cursory appreciation of the level of risk to which infants and children are exposed, one should examine *The Merck Manual.* It acknowledges that vaccines contain toxoids, weakened bacteria that can cause severe, debilitating symptoms—including life-threatening conditions, such as encephalitis.[35]

If vaccines have greatly improved the health of America's children, then it is fair to ask why the United States' ranking among world nations in infant mortality has plummeted from twelfth in 1960 to twenty-ninth in 1990, down to forty-sixth in 2010. The U.S. Centers for Disease Control and Prevention (CDC) acknowledges, "The U.S. infant mortality rate is higher than those in most other developed countries, and the gap between the U.S. infant mortality rate and the rates for the countries with the lowest infant mortality appears to be widening."[36]

While the infant mortality rate in European and Asian countries has decreased, to 2 to 3 deaths per 1,000 live births, more than 6 infants per 1,000 under one year of age will die, and 7.8 children per 1,000 under five years of age will die this year in the United States.[37] The United States lags behind nearly every developed country and even poor countries such as Cuba, whose healthcare expenditure is $230 per capita, whereas the United States' is $6,096 per capita. "Infant mortality and our comparison with the rest of the world continue to be an embarrassment to the United States," said Grace-Marie Turner, president of the Galen Institute, a conservative research organization.[38]

U.S. medicine is dominated by corporate interests—physicians, academic institutions, and government agencies—whose financial interests are intertwined with drug and vaccine manufacturers. Physicians, professional associations, medical institutions, and government agencies are collaborating partners in the business of medicine. The practice of medicine, medical research, and medical information dissemination channels are imbued with the utilitarian business ethics calculus. Thus individuals and communities are viewed as the means by which to increase profits. Neither individuals nor the citizenry are well served when the integrity of medicine is debased, its scientific method is tainted, and its humanitarian and moral principles are violated with impunity.

Conflicts of interest compromise our current vetting system for medical research. The system is designed to bring new products to the market with little regard for their lack of therapeutic value or their potential safety hazards. The system fails to ensure that drugs, vaccines, and medical devices are adequately tested for safety and therapeutic value. The most vulnerable populations—children, the elderly, and the disabled—are unprotected from bad medicine, defective medical products, and unethical physicians.

A government-funded experiment, "fenfluramine challenge,"[39] encapsulates an enduring eugenics mindset within U.S. academic medicine. From 1993 to 1996, senior researchers at the New York State Psychiatric Institute (NYSPI) (including the current chairman of child psychiatry at Columbia University),[40] Queens College, and Mount Sinai School of Medicine,[41] conducted pseudoscientific, wholly non-therapeutic, "predisposition to violence" experiments. State and federal governments funded the experiments. Researchers administered intravenously the neurotoxic drug fenfluramine[42] to one hundred mostly African American and Hispanic boys, aged six to eleven. The researchers sought to prove a eugenic hypothesis,[43] that violent or criminal behavior could be predicted using biological markers, such as levels of brain chemicals.

Neither the researchers, nor the medical ethics reviewers at the institutions and at the National Institute of Mental Health (the federal agency that funded the experiments), nor the journal editorial boards who published the research reports, considered the children's welfare and human rights. They all collaborated in violating the dignity of these children as human beings: None of these academic and governmental physicians had any qualms about exposing young children to risks of harm, causing them to suffer pain and trauma, or labeling them as "biologically predisposed to violence" in an experiment with absolutely no therapeutic rationale. Dr. John Oldham, director of the NYSPI, stated in an interview in the *New York Times* that such studies are "very important to study the biological basis of behavior. . . . Is there or is there not a correlation between certain biological markers and conduct disorders or antisocial behavior? This study was an effort to look at this with a relatively simple method using fenfluramine." Clearly, the medical research community in the United States remains infected with a racist, eugenic ideology that continues to pose a threat to vulnerable individuals, and children in particular.

LESSONS FROM THE HOLOCAUST

The shadows of Auschwitz and the revelations at the Doctors' Trial in Nuremberg are indelible—they cannot be redacted from history. The Holocaust is an object lesson of how the government's usurpation of individual civil liberties and the collapse of professional and institutional integrity—in particular, economic, medical, and legal integrity, which are the underpinnings of a civilized society—led an entire society to be swept into the realm of genocidal murder. The stated rationale for the medical atrocities that were "crimes against humanity" was that they were for "the greater good."

The silence of the medical community—in both Germany and the United States—about its role and responsibility after it turned its back on the Hippocratic tradition in favor of eugenics is an indication that the medical community is conflicted. There has been no disavowal of eugenic healthcare practices carried out "for the greater good," which threaten individual human rights and civilized medicine. Physicians' efforts to distance themselves from the atrocities at Auschwitz—the apex of murderous medicine—by spreading the myth that the Nazi medical atrocities were the work of deranged scientists is an attempt to delude the public and themselves about the profession's criminal culpability.

When medical trends veer from the moral Hippocratic tradition, professionals lose their moral footing and are capable of committing medical atrocities. Mandatory public health policies that trample on personal human rights for the claimed "greater good" threaten individual liberties and the very foundation of a civilized, democratic society. Doctors who practice medicine "for the greater good" participate in a culture that views the individual as a means to an end. Such a culture may set in motion increasingly abusive medical practices culminating in medical atrocities. Unless we confront these moral shortcomings and establish viable safeguards, no one can be safe, and the long shadows of Auschwitz and Nuremberg will continue to haunt us.

Chapter Ten

"THE GREATER GOOD"

Allen Tate

Political philosophy is not a staple of mainstream consciousness. The theories of Thomas Hobbes and John Locke rarely take priority over thoughts of carpools, soccer practice, and grocery shopping in the multitasking mind of the average parent. However, we experience the application of the philosophy of utilitarianism with far greater frequency than we may realize. Whenever we speak of "the greater good," we invoke a utilitarian view. Whenever we take action to value the happiness or well-being of the group over that of the individual, we do so as utilitarians.

As with all philosophical theories applied in the real world, utilitarianism must meet specific criteria for its application to be sound or "good." As utilitarianism necessarily usurps the rights of the individual, the need to meet these criteria is paramount. When utilitarianism is applied to an issue such as vaccination—which involves children, infectious diseases, and the virtual guarantee of death and debilitating side effects for some—it is of the utmost importance that vaccination policy decisions, and specifically vaccination mandates, meet the criteria to satisfy the theory.

When arguing theories, philosophers apply them to ideal worlds, in which they assume all criteria have been met. They do this so they can focus solely on the theory in the purest terms. These theoretical worlds are free of complicating factors such as corruption and human error that might affect one's view of how the theory should be applied. Sadly, we are not afforded this same luxury in the real world, as the existence of corruption in the government cannot be refuted.

For the sake of discussion, however, I will attempt to simplify the discussion as philosophers do.

A sound application of utilitarianism requires that an action or decision must deliver the greatest good for the greatest number. I should note that the greatest good does not necessarily need to be good in the sense of giving happiness or pleasure. The benefit could come in a number of other forms. For example, you might prefer to drive ninety miles per hour on the highway because it would shorten your commute and allow you to get more work done. However, your high-speed driving would also put other drivers at risk of serious injury. Utilitarians would place a higher value on the safety of all drivers on the road than on the pleasure you would obtain from a shortened commute. This is the basis for speed limits on our roads. A speed limit does not give pleasure or happiness to the group; however, it can easily be argued that it provides the greatest good to the greatest number.

Now let us arbitrarily place the speed limit at fifteen miles per hour. While this would most certainly ensure the safety of all drivers, the traffic buildup, the loss of productivity at work, and the total unhappiness of drivers as they crawled to work each day would likely outweigh the benefit of guaranteed safety. Many might reasonably argue that they would prefer to have no speed limit and drive defensively if the only alternative is a painstakingly slow commute. Although the fifteen-miles-per-hour speed limit would aim to provide the greatest good, it would fail to meet the main criterion for sound utilitarianism. An application of utilitarianism is sound only if the decision made actually delivers the greatest good to the greatest number. Any application that fails to meet this criterion necessarily results in the removal of individual rights that cannot be justified within a utilitarian argument.

The justification for mandatory vaccination in the United States is based on utilitarian logic. On many occasions, public health officials have appealed to the American public regarding the necessity of vaccination in order to ensure the safety of the "herd." States deny unvaccinated children admission to day care and public schools not because they are at risk but because they are perceived to pose a risk to other children. Utilitarian philosophers first introduced the concept of the greater good, and it is through this principle of utilitarianism that our government justifies compulsory childhood vaccination. In light of this claim, the burden of proof lies with the government to show that compulsory childhood vaccination does, in fact, provide the greatest good for the greatest number. If vaccination mandates are not truly for the greatest good, and the government has invoked utilitarianism to justify them, then the government would have offered a fraudulent, insufficient justification.

It is indisputable that the vaccination schedule has never been tested for safety in its entirety, or in the way that it is administered. In other words, while the government reviews, licenses, and compels individual vaccines, it does not test—or require vaccine makers to test—the safety and efficacy of vaccines given simultaneously or the cumulative effects of multiple vaccines. For example, while the government reviews manufacturer prelicensure tests and recommends the hepatitis B vaccine for two-month-old infants, it has not tested that vaccine together with the other vaccines it recommends to be administered simultaneously: rotavirus, diphtheria, tetanus, pertussis, *haemophilus influenzae* type b (Hib), pneumococcal, and inactivated poliovirus.[1] Furthermore, despite the public's demand and congressional bills calling for it, there has never been a controlled study comparing long-term health outcomes of vaccinated and unvaccinated populations.[2] Thus it is impossible to determine the true benefits of the vaccination schedule. As it currently stands, it would be impossible to know if mandatory vaccination requirements, in fact, accomplish the greatest good for the greatest number. Because the burden has not been fulfilled, the hypothesis may not be accepted.

To obtain a better perspective of what it would take to justify that compulsory vaccination accomplishes the greatest good, we should consider the risks that the benefits of vaccination would need to outweigh. First, vaccination benefits would need to outweigh the risk of actually contracting infectious diseases, including diseases that have been virtually eradicated such as poliomyelitis, as well as diseases such as rotavirus (better known as childhood diarrhea) and hepatitis B, which is contracted from an infected parent, contaminated needle, or unprotected sexual intercourse. Second, vaccination benefits would need to outweigh the suffering caused by known and documented side effects of vaccination, such as encephalopathy (brain injury), seizure disorders, developmental delay, and death in affected children. Third, vaccination benefits would need to account for the lost productivity and wages of those who die or are harmed by vaccines, and the lost productivity and wages of their caregivers. Finally, vaccination benefits would need to outweigh the cost to the public of compensating the families of vaccine-injured children and purchasing vaccines for children otherwise unable to afford them. Outweighing all of these combined risks is a tall order by any measure. In any case, since the childhood vaccination schedule has not been tested as it is administered, we cannot know the benefits of vaccination, and therefore, whether or not the utilitarian burden can be fulfilled. As it currently stands, utilitarianism cannot justify compulsory vaccination.

At a glance, vaccinations are conventional, prevailing medical procedures that have been standard protocol for over one hundred years and compulsory for our nation's children for over four decades. Nevertheless, it is startling to learn that while the first vaccine was compelled for adults during a smallpox epidemic, we vaccinate most of today's children at least thirty-five to forty times and recommend seventy doses of sixteen vaccines, including one for a sexually transmitted disease within hours of birth. How could we have paid so little attention to such a startling transition?

To explain our seeming acceptance of the dramatically increased number of vaccines children receive today, I will use a famous philosophical device known as the sorites paradox, or the paradox of the heap. This construct is meant to highlight a potential problem with the use of a vague term such as "heap," and in the case of vaccines, it highlights a potential problem with the use of the term "reasonable." I will first illustrate the concept using the term "short" as an example.

We would all likely agree that a fully grown man who is four feet tall is short. Furthermore, we would likely agree that a man who is one millimeter taller than this four-foot-tall man is also short, because a mere millimeter's difference cannot be what separates short from tall. Suppose that we continue this procession of short men, with each man one millimeter taller than the previous man. At the end of our very long line of fully grown men, we would find a nearly eight-foot-tall man who, based on our logic, would also be "short." It is obvious that we must have made some sort of error.

The same is true of our vaccination schedule. Many likely would agree that a single vaccine administered during an infectious disease epidemic is "reasonable." Furthermore, we would likely agree that the addition of one more vaccine would not suddenly make the schedule unreasonable. However, we now find ourselves "reasonably" administering seventy doses of sixteen vaccines to children, and we have the same perplexed feeling we might have when trying to explain why a man who is eight feet tall is "short." We can no longer justify the vaccination schedule in this way. We must not judge a vaccine in relation to previously scheduled vaccines but rather on its merit as an independent entity.

My aim is not to denounce utilitarianism. I seek to examine its application to the United States' compulsory childhood vaccination program. As Harvard Professor Michael Sandel writes in his book *Justice*, "The most glaring weakness of utilitarianism, many argue, is that it fails to respect individual rights."[3] I agree with this statement and, precisely for this reason, I contend that it is essential to ascertain whether the vaccination schedule our nation imposes on its children is truly for the greatest good. To accomplish the utilitarian's greatest good, the benefits of vaccination must verifiably outweigh great costs and suffering. Whether the benefits truly yield a tremendous good is simply unknown—and without this knowledge, the utilitarian's burden of proof is unmet, and utilitarianism cannot justify this program.

As individuals, we will always act in our own best interest to the best of our ability. Without proof that compulsory vaccination is, in fact, what brings the greatest good to the greatest number, there is no utilitarian justification for us to sacrifice our individual rights to do what we discern is best for ourselves and our children.

Part II

BREAKING THE SILENCE OF VACCINE INJURY—PERSONAL NARRATIVES

THE CHILDREN, THE MILITARY, AND THE ADULTS

The Supreme Court ruled 6–2 in favor of federal preemption, making it impossible in the future to claim in a civil court before a jury that a vaccine's design was dangerously defective.

—Russ Bruesewitz

The biggest difference in environmental exposures that happened during those three short years was the increase in their vaccines . . .

—Gay Tate, PhD, LSW, MLSP

What if it were true that the way we now vaccinate our children causes more death, chronic disease, and disability than it prevents in America?

—Judy Converse, MPH, RD, LD

I deeply regret my decision to allow Zeda to get this shot. She has lost all quality of life. I would do anything to get her better, but no one has any answers for me. The doctors still don't believe it was the vaccine.

—Amy Pingel

Unfortunately, these brave defenders of our freedom are defenseless against the mandatory use of dangerous vaccines, such as the anthrax vaccine.

—Capt. Richard Rovet, RN, BSN, B-C (USAF, ret)

Chapter Eleven

JUSTICE DISSERVED:
The Hannah Bruesewitz Odyssey

Russ Bruesewitz

I remember vividly my wife Robie calling me at work, noticeably distressed, and asking if I could come home. She didn't want to tell me what was wrong over the phone, so in the absence of facts, my imagination took over and concocted all kinds of dire scenarios. Once home, I was relieved to see she wasn't hurt, but she had obviously been crying. We hugged for a moment, and Robie told me she was pregnant.

We weren't new to the idea of raising children, as we already had two (daughters, aged eight and twelve), but I was forty-two and Robie was thirty-nine. The term "unexpected" was a colossal understatement when describing the pregnancy. My job required frequent travel. Robie had recently gone back to work part-time, and we were just starting to make some economic gains after our "stay-at-home mom" decision made some twelve years prior. Oh, well, it was just one of those curveballs life throws at you that, in keeping with the sports analogy, might even bring a son I could toss a few to if I wasn't riddled with arthritis by that time. Silver lining? Why not! Kids have a way of keeping you engaged in the present. We just had to get past the nighttime feedings, diapers, and terrible twos, and we'd emerge on the other side as the proud and victorious parents we already perceived ourselves to be, trying to figure out, yet again, a way to pay for one more college experience. Isn't that the American way?

Robie's pregnancy was uneventful and met all the benchmarks of normal progress. Aware of the caveats associated with women having children later in life, we had tests done for Down syndrome, spina bifida, and others. All came back negative. Not that we ever would have considered terminating the pregnancy, but at least we would know. Our Hannah Lee emerged October 20, 1991, full term and healthy.

Over the next six months, Hannah's development progressed normally while we adhered to the prescribed pediatric schedule of checkups and vaccinations. The evening prior to Hannah's six-month visit to the pediatrician, Robie and I talked on the phone since I was away on a business trip for a few days. During that conversation, when I asked about "the babe," Robie expressed nothing more than minor aggravation that Hannah had a slightly runny nose. When Robie called the doctor's office the next day, she mentioned Hannah's symptoms but was told to come in anyway. She did, and after going through the routine physical, Hannah was given her third DPT shot (the whole cell pertussis vaccine was used at that time) and went home. Two hours later, our lives changed dramatically and irrevocably.

While feeding Hannah lunch, Robie noticed a quick jerk followed by a distant stare. Hannah soon recovered and resumed her normal activities, but Robie called the doctor to describe what she had just witnessed. The pediatrician's opinion was that Hannah had had a possible reaction to lingering pain from the shot, and the doctor suggested we give her Tylenol. Several more similar episodes occurred throughout the day. That evening, not long after putting Hannah to bed, Robie suddenly heard Hannah scream, rushed into the room, and found her in the midst of what we now know was a grand mal seizure. Having no familiarity with seizures of any kind but thinking Hannah wasn't breathing and looked blue, Robie quickly got her out of the crib and onto the floor and started to give Hannah mouth-to-mouth resuscitation. Our eight-year old daughter heard the screams, went to Hannah's room as well, and stood in the doorway watching the event. Robie frantically yelled to her to call 911, but our daughter, shocked by what was happening, was frozen in place. The trauma of that experience, coupled with the guilt she felt over her inability to do as asked in that crisis, eventually resulted in her getting therapy to address some behavioral issues that later surfaced. Once Hannah came out of the seizure, Robie called our pediatrician, then 911, and was soon on her way in an ambulance to the first of many hospitalizations for Hannah.

I was in Detroit on business at the time and couldn't be reached until the next day (cell phones were not yet standard issue), leaving Robie alone to deal with the trauma as well as juggling care for our other two children. We had no immediate family nearby, so we relied on neighbors and friends for help. Fortunately, we had a pediatric neurosurgeon friend at Children's Hospital of Pittsburgh who met Robie in the ER and expedited the admission process while Hannah continued to seize. In triage, a resident administered a sedative to stabilize Hannah's condition, and she subsequently was moved to the ICU, where she remained for the first few days. After catching the first available flight back, I arrived the following afternoon dumbfounded, scared, and angry, given the obvious connection to the DPT shot. Sure, we were probably advised of the percentage risks for possible adverse effects, but my recollection was that the risks were described as so minute, it wasn't even a consideration when weighing the vaccine benefits. We had never heard of anyone having a problem! Was Hannah now one of those obscure statistics no one really thinks could be them? *Let's just get her back to normal, and we'll deal with those issues later,* I thought.

Our initial hospital stay was for fifteen days, during which Hannah experienced more than 125 seizures—some as frequent as every ten minutes and lasting up to five minutes in duration. The anguish we experienced as parents was indescribable as we watched our previously healthy baby become catatonic, seeing her lips turn blue from lack of oxygen, hearing the alarms go off to notify the nurses she had reached critical levels, then helplessly watching the flurry of medical activity to address her condition. In a somewhat naïve attempt to feel more useful, we recorded each seizure, noting the time of day, duration, body movements, medications, and recovery. Subsequently, and for a number of years after, Hannah's medical team would occasionally review our notes to supplement their clinical records, hoping they, too, might uncover some clues for her treatment. Phenobarbital was the anticonvulsant of choice at that time, used with marginal success. It left Hannah's demeanor between seizures significantly altered from that bright-eyed personality of the previous six months. To this day, I look back on that time as one of watching the light go out of my daughter's eyes, a light that has never returned.

While Hannah's return home was welcome, the seizures continued. Over the next eighteen months or so, hospitalizations were required every four to six weeks when a succession of seizures wouldn't stop. Hannah progressed physically, but she was regressing cognitively. When it was necessary to administer Valium or Dilantin to stop the seizures, she became extremely agitated afterward, screaming and flailing to the point where we had to put a cage top on her crib for her own safety. When not seizing, her demeanor became mostly flat and expressionless, seldom smiling at things she liked in the past. Any vocabulary she had previously acquired was lost. Her only words were an echo to a prompt, as Hannah expressed nothing on her own initiative.

We were desperate to help our child, spending much of our time investigating alternative doctors, treatments, therapies, drugs, and play groups in an attempt to find a path to recovery. Some of them initially gave us hope, but ultimately we found them ineffective. In that investigative process, we were surprised to discover a source of financial help for those injured by vaccines. While I can't recall specifically how I learned of the National Vaccine Injury Compensation Program (VICP), I can assure you it was never mentioned to us by anyone in the medical profession. In fact, in numerous discussions with Hannah's medical team of neurologists, pediatricians, and social workers, no one ever acknowledged the obvious connection between Hannah's profound injuries and her DPT shot. The closest we came was a comment from a pediatric neurologist who said that Hannah's case was a major topic of discussion at a neurological conference he had just attended. The conference considered her situation unusual, with the close temporal relationship between the shot and her reaction.

Every time we were asked for Hannah's medical history (and you can't imagine how often that occurred), Hannah's reaction to her DPT vaccine never seemed to prompt further questions from doctors seeing Hannah for the first time. I was always struck by that lack of curiosity. Were we being dismissed as just reactionary parents looking for someone or something to blame for their child's condition? Were doctors afraid to open that box for fear of impugning the reputation of a colleague? Or worse, was there an implied code of silence in the medical profession surrounding adverse vaccine reactions based on the fear that, if acknowledged, these reactions might lead to reduced compliance? Only years later, during our legal pursuits, did we learn of the Vaccine Adverse Event Reporting System (VAERS) and were amazed by its obscurity. The professionals most likely to benefit from the information it contained didn't seem to know about this database. If VAERS isn't even used by medical professionals, who is it for? Hannah's own pediatrician who administered the vaccine that caused Hannah's injury later stated that had she known of the number of incidences surrounding the "hot lot" attributes of the vaccine batch Hannah received, she never would have given her the shot that day.[1]

Being a die-hard pragmatist who supports the concept of "two sides to every story," I have a hard time subscribing to most conspiracy theories. But after twenty years of navigating this sole, marginal system of recourse for vaccine injury, one created by the National Childhood Vaccine Injury Act of 1986, I believe it is failing precisely those families Congress intended to help.

In the three years following Hannah's vaccine reaction and subsequent injuries, we continued to focus on her care and development, but it became increasingly evident that her prospects for recovery were not good. That reality was, and still is, hard to accept. It forced us, however, to start looking at the practical aspects of providing for her long-term care. While Robie continued her almost daily dealing with doctors, insurance companies, pharmacists, social agencies, parent support groups, caregivers, and school districts, I started to look into the VICP. With the help of my brother, who is an attorney, we started due diligence and were shocked to learn that we were up against a three-year filing deadline that expired the next day. No time to explore options—we were out of time. Using his legal resources, my brother ended up speaking that day with Cliff Shoemaker of Shoemaker & Horn Associates, a two-man Washington, DC, firm specializing in vaccine injury cases. After frantically gathering the information he needed from us over the phone, Cliff Shoemaker prepared the petition and literally walked it over to the U.S. Court of Federal Claims. The paperwork was stamped to verify compliance with the April 3, 1995, filing deadline. Had we not met that deadline, the VICP would have denied our petition, and all other possible civil court actions would have been unavailable. If we made the deadline by one day, how many other deserving families probably missed their opportunity for help because of the underpublicized existence of the program, intentional or otherwise?

From our initial review of the compensation requirements, we had met all the conditions and, with the DPT vaccination as the obvious catalyst, everything seemed to point to a quick and favorable compensation decision. That expectation soon proved to be misguided. Contrary to the stated intent of the legislation to "provide for a less adversarial, expeditious and informal proceeding for the resolution of petitions," our petition was assigned, then reassigned, then delayed during the chaotic government shutdown over budget approval in 1995 and 1996. Finally, our petition was transferred to Special Master Laura Millman on February 29, 2000, almost five years after we had filed. Specifically, our petition claimed that Hannah had "suffered an on-table, residual seizure disorder and encephalopathy after receipt of her third DPT vaccination." That turned out to be only partially accurate. We soon discovered, effective March 10, 1995, and just twenty-four days before we filed our petition, that U.S. Department of Health and Human Services (HHS) Secretary Donna Shalala had removed "residual seizure disorders" from the Vaccine Injury Table in the National Childhood Vaccine

Injury Act.[2] Had residual seizure disorder remained on-table as originally established in the Act, Hannah's DPT shot would have been legally presumed to have been the de facto cause of injury, and we would most likely have received an award quickly.

This stroke of an administrator's pen was a game changer. Without a single vote cast, Shalala significantly altered the program that Congress had painstakingly crafted as a compromise between a threatened vaccine supply and those children injured from receiving the vaccine. We were now compelled to prove that Hannah's shot was the cause of injury. Proof instead of presumption. This posed a far more difficult and costly burden for our family, and as we'd learn, for almost any petitioner.

Given its timing, a cynic might wonder if this administrative tweak to the table of injuries was the result of a brokered concession that emerged from those budget battles in the mid-1990s between President Clinton and House Speaker Newt Gingrich surrounding Medicare, education, and public health. Regardless of the reason, proceedings in the VICP were about to become more adversarial, more costly to pursue, and less likely to succeed for the many families unfortunate enough to be seeking justice for their children's vaccine-induced injuries and deaths. Is it possible that the intent of Shalala's action was to reduce the number and amount of awards?

As we approached our hearing date, one of the experts slated to testify on Hannah's behalf suddenly became unavailable, with suspicious timing that necessitated an eleventh-hour request to the court to reschedule so that we could try to find a replacement with a similar pedigree. This turned out to be a critical blow in our effort to prove "causation." Additionally, as noted later in the Special Master's written decision, our attorneys were accused of using "high-handed" tactics in creating the delay we needed, leaving in question whether this had an effect on her judgment.[3] Discussions with our attorneys also revealed that Special Master Millman's past decisions exposed a bias; she believed that to implicate the DPT shot as the cause, a fever must be present at the onset of the seizure.[4] Although arguments were made to the contrary, she ruled Hannah did not have a fever at the time of her initial hospitalization. The Special Master's bias about DPT causation combined with her already expressed aggravation about the previous delays did not bode well. Hannah's hearing finally went forward on July 23, 2002, more than seven years from the time we filed our petition with the court. These years of delay to get a hearing were hardly consistent with the "expeditious" intent of the 1986 Vaccine Act.

The actual proceedings went pretty much as anticipated, but with a few surprises. The hearing lasted one day and involved testimony from our substitute expert, an adult neurologist, and Hannah's treating neurologist with Children's Hospital of Pittsburgh, as well as Robie and me. The government's expert was a pediatric neurologist and author with twenty years' experience. That we now needed to prove causation made our path much more difficult. In addition, our lead attorney was presenting her first case in a trial setting and, although she did a good job, she did not project the confidence of a seasoned trial attorney.

The Special Master, on the other hand, seemed to have an irritable demeanor, constantly probing the experts on both sides for their opinions on Hannah's seizures and whether there had been fever or not. Both Robie and I were surprised by the dismissive and arrogant attitude the Special Master seemed to exhibit, typified in part by the several meals she ate while on the bench. If this is what was meant by a less formal legal setting under the 1986 Vaccine Act, then we would have vastly preferred a more formal legal environment. Had we been permitted to introduce the drug manufacturer's incriminating evidence of a "hot lot," Hannah would likely have won, but that was not the case. Special Master Millman denied our petition on December 20, 2002, seven and a half years after we filed.

After reading her decision and then reflecting on the big picture, the outcome should not have been surprising. Although the two-hour connection of cause and effect between Hannah's vaccination and her tailspin into lifelong dependency seemed obvious to us, Hannah's claim was rejected. Our expert's testimony was dismissed as not credible, and the testimony of Hannah's treating neurologist was discounted as merely "attempting to help the petitioners in a very sympathetic case." The close temporal relationship of vaccine to reaction was not enough to prove causation. The absence of any other definitive cause for Hannah's condition could not be used as evidence that the shot was the cause. And Special Master Millman's decision was peppered with her contention that Hannah's seizures were afebrile and therefore the shot could not have been the cause.

What were we supposed to take away from this verdict? Did the Special Master believe that Hannah's seizures and developmental regression were an unfortunate coincidence of events, like being hit by a meteorite? Did she think that Hannah was predestined for the life she is living regardless of the shot?

The VICP has morphed from its original mission as a noble effort to do the right thing into a self-serving pretense of justice. Parents who believe this forum offers legitimate recourse for their vaccine-injured child quickly realize their interests are second to the alleged "greater good." If you add in the specter of a vaccine-autism link, you have the proverbial gorilla in the room that no one wants to let out for fear of the damage it could do. Let the gorilla wreak havoc on those in the room—just don't open the door!

Disappointed and now even more angry, we were soon contacted by several law firms that had watched our case closely as it sputtered through the VICP. We learned that we had the option of appealing the Special Master's decision or bringing suit against the vaccine manufacturer in civil court. Hannah's long-term needs had become even more certain now that it was apparent that there would likely be no improvement in her abilities. Faced with the highly unlikely prospect of favorable rulings on appeal, we chose to pursue the suit in state civil court. Ironically, had we been successful in the VICP, one of the Act's stated objectives would have been served: we would never have gone the civil route, and the manufacturer would have been spared the legal challenge we were about to initiate. Our legal efforts were never about chasing a windfall, but solely about providing a safe and supportive environment for Hannah when Robie and I were no longer able to do so.

We decided on the Houston, Texas, firm of Williams Bailey (now Williams Kherkher) for a number of reasons, including our sense that they had the financial staying power to endure for the long-term, never imagining that it would take so long. Over *the next eight years*, we worked with this firm to advance our case through the courts. While I'm sure the firm held internal discussions about the costs incurred and the prospects of continuing, they always pursued a successful outcome with the aggressiveness they showed from day one. Years of litigation wore on, and in spring 2010, the Supreme Court of the United States agreed to hear our case.

After getting the call from our attorney that our petition had been granted by the Supreme Court, we were both exhilarated and relieved. For the first time in nineteen years, we allowed ourselves to feel validated. The highest court in the land was going to grant us this one last dance. We weren't naïve enough to believe that after hearing the merits of our case, these nine compassionate people would acknowledge the broken system through which we had come and correct the wrongs done. Still, we were at least being given a large stage and an engaged audience to which we could cry "foul." The reality was that even if we were fortunate enough to get a favorable decision from the Supreme Court, we would merely have won the right to start all over again in civil court.

When questioned what Hannah's case was about, I offered a condensed overview of preemption—in other words, whether federal law takes precedence over a state's traditional jurisdiction in product liability cases. Pretty dry stuff; hardly a subject that they write songs about, right? But if the case were decided in favor of federal preemption, any child suspected of having a vaccine injury from a defectively designed vaccine would have her financial fate sealed with a thumbs up or down decision from just one individual, a Special Master in the VICP. Regardless, the 1986 Vaccine Act absolves the vaccine manufacturer of any financial liability, and by default, there is then a presumption that the manufacturer committed no fault. I suspect most companies might abuse the privilege of no liability if given that free pass; fortunately, almost all companies are financially liable for harms they cause.

The Supreme Court ruled 6–2 in favor of federal preemption, making it impossible in the future to claim in a civil court before a jury that a vaccine's design was dangerously defective. In practice, all vaccine injury claims must now be decided exclusively in the VICP.

Disappointed, but not surprised, we recognized that the vaccine manufacturers' argument of a threatened vaccine supply and future infectious disease outbreaks won the day. The drug industry had mobilized their forces and had successfully linked our case to the looming storm of five thousand potential lawsuits in the Omnibus Autism Proceeding, in which parents claim that vaccines caused their children's autism. Wyeth's lawyers argued that if these cases were allowed to proceed in civil court, devastation would certainly ensue and, once again, children might be deprived of the benefits of vaccines. Leaving aside whether that threat is real or contrived, I found it interesting that Chief Justice John Roberts owned stock in Pfizer, the parent company of Wyeth. He recused himself, subsequently sold his stock, then withdrew his recusal to resume his leadership position on the bench. I was surprised that these questionable actions drew little scrutiny.

Before the commencement of the Supreme Court hearing, one of our attorneys sent us a card with a bracelet for Hannah. She described Hannah as courageous for her perseverance and noted that Hannah, like all Supreme Court petitioners, was now tied to history: she "walks with giants." We couldn't agree more, but the pride we all felt was ours alone as Hannah was unaware of her legal participation and will be forever locked in a world without ordinary cognition and language. She endures pain without expression, unlike those who love her most of all and know her struggle. It has been twenty years since that fateful vaccine and eighteen years in our search for justice. Through it all, we learned a simple truth: Our system of judicial compensation has ruled repeatedly that Hannah is not entitled to any remedy. What of all the others now affected with even less recourse?

As a citizen unwilling to give up on the legitimacy of our legal system, I have to believe that the blindfolded lady with the scales of justice in her hand can feel the tugs at her robe. Robie and I remain hopeful that she will do the right thing, if not for Hannah, then for the vaccine-injured children who follow.

Editors' note: On February 22, 2011, in a 6–2 decision by Justice Scalia, the Supreme Court held that the federal vaccine act preempted all design defect claims based on the text of the statute. In chapter 5, "The Right to Legal Redress," the authors Holland and Krakow provide excerpts from the forceful dissent by Justices Sotomayor and Ginsburg. In our opinion, the dissent incorporates a clearer understanding of the legislative balance struck by Congress between the rights of injured children and protection of vaccine manufacturers and is therefore the more compellingly reasoned decision.

Chapter Twelve

THREE SHORT YEARS

Gay Tate, PhD, LSW, MLSP

Not many things can compare to the heartache of helplessly watching your young child lose developmental skills and regress into a world of mental confusion and physical pain. This is something you would never wish on any parent. Sadly, I have had to face this challenge not once, but twice. My name is Gay Tate, and I am a mother, former scientist, and practicing psychotherapist. My husband Allen and I have three children, ages twenty-one, eighteen, and fourteen. Our oldest was born in 1989 and was vaccinated according to the recommended schedule of the 1980s. He has always been strong and healthy. Our two younger children, born in 1992 and 1996, received their vaccinations according to a sharply expanded vaccination schedule, beginning at twelve hours old—it was the schedule typically followed for American children born in the 1990s. They have both been diagnosed with autism.

They are not alone.

This series of events and outcomes has repeated with thousands of other children and families, and autistic spectrum disorders have reached epidemic levels in this country. My family has experienced, firsthand and from the beginning, the unfolding of what I believe to be a man-made catastrophe. In less than two decades, we have moved from needing to explain what "autism" is, since so few had ever seen a person with autism, to seeing autism studies offered as a major in college.

This is our story.

Young Allen, my firstborn, was almost three years old when I was expecting my second child. I had left a rather intense and demanding life as a research scientist to experience motherhood full-time. Allen was sweet, precociously verbal, and very excited to be getting a new brother or sister. My pregnancy was smooth and uneventful, and Kenny was born in June, big and healthy, the day before my fourth wedding anniversary. Life was hectic, but good.

During his first year, Kenny hit his developmental milestones. He was good-natured, had a great appetite, and grew well. At his four-month well visit, he had developed a wheeze following a cold. His doctor prescribed a wide-spectrum antibiotic and a nebulizer, and because he was running a fever, she agreed to postpone his four-month round of vaccines for a week or two. Ten days later, although he still had a wheeze, he was not running a fever, and we went ahead with his shots. With time and a couple more rounds of antibiotics, his wheeze eventually resolved and he continued to grow and progress.

I remember that Kenny spoke fifteen words at fifteen months old, just like his big brother. However, I was concerned that he was not walking yet and even seemed oddly fearful of cruising at times. When I reported this to his pediatrician, I was told not to worry: "Big babies are lazy, he'd rather talk than walk, the intellectual type." The doctor was mildly concerned, however, because his head size had increased and veered off his growth curve slightly. Still, he gave Kenny his MMR vaccine. Just to be sure, off we went for a skull X-ray and ultrasound. Both were normal, no problem; but by eighteen months I was aware of a significant change. Kenny had slowed down. He spoke less and tended to sit in one spot, often with his head tilted to one side as if it was too heavy. He still was not walking on his own. Again, the doctor tried to reassure me that boys are slower to develop and that we should give him until twenty months to walk on his own. In my gut, I knew something was very wrong, but the people around me dismissed and minimized my concerns. It was a lonely and dreadful feeling.

Then Kenny's disposition, which always had been sunny and affectionate, began to take on a quality of misery. Kenny often seemed disconnected, hypersensitive, and fearful. He was obsessed with ceiling fans and lights, and started to wake up many nights either laughing uncontrollably or screaming in pain from diarrhea. By now, my husband was very concerned. At age two and a half, the pediatrician finally agreed that Kenny showed developmental delays in all areas and recommended a specialist's evaluation. I was desperate to know what this meant for his future, but doctors said that it would take a couple of years before anyone could say for sure. We would just have to wait and see. In the meantime, Kenny qualified for early intervention services. I watched Kenny constantly, calculating his every move in terms of the developmental charts. It sometimes took conscious effort to pull my eyes off him to pay attention to Allen, which added more guilt to my emotional load. We were also now on the treadmill of endless speech, physical, and occupational therapy sessions, usually with Kenny screaming, me sweating, and Allen trudging along with a toy and a book.

As Kenny made progress in some areas, new behaviors emerged. For example, as his walking improved, he went up on his toes and flapped his hands. As we got some language back, he began to verbalize with endless, loud *eeeeeeeee* sounds. Moreover, he was obsessed with videos.

Discontent to wait years for an answer, I did my own research and concluded that Kenny's behaviors fit the criteria for autism. As surprising as it might sound in today's world, my doctors and therapists again minimized my concerns and tried to talk me out of it. Developmental delays, yes; some sensory issues, maybe; but not autism. It was early 1995, and the wave of regressive autism that is so prevalent today was just beginning. It was not until Kenny was nearly four and I was pregnant with my third child that I was gently informed that Kenny was "moderately to severely autistic," as though it would be a terrible shock. I was past shock and it was certainly not news to me—I knew well before the "professionals."

I also could not get satisfactory answers about all the physical maladies afflicting Kenny. His nose was constantly stuffy. He had crust behind his ears and nearly nonstop diaper rash. When I asked about his eyes rolling back in his head, the doctor said that as long as Kenny could respond to my voice when it happened, it was not a seizure (good . . . I think . . . but what was it?). The diarrhea was the worst, though—foul and discolored. It was obviously so painful for him. He would burst into tears and the diarrhea would start, sometimes so often that his bottom would bleed. It was heart wrenching to see, and it just didn't make sense. He had been fine and was born healthy and beautiful. I was from a medical family with a strong background in science and was not aware of any disorders with this timing and this particular array of symptoms. He just seemed so sick.

I bounced for a while between anger and grief distorted into a form of aggressive industry. My husband Allen and I found a program that favored dropping diagnostic labels and treating problems like Kenny's as a brain injury. We hired a caregiver, attended a weeklong intensive training, and put together an extensive physical, behavioral, and academic program for Kenny. We tried a different diet, sound and smell therapies, set up a special gym, and looked at endless flashcards and photographs. My husband Allen, our caregiver, and I spent many hours doing creeping and crawling strengthening exercises with Kenny through the house, and we did patterning therapy most mornings with another family whose daughter had developmental delays. I can still remember young Allen demonstrating the patterning technique and coaxing the younger kids up onto the training table. It felt better for all of us to be doing something, and it did help—Kenny was getting stronger.

In the middle of putting this all together, my daughter Olivia was born. She was a joy, bright and curious, and always engaged with the people and activities around her. By now, Allen was seven, doing well in school and playing soccer. I was doing OK. When I started to feel sorry for myself, and overwhelmed by Kenny's disabilities, I told myself that I could handle it. I tried not to think, "Why my child?" but rather "Why not? There are no guarantees." With intensive work, he would get better. I had struck a fragile balance and life went on.

Olivia continued to flourish, and was more active and social than even young Allen had been at her age. She was physically precocious and had easily hit all her developmental milestones by her eighteen-month checkup, when she was given the MMR vaccine. (I can still remember the nurse practitioner's panic that Olivia, for some reason, had not been given the MMR at an earlier visit.) At two years old, she knew colors and shapes and loved the alphabet. However, I noticed that a change was occurring. In stores, she began rubbing the rugs instead of focusing on shopping, and although she knew many words, she wasn't stringing them together. When I expressed my uneasiness, friends and family told me I was understandably a little paranoid after dealing with Kenny's development. After all, the pediatrician had never found any cause for concern at her well visits. I latched onto this and managed to stay in denial for a few more months.

In late 1998, I was in New York City attending a conference on autism, hoping to gather the latest information for Kenny's program. Parent groups, along with a small but growing group of medical researchers, were mobilizing, trying to figure out what was happening to so many children. At the end of the day, I remember listening to a talk on a new developmental screening called a CHAT test (Checklist for Autism in Toddlers). Apparently, a lack of forefinger pointing at objects in a young toddler's behavior was the single most telling sign in predicting autism. It hit me with such force—Olivia had stopped pointing. Within seconds, my denial disappeared, and I felt as if someone had pushed me into an empty elevator shaft. To this day, I can still remember the suddenness of the realization and physical shock that Olivia was regressing toward autism. I am not sure how I made it home to Pennsylvania.

Over the next year, despite private speech therapy, my daughter's language slipped away and was replaced by occasional inappropriate words and phrases from Barney videos. Her behavior changed from outgoing and curious to fearful and rigid. She cried a lot, and her facial expressions took on a distant, even bewildered, look. At three and a half, she was officially diagnosed with autism, and I was utterly heartbroken. I was no longer even superficially OK. I could not fathom how I would get through this again.

For a while, I struggled with feeling depressed and overwhelmed. Everyone has a limit, and I had reached mine. Nevertheless, I tried to keep my game face on and keep moving forward with interventions for Olivia. At this point, we scaled back Kenny's home program because he had entered school full-time as one of four boys in the first autism support classroom in our district.

We were still exhausted from the many nights Kenny would wake and stay up. From age three to almost nine, he slept on the couch in our den, often not more than four or five hours. In addition, our caregiver, only nineteen when we hired her, had understandably moved on with a new career, and finding a reliable replacement had been challenging. It was a difficult period. Looking back, I realize that I was also incredibly sad. Grief after an autism diagnosis is difficult to process. For me, it felt like there was no time, as if the window of opportunity would close for turning things around if I relaxed or got lost in emotions. There was a sense that I was never finished, that I hadn't done enough or had missed something crucial. There was a lot of guilt.

Eventually, though, things began to look up a bit. After a failed attempt at regular preschool, Olivia began attending a full-time autism support class. By this time, children with a PDD (pervasive developmental disorder) or autism diagnosis were not so rare and services had to adjust. I worried that a full-time preschool program would be too much for her, but she seemed to enjoy it, bus ride and all. Around the same time, we found a wonderful new caregiver. Young Allen had become quite involved with traveling soccer and basketball teams, guitar and dance lessons, and this gave his dad (coach) and me more time to support him and attend his games and performances together. Also, with the help of a good Defeat Autism Now! (DAN) doctor and developmental pediatrician, we were able to calm Kenny's raging intestines. Improving his intestinal health, along with a low dose of Risperdal, helped him finally sleep though the night and tolerate sleeping upstairs in his own room in a regular bed.

Over the next few years, we chiseled away at Kenny and Olivia's physical symptoms and behaviors with various prescription medications, supplements, diets, and therapies. Nothing dramatic happened with either child, just slow, steady progress. Both children had developed some very basic functional language. Although still in pull-ups, Kenny's muscle tone had improved, and he seemed happier and more comfortable in his own skin. Olivia sometimes struggled with anxiety and obsessive-compulsive tendencies, but she loved the structure of school and asked to go every day. Our district had developed a good autism support program and both kids had excellent, committed teachers and therapists. However, what seemed to be lacking there, and everywhere, was alarm—recognition that something very wrong was happening and causing autism diagnoses to skyrocket. In one school district, in just six years, the need for autism support had ballooned from Kenny's original class of four boys to seven full-time *classrooms* at the elementary school level!

In 2005, after sixteen years of being home full-time, I decided to venture out to go to graduate school, part-time, for a master's degree in clinical social work. Yes, it was still true that our family's needs and complexities were daunting, but I felt we had reached a comfortable equilibrium that could accommodate my new goals. However, although I finished my master's program, a challenge we could not have fully anticipated was the powerful impact of puberty on Kenny and Olivia. Again, this was uncharted territory with this new wave of sick children.

During the first year of middle school, Kenny, usually sweet and cooperative, had some episodes of aggression with teachers and aides who were not familiar with his signals and behaviors. His school called Allen and me into several meetings with the rest of his team and the situation was easily resolved. Then, at age fifteen, the day after Christmas, Kenny had a massive grand mal seizure. Allen was coming back from driving young Allen to the airport on his way to a basketball tournament, and I was home alone with Kenny and Olivia. I had taken Kenny to sit on the toilet when he started to seize. He was fully grown by then and in the small bathroom, there was no place to get him down to the floor without one of us being injured. I don't remember how I dialed 911, but I did. Inexplicably, they did not arrive for fifteen minutes. In that time, his breathing stopped and started several times, and I had trouble finding a consistent pulse. When the paramedics finally arrived, they hustled me out of the house and began working on Kenny. I remember sobbing in our driveway while Olivia smoothed my hair and wiped tears off my cheeks. The ambulance transported Kenny to our local hospital, and then quickly airlifted him to Children's Hospital in Philadelphia. He was in intensive care and neurology for a week and has been on seizure medication ever since. I believe that if this had happened during the night or I had not been nearby, he would have died.

Olivia began middle school and her menstrual period at about the same time, in the fall of 2008. Her anxiety and obsessive-compulsive disorder (OCD) symptoms had already been escalating and the erratic cycle of hormone activity put them over the top. The girl who had loved to go to school now had more and more days of either lashing out physically at teachers and aides or curling up in a ball in the foyer of her classroom. Several times, her school asked me to come pick her up because she had her period and refused help in taking care of her hygiene needs. In January, our developmental pediatrician directed us to an adolescent specialist who suggested putting her on birth control pills to regulate her periods to four times a year. The doctor encouraged us to weather any early, unpleasant side effects, and stick with the pills for at least six months. She was miserable and we were desperate, so we tried it. Olivia went downhill fast, and her behavior deteriorated even more. In May of that year, a boy in her class had his first seizure. Olivia apparently became very agitated by the unusual activity in the classroom and began throwing things. When the paramedics arrived, she went after them, trying to shove them away from her classmate. That afternoon, her assistant principal informed me that she could not return to school because she was a threat to the safety of other students.

During the first few weeks at home, she spent most of the time in her bed, under covers. She looked haunted and miserable. We brought food to her and had to encourage her to eat. I could not get her in the bathtub for two weeks. It was so painful to see her suffering so much and unable to explain how she felt. When she did come downstairs occasionally, she sat in one spot on a wooden chest, but not on couches or chairs. This went on for several months. When her homebound tutoring began, her teacher sometimes had to work with her from the side of her bed. Olivia had always been physically talented and inclined. She loved to swim in our backyard pool, roller skate, and do yard work with Dad. All of that stopped. Olivia did not go outside for four months, and it felt like we had lost her all over again. It was devastating.

Over the past year, we have weaned her off hormones and worked more with supplements to restore normal neurotransmitter levels. Five months ago, Olivia was able to tolerate attending a specialized autism support class in an alternative school. It took two hours to coax her into the building, but she has been able to build up slowly, in half hour increments, to a full school day. We are very proud of her courage and progress. However, her OCD symptoms remain a challenge for her and for others in the classroom, and these symptoms often rule our home.

Just as we were rejoicing over her school achievements, Olivia had her first seizure. Although it was not as life threatening as Kenny's, it brought her patient and unflappable father and me to tears.

And so our story goes. Just as we begin to exhale, or think we cannot possibly dig any deeper for strength, another scary challenge comes along that needs to be added to our pile of worries and uncertainties. At this stage, however, what is not uncertain for us anymore is that Kenny and Olivia's symptoms, which are labeled as "autism," are really symptoms of vaccine damage and mercury toxicity. The full realization of this came slowly, with times of resistance and disbelief, probably for several reasons. First, neither child had a sudden and obvious adverse reaction. In hindsight, Kenny had shown more of a slow deterioration with each round of vaccinations. Olivia's regression had been a little steeper, and seemed to begin around the time of her MMR, at around eighteen months. Second, they were born at the beginning of the autism epidemic, particularly Kenny. Looking back, I remember monitoring both children, as instructed, immediately following each round of shots, for symptoms such as "extremely rare" high-pitched screaming. However, I was not looking for any delayed signs of an adverse reaction, and therefore, did not connect later problems to their vaccines. I doubt if many parents did, especially back then. In the early 1990s, parents and most pediatricians had never seen this kind of physical deterioration and developmental regression in formerly healthy, normal toddlers. Today, less than twenty years later, with 1 percent of American children developing autism, the appearance of autistic symptoms is a major concern among new parents, and pediatricians now routinely screen toddlers at well visits.

Another source for my disbelief that vaccines could have caused such harm to my children is my age and background. I was born in the 1950s and was very aware of the excitement over the polio vaccine. My mother was a physical therapist, trained at a time when the career was in its early stages. In our home, I remember that she prominently displayed a framed picture from a magazine article about her working with a child with polio. My father was a family doctor and we lived in the country around horses and cows. We were not concerned if I fell and scraped my knees because a tetanus shot would save me. For several years, my older sister was one of the pediatric nurse practitioners doing well baby visits at the large pediatric practice where I eventually took my children. The belief system was (and for many still is) that vaccines were one of the greatest achievements of modern medicine, and it was irresponsible and very risky not to vaccinate your child. It was a powerful message all around me that eventually clouded my common sense and contributed to a disconnect between what I learned about the development of the immune system in graduate school and research, and the delivery of so many vaccines so early, beginning in the first twenty-four hours of life.

Although not from a medical family, my husband's parents raised him with a powerful belief system as well. He was a molecular biologist by training and we had met and briefly worked together in a research lab at New York University Medical Center. After a couple of years, he returned to graduate school for an MBA. With that combination of skills, he was quickly, and suitably, snapped up by the pharmaceutical industry. Even though he was never directly involved with vaccines, his work culture, understandably, had an impact on the process of piecing together what had happened to our children.

To be clear, I am not anti-vaccine. I am for safer vaccines and vaccination choice. Vaccines serve as important tools for controlling certain infectious diseases. Nevertheless, the reality is that they are not miracles; they are medicines, with potentially negative side effects expressed differently in different people and under different conditions. The side effects are additive with each individual vaccine as well as from combinations. From my graduate studies, I came to appreciate how highly evolved and intricately balanced the immune response is, and that one of the main tasks of the developing immune system in the first months of a child's life is to learn to differentiate between what is self and what is foreign. This is exactly when our vaccination schedule is applied most aggressively, artificially stimulating an immune response via an unnatural route of infection. It follows that some children, perhaps because they are sick or have a family history of autoimmune disorders (like mine do), will simply not be able to tolerate this bombardment, and chronic, systemic damage will likely ensue. All of this I knew intellectually (pangs of guilt). Yet when our government suddenly increased the number of vaccines in the early 1990s, the "vaccines are good and necessary" message somehow superseded my ability to make the connections. I put my faith in the pediatric experts to keep me up to date on the best and safest infant care.

The difference in the vaccination schedule in the three years between Allen and Kenny was dramatic. Our government had more than doubled the number of shots in the first six months, and they were also given earlier, beginning on the day of birth. However, I believe the most damaging aspect of the change was the sharp increase in the amount of mercury that Kenny, and later Olivia, received. Their pediatricians gave Kenny and Olivia the additional hepatitis B and *haemophilus influenzae* type b (Hib) vaccines, both containing the preservative thimerosal, which is 50 percent ethylmercury by weight. Mercury is a potent neurotoxin, and very damaging to the immune system. Kenny and Olivia each received three times the amount of mercury, beginning just days after birth, as Allen. Those in charge of the vaccination schedule apparently forgot to add up the amount of mercury that American children would receive because of the added shots. How careless! Even worse, how reprehensible to continue, to this day, to publicly deny the damage that was done.

How do I know Kenny and Olivia suffered from mercury toxicity? I have test results. We had each child's porphyrins tested. Porphyrins are derivatives of the heme synthesis pathway and are normally found in urine. Variations in urinary porphyrin metabolite patterns can be used to measure the overall body burden and toxicity of mercury, as well as other heavy metals and toxins. I can still remember staring, through angry tears, at the results of my three children's porphyrin profile tests. Kenny, cognitively and physically the most affected child, showed the highest level of mercury toxicity. Olivia, less affected, showed clear but less toxicity, and Allen's results showed no evidence of mercury toxicity at all. All three children shared the same parents, food, water, air, time spent nursing, and even the time of year they were born. The biggest difference in environmental exposures that happened during those three short years was the increase in their vaccines. There it was, clear data from my own family and a microcosm of what had happened to so many children. A sudden increase in autism, beginning in the 1990s, that coincided with a change in a single set of exposures—vaccines, received by all children in our country without regard to genetics, diet, socioeconomic status, geography, or quality of medical care.

Yes, my family has ridden the wave of this tragic, man-made disaster from the beginning. Autism, practically unknown when young Allen was born, now affects 1 percent of our population and one in seventy boys. The pain and sickness it has caused the affected children and the toll it has taken on families is incalculable. Marriages have been hard hit, and the divorce rate is high. Fortunately, Allen and I have become tight partners during the last twenty years, but we struggle to find time for each other and respite from the full-time care needed for Kenny and Olivia.

Siblings are tremendously affected, not only as children, but also as adults who will carry the burden and face the challenges of the next generation. Young Allen has been an incredible older brother, from the beginning, in so many ways. As a young boy, he would carefully practice the best and most informative ways to explain Kenny's behaviors before a new friend came over. I can still remember his sad and stoic face when I told him that Olivia was going to need speech and some other therapies, like Kenny. He dug in and helped with her therapy. In high school, he would announce autism fundraisers, and urge other students to participate in autism walks and races. As a young man in college, he has expanded his advocacy to writing and speaking, as part of his involvement with such groups as the Elizabeth Birt Center for Autism Law and Advocacy (EBCALA) and the Center for Personal Rights. You will again be introduced to young Allen, as he is the author of chapter 10 of this book, "The Greater Good." Watching Allen with his brother and sister through the years and seeing his advocacy for children like them, who may be susceptible to vaccine regression, has been a quiet, steady lesson in unconditional love.

More than a few times, people have told me what an amazing kid Allen is. Then with the very best intentions, they mention that growing up with Kenny and Olivia must have given him extra compassion and maturity. Alternatively, people will comment on how very patient (my husband) Allen and I are, that we are the right people to be raising Kenny and Olivia. I thank them, and there is truth to what they say. However, a part of me bristles. Allen was an amazing four-year-old long before the word autism came into his life; my husband and I have always been patient people. One thing did not have to happen for the other to be true. The vaccine damage to Kenny and Olivia should not have happened at all. They and thousands of other children have been denied the opportunity to achieve their full potential at a great cost to our society, both financially and in terms of human resources. I don't want people to neatly arrange any part of this unnecessary tragedy in their heads. I don't want this to happen to anyone else.

With that in mind, I urge those who know the story of our family to pay close attention to the vaccine injury debate, to learn about the autism epidemic and the growing epidemics of neurological and autoimmune disorders happening today, and to demand both informed consent and a safer, individualized vaccination schedule from our public health officials.

Chapter Thirteen

PEDIATRICS: SICK IS THE NEW HEALTHY

Judy Converse, MPH, RD, LD

I am struck by the contrasts in a certain three doctors I know. Two are pediatricians and the third is a renowned surgeon with expertise in immunology.

The first pediatrician's career included—in addition to a busy practice spanning three decades—defining and promoting health and immunization policy in conjunction with his state health department and an Ivy League university. It also included years of volunteer work for the autism community. He is a recognized expert witness in litigation related to vaccine injury. He is happily married to his first wife and has four children and twelve grandchildren.

The other pediatrician received his medical degree from the same Ivy League university, served in the National Health Service, has a master's degree in public health in addition to his medical degree, and has practiced with no particularly distinguishing achievements. He is married and has two children.

One unequivocally supports vaccination to the fullest, believes all vaccines are safe and effective, and vaccinates children daily without the slightest hesitation. The other now accepts the possibility that vaccines can trigger autism in susceptible children, supports findings that show a link between inflammatory bowel disease and MMR vaccine, and questions the current pediatric vaccination schedule.

One is knowledgeable about food allergies and sensitivities, and encourages special diets to mitigate them. The other is not informed about these issues, and has advocated the use of asthma medication for children with multiple food allergies but no asthma symptoms.

One has a grandchild with autism. The other has a child who committed suicide.

One operates within the status quo of pediatrics, assiduously following the guidelines of the Centers for Disease Control and Prevention (CDC) or American Academy of Pediatrics (AAP) in practice. The other is aghast by the degree of chronic illness and disability that has become normal in children in the new millennium, is relieved to be retired from practice because of this, and hopes for a dramatic change in his field.

For years, we were patrons of the pediatrician who seemed unable to answer my questions with anything other than broad-stroke missives from the CDC or the AAP. My son didn't particularly like or trust him. He once told us that rice was not allowed on gluten-free diets (not true), and his head nurse did not know what a gliadin antibody test was (it checks gluten sensitivity; my son's test results are highly positive). Like many of the parents I've met in my pediatric nutrition care practice, I've felt frustrated by having a pediatrician who not only didn't understand our healthcare needs but was not interested in them, and much to our dismay, neither was anyone on his staff. He continues to practice, unfettered by controversy.

The other pediatrician described here is a former neighbor of mine, who is retired. In spite of his spotless record, expertise, and tireless service, he has been marginalized for investigating vaccine safety. He is trivialized for speaking up about the possible role of vaccines as a trigger for autism, seizures, sudden infant death syndrome (SIDS), autoimmune disorders and allergies in children, and shaken baby syndrome (SBS)—even though any vaccine package insert states that the product inside may trigger seizures, death, or allergic reactions, among many other adverse events.

I would like to have a pediatrician like him. He has an impeccable reputation among his patients, is kind and caring, and is an independent thinker who listens and then answers questions thoughtfully. My son knows him, likes him, trusts him, and always has—even when he was a young toddler with sensory issues that made him anxious around almost anybody. He felt instantly comforted by this man's reassuring manner with kids. When this doctor has no answer, he says so and directs me to someone who might know more. Unfortunately, I can't have him or a younger, unretired version of him as my pediatrician. A pediatrician who questions vaccines is not only unwelcome among his peers and professional organizations but is cast aside as unscientific, intellectually feeble, and even unpatriotic—a knee-jerk reaction that smacks of bias and zealotry.

I use the other pediatrician only because he is available. He is all that our beleaguered healthcare system sanctions. Insurance pays for doctors like him, but not for the ones who think independently. This pediatrician's scope of practice is in lockstep with what the health insurance industry dictates, which, in turn, is in lockstep with what the pharmaceutical industry dictates. If there is no need for prescription drugs or vaccines, there is no need for this pediatrician; indeed, we rarely see him. I am sure he is a good, kind, and even smart man. Why did he stop thinking independently and critically?

Imagine your child off drugs. All drugs. No antibiotics, vaccines, reflux medication, laxatives, statins, antipsychotics, sleep medication, stimulants, antidepressants, seizure meds, insulin, steroids, nebulizers, antiinflammatories, suppositories, or anything else I've missed. Imagine your child being so healthy that perhaps just one thing from this list is needed every five years or so. If you were born before 1970, you grew up with few drugs and vaccines. Odds favor polypharmacy for those born after 1990 or 1995.[1] If your child never used any of these things, would there be any reason left to go to the pediatrician? What else would be offered for care if not a medication or a shot? Incredibly, it is now the norm to turn the pharmaceutical spigot on children, starting at birth—and nobody blinks. I often ponder the colossal profitability of this. Through a persistent, pervasive hypnosis from pediatricians, news outlets, print ads, web content, governmental press, TV programs, and magazines, parents are being taught that it is normal to have a child who can't talk, poop, walk, eat, digest food, grow, attend school, or perhaps even breathe without the benefit of a litany of medications and vaccines. Worse, we are conditioned to believe that children who *don't* join in are somehow *not* normal, or—perhaps more inexplicable—that they are a danger to children who do submit to the dozens of vaccines, antibiotics, and drugs. It is arguably the other way around, but that's another conversation.

In the same time frame in which giving babies and children dozens of drugs and vaccines became normal, our children have gotten sicker, not healthier. It cannot be claimed that this style of healthcare works. My vantage point on this is different from your pediatrician's, by virtue of a different training in child health. I am a registered dietitian, and I have an undergraduate degree in nutrition sciences and a master's degree in public health nutrition. This training taught me that nutrition status is what drives a child's ability to fight infections. Low nutrition potentiates susceptibility to infection, and nutrition status declines during infection.[2] Ample and strong data agree—data that show that one-third to one-half of the millions of deaths in children across the world, caused by measles, diarrheal diseases, pneumonia, AIDS, and malaria, are due to undernutrition, not undervaccination.[3] Data show that even mild malnutrition (the kind I see in practice here in the United States, especially in kids with autism) changes how often kids get sick and whether they live or die from an infectious disease.[4]

I was taught about vaccines too. When I finished a graduate degree in public health in 1988, nutrition and vaccines were not presented as mutually exclusive tools for maternal, infant, and child health. Now it seems that data on child nutrition are ignored—and parents are told that the only way to have healthy children is to fully vaccinate them. This is a myth that the body of evidence in child nutrition does not support. What *is* well supported is that nutrition drives learning, growth, development, and immune function in children. Look at it this way: For your child's first year of life, if you could afford only a box full of the thirty-odd vaccine doses now recommended for zero-to-twelve-month-old infants or a year's supply of nutritious food, which would you rather have?

I was trained to uphold the CDC's goals and objectives for public health. In 1988, its "goal number 1" for the year 2000 was to "increase span of health life."[5] It is not difficult to see that American public health policy has failed at this: Data from several sources show that, for the first time in U.S. history, children's life expectancies may be shorter than their parents'.[6] Despite the rapidly growing vaccination schedule, American children are more chronically ill and disabled than ever. They are more likely to die as babies than the children in more than forty other countries. We have an abysmal infant mortality ranking that has done nothing but slide from bad to worse since the 1950s. Children in the United States today have much higher rates of asthma, autism, diabetes, Crohn's disease, epilepsy, life-threatening food allergies, obesity, ADHD, learning and behavioral disorders, and suicide than kids did in the 1980s, when fewer vaccinations were given.

Were these genetic phenomena unnoticed before? Martha Herbert, a Harvard neurologist who also holds a PhD, put it this way: "Genes load the gun, but the environment pulls the trigger." The one trigger we pull dozens more times than ever before is vaccination. Children now get seventy doses of sixteen different vaccines through age eighteen. Among possible causes of the health problems plaguing our children are immune dysregulation and toxic exposures, so we must examine the role of vaccines as possible triggers.

What if it were true that the way we now vaccinate children causes more death, chronic disease, and disability than it prevents in America? Blanket assurances that this can't happen are naïve. Pharma-funded studies to diffuse concern—whether the money flows first through university foundations, hospitals, or nonprofits—are obviously flawed, inherently biased, and can't yield clear information. I am disturbed by how sick the children are who I meet each week in practice, but I am more concerned that their pediatricians seem nonchalant. They are like my pediatrician—not well informed beyond pharma-driven information, not thinking critically, not using their training and skills to the fullest, and bending to the dictates of the insurance and pharmaceutical industries, who ultimately underwrite what they do every day. Is this the best we can offer?

My graduate training fully indoctrinated me in the reasons for vaccination. I need no instruction on its touted benefits. I was also privileged to have classmates in graduate school at the University of Hawaii who were doctors, nurses, and health workers from all over the world. As a *haole* (white) U.S. mainlander, I was in the minority. They had lived or worked in places like Vanuatu, Papua New Guinea, or Karachi, Pakistan, where they had seen children die of measles. They were from impoverished regions in Africa, Indonesia, or Chinese Taiwan, where children die of infectious diseases more often than in the United States. I understand what it means to lose children to preventable deaths. I also watched my brother die a slow, agonizing death from AIDS—a disease whose very existence some trace to unabashed negligence in the vaccine industry,[7] and from which the same industry hopes to profit with a new product.

Any public health training teaches that vaccines do kill some babies and children. I was taught that infants who die from vaccine injuries are probably defective and would have died anyway, that these are necessary deaths, and that it's best to weed these children from the gene pool. Even then, years before I became a parent of a healthy full-term baby who would nearly die from a vaccine injury, this shocked me. Is this eugenics? Isn't public health for *all* of the public? How many parents would like their children's lives to become necessary collateral damage? Is it constitutional to kill or injure some children for the supposed good of others? Is it moral?

Is it good for others? Read the year-end summaries of reportable diseases in *Morbidity Mortality Weekly Report*, the CDC publication paid for with your tax dollars. Look historically—go back and view trends. You will see that infectious diseases wax and wane independent of vaccination, which is proof that it takes more than a vaccine to prevent a disease. You'll see discussions over the years about risks and benefits of preventive and treatment strategies, and research on how infectious diseases move through populations, what makes some of us die, what lets some of us live, and why some of us get no illness at all. It is never just about a vaccine.

The last physician in my story knew this. Francis Moore, MD, was the late-in-life spouse of my husband's grandmother, Kathryn Saltonstall. Both knew well what my son had endured from an adverse reaction at birth to the newborn hepatitis B vaccine; both knew of my testimony before a congressional subcommittee on this in 1999, and both knew I wrote a searing personal memoir about this family debacle.[8]

Some readers might recognize Dr. Moore's name. He was a Moseley Professor of Surgery, emeritus, at Harvard Medical School. He maintained an address on campus well into his eighties. He was surgeon-in-chief at Brigham and Women's Hospital in Boston. His accomplishments are too numerous to list. Hailed as one of the greatest surgeons of the twentieth century, he pioneered techniques in organ transplantation and the metabolic support of surgical patients. He led the team that performed the first successful human organ transplant—a kidney transplant in identical twins—in 1954. Dr. Moore authored hundreds of medical articles and six books, including a text that became a standard in the field. By the time I met him, through family gatherings, he was mostly busy traveling the globe to receive awards and honoraria, and having audiences of dignitaries the world over. Out of sheer respect for his stature in the medical community, I never told him my opinions of vaccines—but once I testified before the congressional subcommittee reviewing hepatitis B vaccine safety in 1999, of course, he found me out.

The conversation I avoided for years took place when he phoned me to discuss this. As one might think, Dr. Moore had a very commanding manner and tone. I was more terrified to take this call than I was when I faced off with Henry Waxman (D-CA) at the subcommittee meeting, with news cameras rolling. I did not want to disrespect this man who had contributed and accomplished so much in his field, or my husband's grandmother, the kind, gentle matriarch of the Marion, Massachusetts Saltonstalls. I was certain Dr. Moore was about to crush me, using that signature booming voice, with a rant that would effectively bar me, my husband, and my son for good from further family discourse.

That never happened. Dr. Moore was kind. He listened. He was appalled and puzzled that newborns were being given hepatitis B vaccine, a shot he felt they could not possibly need. He did not know of this policy, though, at the time of our conversation. Babies nationwide had been receiving hepatitis B vaccine at birth for about ten years. He described a "window of vulnerability" for newborns, from ages zero to two months, in which he felt the immune system was too immature to receive a vaccine of any kind, lest it trigger damage. He thanked me for contributing my testimony. I hung up and dropped my face in my hands, overwhelmed by this unexpected reception from a physician at the zenith of accomplishment in his field, after being scoffed at for years by my son's doctors.

Dr. Moore's letter to me is in the appendix. He sent it after our chat on the phone. He calls for "absolute avoidance of neonatal vaccination" in this letter. Read it, and wonder: If one of the world's most respected medical achievers had this opinion, but has now passed on, who else of his stature can continue this conversation? Where are the sage experts to guide us? If he were alive and publicly voicing this point, would he too be dragged through a spurious trial and stripped of credentials and authority, as the British General Medical Council did to Dr. Andrew Wakefield for his research on vaccines and autism? When our doctors are no longer free to think independently, pioneer, investigate, and lead—as Dr. Moore did throughout his career and, as I believe, Dr. Wakefield endeavored to do—medicine is truly dead. Our children will suffer the most. It then falls upon you, the individual parent, to uphold your right to choose with your doctor how you vaccinate your children. It falls on you to investigate, read, explore, defer, or simply stop going to pediatricians if all they can offer is rote information and new needles at each visit. Support the independent thinkers. As Gandhi said, "First they ignore you, then they laugh at you, then they fight you, then you win."

Chapter Fourteen

MY DAUGHTER IS "ONE LESS"

Amy Pingel

I am a single mother of four children, two boys and two girls. We live in Lake Station, Indiana. Before my oldest daughter Zeda got sick, we had normal lives. I worked a full-time job. Thirteen-year-old Zeda was a cheerleader and straight-A student. She was also a big help to me with my other children. She was the daughter every mother dreams of having and her future was bright.

I have always done my best to keep my kids healthy. I made sure they received their wellness checkups and they always got whatever vaccines the doctor suggested. All my kids were completely healthy aside from simple colds here and there—completely healthy, that is, until it was time for Zeda's well visit on November 5, 2008. Then her life, as she knew it, completely changed along with the lives of the rest of our family.

On November 5, 2008, at her annual checkup, Zeda's pediatrician suggested that she receive the human papillomavirus (HPV) vaccine or, as I know it now, the Gardasil vaccine. I thought, "OK, this is what I am supposed to do to keep her healthy." My doctor said to do it and so we did it. I knew absolutely nothing about the vaccine, other than it was for girls her age. The doctor did not advise me on any side effects, other than the potential for soreness around the injection site.

A week after Zeda got the shot, she started to complain that she wasn't feeling well. She had headaches and felt sick to her stomach. Perhaps like other girls at her age, Zeda could be a drama queen. So when she told me these things, I would tell her to lie down, murmur to her that she had probably had a long day, or give her some Tylenol. It would have never occurred to me that those little signs could have been red flags until November 28, three weeks after she received her Gardasil vaccine. We were on our way to pick up her little sister from a friend's house. While in the car, Zeda kept dropping her phone. Suddenly, my son said, "Mom, I think something is wrong with Zeda!" Zeda was crying, drooling, and her eyes were not looking right. She looked to me as if she had just had a seizure, which I recognized because my brother had seizures. I rushed her to the nearest emergency room at St. Margaret's Hospital in Hammond, Indiana.

Little did I know that this was the beginning of an unbelievable nightmare for Zeda, for me, and for our entire family. The ER staff did not even know what was wrong with her. First, they accused Zeda of overdosing on drugs. They yelled right in her face, as if she was hard of hearing, that she needed to tell them what she had taken. I was scared out of my mind. I had no clue what was going on, and these doctors and nurses were screaming at my daughter. Zeda was as scared as I was. I believe she knew that something was seriously wrong. She was confused and unable to say what she wanted. It wasn't until she had another major seizure, right in the ER, that the staff finally started to take us seriously. Of course, by then, they also had received her drug test results, which were negative for any illegal drugs.

After those first few horrible hours, they transferred Zeda to our local hospital where a pediatric neurologist and Zeda's pediatrician could see her. I thought we would surely get some answers there. The doctors started running all kinds of tests. At this point, Zeda was unable to say a whole sentence and could only say a few words at a time. She was terrified and was crying very hard. Zeda had an MRI, a CAT scan, an EEG, and a spinal tap. Whatever they were looking for, they did not find it. Every single test was negative. The MRI did show a "shadow" on the right side of her brain, which they treated as viral encephalitis.

During those four very long days at our local hospital, I began to lose my precious daughter. Zeda stopped talking, stopped eating, stopped walking, and started urinating on herself as she lost control of her bladder. Even as Zeda's health deteriorated rapidly, her pediatrician continued to say—unbelievably—that she was doing this to herself, that Zeda was "faking" it.

Words of reason finally came, not from the doctors, but from the nurses who agreed with me that something was terribly wrong. Zeda's symptoms were not taken seriously until a psychiatrist came in, tested her, and issued a professional opinion that she wasn't faking it. One nurse suggested we go to a specialty hospital, because we were getting nowhere. We agreed.

Within forty-five minutes, we were en route to Riley Hospital for Children in Indianapolis. Once again, I thought that we would finally get answers. We were not so lucky. Even there, the doctors accused Zeda of doing this to herself. They installed twenty-four-hour video surveillance on Zeda and monitored her for about two weeks to try to "catch" her in the act of faking it. They continued to perform test after test, all of which came back negative. Zeda developed uncontrollable movements, an extremely high heart rate (up to 180 beats per minute), and high fevers, including one that was 108.[7] degrees. At one point, her doctors placed her in a "medically-induced coma." Through all this, they still couldn't tell us what was wrong or why this was happening to Zeda. They opted instead to spend most of their time blaming us and trying to prove that Zeda was faking her serious medical problems. Although I repeatedly asked if it could be vaccine injury, no one would touch that with a ten-foot pole. Not one medical professional would go on the record saying that the Gardasil vaccine did this to Zeda. They wouldn't even say that a vaccine *could* do this to her.

Zeda eventually lost lung function and her doctors placed her on a ventilator for several months. They also inserted a tracheotomy and feeding tube because she could no longer breathe or eat on her own. We were at Riley Hospital for four months and her doctors still could not tell us what happened to her, only that they were sure it was not a vaccine reaction. They refused to acknowledge Zeda's vaccine injury despite dozens of cases of similar reactions in previously healthy girls following the HPV vaccine.

Today, Zeda still breathes through her trach and eats with a tube. She is mostly non-responsive and lives her life in a hospital bed in our living room with round-the-clock care and daily nursing visits. This has become the defining struggle of my life. I struggle every single day to do what I can to get Zeda better.

I deeply regret my decision to allow Zeda to get this shot. She has lost all quality of life. I would do anything to get her better, but no one has any answers for me. The doctors still don't believe it was the vaccine. Now that she's sick, no one knows how to help my baby girl.

Tragically, the number of "Gardasil Girls" like my daughter continues to grow. The *Truth About Gardasil* website (http://truthaboutgardasil.org) posts many of their stories. The federal government added Gardasil as a recommended shot for sixth grade girls in 2008, the same year that Zeda received this brand-new vaccine. I can't help but think, if only . . . if only I had read the editorial in the *New England Journal of Medicine* that someone recently showed me—"Human Papillomavirus—Reasons for Caution."[1] It was published two months before Zeda's doctor gave her the Gardasil shot. I now know that pap smears are just as effective, if not more effective in preventing cervical cancer. If this isn't already too much to bear, in the summer of 2010, the Centers for Disease Control and Prevention's (CDC) Advisory Committee on Immunization Practices added the recommendation that sixth grade boys should now receive the HPV vaccine. My sons and my younger daughter will never get this vaccination. We had to sacrifice Zeda to learn this hard lesson.

Even as I have been writing this chapter, another family contacted me to ask for help. They said their daughter had the same experience as Zeda. Our heartbreaking reality is now theirs and they too are being offered no diagnosis or help from their doctors, but only accusations that their daughter is "faking it." It is heartbreaking to watch this play out for them and to have no help to offer.

Judicial Watch, a nonprofit, nonpartisan watchdog organization, began investigating reports of adverse events from the HPV vaccine in 2007 and issued two critical reports detailing the adverse outcomes of the vaccine. After reviewing government reports on Gardasil, Tom Fitton, president of Judicial Watch, remarked, "The FDA adverse event reports on the HPV vaccine read like a catalog of horrors. Any state or local government now beset by Merck's lobbying campaigns to mandate this HPV vaccine for young girls ought to take a look at these adverse health reports."[2]

As of November 17, 2010, there have been 20,978 adverse HPV vaccine reactions reported. There are eighty-nine deaths associated with the vaccine. The U.S. Food and Drug Administration (FDA) has yet to issue a position on these cases. The Department of Health and Human Services (HHS) conceded an HPV vaccine injury-related case on October 5, 2010.[3]

I am here to tell you that I didn't know the harm vaccines could do. I think many parents are like me—they don't know either. I thought vaccines would keep my children safe and healthy. That's what the doctor told me. I never imagined that a vaccine could do this to my daughter. My doctor never told me the risks. No one else did either. I was never shown anything describing potential side effects of vaccines. If you do choose to vaccinate, you had better be very comfortable about the need for each vaccine, because every time you vaccinate your child, there is a risk of severe injury and death. It is crucial that parents understand what is at stake and that the choice is theirs to make. I am not telling you not to vaccinate. I am telling you that people who pressure you to vaccinate don't own the consequences. Only you, as parents, do.

Chapter Fifteen

A MOTHER-SON STORY

Sonja Hintz, RN, and Alexander Hintz

On May 26, 2010, my son Alexander Hintz, acting on his own decision to participate, spoke at the American Rally for Personal Rights about his life story and the need for vaccination choice. Alex regressed into autism following his childhood vaccinations and recovered from autism through biomedical intervention. In his own words, Alex can explain how he developed a personal conviction that all parents must have the right to make individual medical decisions for their children.

ALEX:

I am thirteen years old and in the eighth grade at St. Peter's Catholic School.

As a preschooler, I went to Lakeland School in Elkhorn, Wisconsin, where I received special education and help in learning important new communication, sensory, and social skills. I received physical, occupational, and speech therapy. I met many kids like me and I never thought much about it. I could see that other children had challenges. I didn't think I had any myself. My parents put me on a gluten-free, casein-free diet. It made me feel different from the other kids.

When I was in kindergarten, I attended Lakeland two days and St. Peter's three half-days each week. I felt lonely at St. Peter's. I wasn't interested in what the other children were doing. I didn't like loud noises and avoided the gym. At recess, I watched the kids play. My teacher helped me a lot at St. Peter's. She taught me how to be part of a group. She helped me learn my letters and colors.

In kindergarten, I started working with a tutor. I kept up my speech therapy. In addition, this was the year I started to play my first sport—soccer. I remember being scared of the coach and the other kids. I remember having no idea how to play this sport.

My first grade teacher taught me to read my first book. She even noticed I was growing a lot more. I think it was because of my special diet. I had become less picky and starting eating new foods. In second grade, I continued to receive tutoring and speech therapy. I pretended to be sick because I had no friends and wanted to stay at home.

In third grade, I had to take my tests in a different way. The teacher had to read the test questions to me because I had "receptive speech delays." That meant that I had difficulty understanding the meaning of the words and what the questions meant. By the end of third grade, I had made some friends and enjoyed school a lot more. My friends were able to write in cursive. I still had to print because my hand cramped and hurt.

The summer after third grade, my parents started taking me for chelation therapy. Chelation therapy is the medicine that I received in my arm to drain out the heavy metals in my body. It helped me think more clearly and be more active in sports. In just a few weeks, my stick figure drawings developed into detailed pictures. I remember drawing myself walking my dog! I could also write in cursive.

Fourth grade was the biggest year of change for me. Chelation helped me with my schoolwork. Because of the changes over the summer with chelation, I was able to take written tests on my own. It was easier to write out my answers. I also became involved in new sports, including flag football. During fifth grade, I started to get better grades. I continued receiving tutoring.

In sixth grade, I was able to do all my schoolwork by myself. I didn't need tutoring. My parents and older sister didn't need to help me with my homework. I began to receive As and Bs on my report card. I started playing tackle football and baseball. In seventh grade, I made a lot of friends. I added a new sport—basketball. And I got almost all As.

I no longer follow a gluten-free diet, and I eat some dairy. I know that the therapies and biomedical treatments my parents gave me helped me to get better. I have made a lot of progress in the nine years since I was in kindergarten. My kindergarten teacher recently said to me, "Wow! Do you remember how we used to have to put cotton balls in your ears so you would enter the gym? And today, you play the drums in the gym!"

My mom tells me that many people don't understand how children with autism can be helped. She says that doctors offering biomedical treatments to children like me are criticized and threatened. I just want people to know that there's another side to this story. I am very lucky that my parents made these decisions for me. I am very lucky that my parents had the right to make these decisions for me.

SONJA:

My son's medical history is a complicated one and includes two severe vaccine reactions from shots that we—my husband and I—were pushed into giving him.

Alex showed signs of motor delays from birth. He was floppy when lying down and stiff when held upright. He did not gain weight normally; he had colic and severe reflux, which influenced his ability to suck, swallow, and breathe. Alex refluxed during his feeds and had to coordinate what he refluxed up as well as what was going down (double volume of fluids). In addition, doctors observed during a swallow study that he had silent aspiration with his reflux. I had to nurse him every hour and a half because he could not properly coordinate his feeding and became cyanotic around his mouth (his lips turned bluish with every feed).

After discussing his colic and reflux with his doctor, my husband, Gregg, and I made the decision to delay his vaccinations because of our concerns for Alex's health.

By six months of age, Alex had recovered from a severe case of chickenpox and continued to have serious reflux that caused him to stop breathing (apnea). One night I awoke at three o'clock (from six months of habitual night feeds) to find Alex cold and lifeless. I was so shocked at first that I screamed to my husband and forgot that I knew CPR. I actually had to slap myself in the face, remind myself that I was a registered nurse and tell myself, "You know CPR!" I started working on him and Alex came around. I had never been more afraid.

When we took Alex to the doctor that day, my husband and I were under a great deal of stress after seeing our son so close to death. We don't feel that we were properly prepared to make a good decision on the recommendation that came from the doctor. She wanted to vaccinate Alex during the visit.

The doctor pointedly asked, "You wouldn't want to lose him to pertussis, would you?" I felt a combination of pressure and guilt—pressure because even as a nurse, my training did not educate me to question the safety of vaccinations, and guilt because I had almost lost my son once and the doctor implied that not vaccinating him at that time would put Alex back in that terrifying place. Despite the fact that Alex's doctor did not take his risk factors and health status into account before making her vaccine recommendation, the not-so-subtle point hit its intended mark and we vaccinated Alex for the first time. The doctor gave him the combination diphtheria-tetanus-acellular pertussis (DTaP) vaccine and my baby had a severe reaction to his shot. That evening, Alex began to arch his back and then screamed for two days, nonstop. His doctor claimed that this was normal and failed to recognize this as an adverse vaccine reaction.

Alex had been receiving speech and occupational therapies since birth. He benefited from the sensory work from his occupational therapist and the oral motor therapy from his speech therapist. By two years of age, he had a few words, and would even appropriately use "me do" and "sorry" when he did something he was not supposed to do. He was improving in his development. However, when Alex received another series of vaccinations at twenty-five months, he lost those phrases, and everything changed. Regressing yet again, he lost weight, his hair thinned, and he exhibited a weakening muscle tone. He perspired from his head, had a pale complexion, and had severe constipation. He exhibited behaviors typically associated with autism, such as hand posturing, difficulty sleeping, and picky eating. He would only eat dry, crunchy foods, stopped repeating the few words he had acquired, and began to drift away from us. By thirty-seven months, he was extremely difficult to engage and spent much of the time obsessively lining up toy cars. This is when his pediatrician told us that she thought Alex had autism.

Alex was also evaluated by a mitochondrial specialist. After reviewing the test results, he advised us to get Alex's affairs in order because he would always need support and care. I will never forget his parting words: "He will be low functioning. Don't expect much." We were further discouraged by the nutritionist at the hospital who recommended that Alex have a feeding tube inserted: "It would be easier to place a G-tube than to try and feed him."

During his medical workup, we also found that Alex had insufficient enzymes to break down carbohydrates, resulting in poor absorption of nutrients. The word "insufficient" was putting it mildly. There were no enzymes detected when they did the study.

Fortunately, our neurologist was supportive when I mentioned dietary intervention. During his initial workup, Alex had an abnormal electroencephalogram (EEG). After a six-month trial of the gluten-free, casein-free (GFCF) diet, Alex's repeated EEG results were normal.

The word "autism" had little meaning for me. Alex's body was struggling medically. Where others saw behaviors called "autism," I saw a sick child. I refused to settle for this diagnosis and instead pursued a biomedical route. The lab work showed that Alex had nutritional deficiencies, mitochondrial abnormality, severe gastrointestinal reflux, and lymphoid hyperplasia of the bowel. As devastating as it was to receive his lab results, it was also a relief. I now had something to work with. My intuition was right. While the diagnosis of *autism* offered no solutions, Alex had very real medical problems that we could address.

For ten years we treated Alex's medical problems and the result is a healthy, verbal, social young man. The conventional medical community does not think that it's possible that vaccines either caused or exacerbated his health challenges. I am highly skeptical of their view. My son's biology did not match up with the "one-size-fits-all" vaccination schedule. He suffered two clear vaccine adverse reactions. His behavioral symptoms and decline in health coincided with his shots, and the tests showed he was sick. As we addressed the medical problems, his "autism" began to get better and eventually went away.

It is difficult for me to understand how members of the medical community can see the thousands of cases like my son's and continue to deny that vaccine-induced autism—and recovery from autism through medical treatment—is a reality. My experiences have taught me that we must pay more attention to the individual children receiving vaccines because some are inherently more vulnerable than others. Unfortunately, my family—and others—had to learn this the hard way.

If the doctor had been compelled to offer information that constituted true informed consent prior to vaccination, Gregg and I never would have vaccinated Alex after his apnea event. True informed medical consent stems from the legal and ethical right to decide what medical interventions your child will undergo based on a complete understanding of the known benefits and risks of any procedure or drug. Our doctor did not give us the information we needed to make wise choices for our son. After his first adverse vaccine reaction, I wish our doctor had given us the option to check Alex's titers to see how his body responded immunologically to the vaccinations already given.

We were pressured into making the wrong choices for Alex.

This is true to varying degrees for nearly all families, because there is no true informed consent in vaccination. Doctors are not required to fully disclose the risks of vaccination and, in many cases, have not been properly informed by the companies, agencies, schools, and professional organizations tasked with educating medical professionals. I know this because I was one of those uninformed medical professionals.

Eighteen years ago, I was a nurse who administered vaccines for the Milwaukee Public Health Department when the court charged families to vaccinate their children. In this program, parents were required to appear in court after the school had identified their children as not being up to date with the recommended vaccination schedule. The city government sent letters informing these families that school could deny their children admission because of vaccine noncompliance. Following the court hearing, I administered those vaccines.

What strikes me now is that, just as no one told me my options, no one told these parents their options either. I don't know if public health officials ever offered these families any information on vaccine exemptions. I did not evaluate or review the children's health status before I vaccinated them. I had the parents sign a consent form, but in some cases they may not have understood or have even been able to read these forms. I assumed they understood, and then I gave the vaccinations. That was it.

Even at that point, I began to ask questions about vaccine policy: Why do we vaccinate the way we do? How do we know the body needs the number of vaccinations we give? Can we measure the child's immune response before administering booster shots? The answers I received from health department officials did not seem to justify these policies:

> "It is cheaper to vaccinate repeatedly than to check titers."
> "Moms are more willing to bring their babies than their older children for vaccination."
> "Herd immunity would not be achieved."
> "This is the way it is done."

Little did I know that the birth of my son, Alex, would prompt me to ask the same questions again.

Vaccination is a serious medical intervention and only parents can assess the risk-benefit tradeoffs for their individual children. One-size-fits-all vaccination schedules and mandatory vaccination policies represent an outright challenge to parental rights and personal bodily autonomy. In Alex's case, vaccinations compounded his already complicated heath challenges. The belief by the medical community that vaccinations are benign interventions robbed our son of his basic functioning skills in the early years of his life.

Fortunately, Alex has recovered from autism. On many levels, it would be easier for our family to pretend that it never happened. There is no social benefit to Alex for his friends to know about his old diagnosis. I think it's pretty extraordinary for a thirteen-year-old boy to make the decision to share his story. Alex and I hope our story inspires others and allows for an open discussion of the medical problems that these children face.

We are so grateful for Alex's healing. If not for doctors who have willingly become a part of this open discussion—like Dr. Anju Usman, who oversaw Alex's chelation and other therapies—we do not believe that he would have recovered from autism. His improvement all started with our belief in the possibility of Alex's recovery.

Alex is living, thriving proof that autism is treatable and children can recover. I bear witness to this, as a parent and a nurse. I am opposed to vaccine intimidation. Vaccination choice is every parent's right.

Chapter Sixteen

WHO WILL DEFEND THE DEFENDERS?

Captain Richard Rovet, USAF (Ret.), with afterword by Colonel Felix M. Grieder, USAF (Ret.)

I write about some of the most vulnerable of society's members: our men and women in uniform. Right now, many are protecting our freedom in austere conditions and dangerous places around the globe. Some may have given their lives today, in support of that freedom to do exactly what this book aims to do—to protect the principles we hold dear.

Unfortunately, these brave defenders of our freedom are defenseless against the mandatory use of dangerous vaccines, such as the anthrax vaccine. For the past sixty-four years, the United States military and other government agencies have used our servicemen and women as test subjects, often in secret and without informed consent. Moreover, there is no end in sight; there are additional biodefense vaccines waiting in a pipeline to be tested on our soldiers.

Sadly, this is not a conspiracy theory. This is documented fact. In December 1994, right after the first Gulf War, the United States Senate released a report titled, "Is Military Research Hazardous to a Veteran's Health?"[1] This report outlined the unethical use of servicemen and women as test subjects. The report revealed that the Pentagon had quietly used soldiers in clinical trials and did not record the resulting information in their medical records, preventing the soldiers from receiving appropriate follow-up care. Many were simply left to die.

Witness the well-illustrated connection between Gulf War Illness and the anthrax vaccine. "Gulf War Illness" encompasses a wide range of acute and chronic health disorders, including chronic fatigue, severe joint pain, neurological problems, memory loss, and unexplained persistent rashes and sores, to name a few. Many illnesses are autoimmune in nature, meaning that a person's own immune system has turned on certain tissues and organs in his or her own body. Roughly one in four of the 697,000 veterans—my fellow servicemen and women, who served in the first Gulf War—are afflicted with Gulf War Illness. Multiple studies and governmental reports demonstrate a higher rate of Gulf War Illness in vaccinated veterans.[2] At the time that those fighting in the first Gulf War were vaccinated, the Department of Defense ordered medical staff not to annotate or identify the anthrax vaccine and the botulinum toxoid vaccine in soldiers' vaccination records. Instead, they mysteriously recorded these vaccines as "Vac A" or "Vac B." [3]

While investigating potential causes of Gulf War Illness, Vancouver neuroscientist Chris Shaw and his research team were shocked to find that the aluminum hydroxide adjuvant in the military's anthrax vaccines caused massive brain-cell death in laboratory mice.[4] Vaccine makers now routinely use this adjuvant in childhood vaccines.

There is no doubt that the first Gulf War was a toxic war, and that there may be other contributory factors to Gulf War Illness besides the anthrax vaccine. One of the most damning pieces of evidence implicating the anthrax vaccine, however, is that the coalition forces from other countries that fought alongside United States troops in the Gulf War, but did not receive the anthrax vaccine, have relatively few incidents of Gulf War Illness.[5] Meanwhile, other coalition troops, such as British forces, who were also vaccinated against anthrax, do have significant incidence of Gulf War Illness, like their U.S. counterparts. The French report a small number of cases, which are primarily associated with the French soldiers who served with U.S. soldiers, and were given the anthrax vaccine.

General Douglas MacArthur outlined the damage done to a fighting force when such treatment of soldiers is undertaken:

> The unfailing formula for production of morale is patriotism, self-respect, discipline, and self-confidence within a military unit, joined with fair treatment and merited appreciation from without. It cannot be produced by pampering or coddling an army, and is not necessarily destroyed by hardship, danger, or even calamity . . . It will quickly wither and die if soldiers come to believe themselves

the victims of indifference or injustice on the part of their government, or ignorance, personal ambition, or ineptitude on the part of their leaders.

—General Douglas MacArthur, 1933 Annual Report of the Chief of Staff, U.S. Army http://www.lanl.gov/external/news/reflections/0900.pdf (pg 2 of 12)

Our soldiers' calls for help have not only been ignored, but their own government, the one they swore to serve and protect, has tried to discredit them. For many years, veterans of the first Gulf War and their families have begged for help and answers. Meanwhile, in a misguided effort to mislead Congress, the press, and the American people about the extent of the damage done to personnel during the conflict, the Pentagon launched Operation Bronze Anvil,[6] a propaganda program designed to deflect any inquiries into the Gulf War Illness–anthrax vaccine connection and to harm the reputations of those who spoke out about the connection. This effort has branded honorable U.S. servicemen and women complaining of anthrax vaccine reactions as malingerers, liars, whiners, and malcontents.

This misuse of our military personnel did not stop after the Gulf War or after the Senate's 1994 report. Three key areas continue to provide fertile ground for the abuse of our servicemen and women as guinea pigs by the pharmaceutical industry in cooperation with senior military leaders.

First, the military may order soldiers to take medicines or receive vaccines against their will. If they refuse medications, these proud soldiers risk incarceration, forfeiture of pay, and a dishonorable or bad conduct discharge, which would lump them together with murderers and rapists.

Second, the military may waive informed consent for the use of untested or unstudied medications and vaccines in two possible ways: when there is a Presidential Executive Order[7] or if the defense or intelligence establishment deems that there is a credible military threat.

Third, servicemen and women are barred from suing the federal government under an archaic and arguably unconstitutional law from 1950 called the Feres Doctrine.[8]

Sadly, recent history has taught us nothing, and our soldiers continue to pay with their lives and their health.

Despite mounting concerns over anthrax vaccination as a contributing factor to Gulf War Illness, the U.S. military launched a mandatory anthrax immunization program in the late 1990s. I was at Dover Air Force Base in Delaware at the time and I witnessed a hotbed of illnesses and controversy surrounding the anthrax vaccine. Among the documented illnesses witnessed, many were autoimmune in nature, such as multiple sclerosis and rheumatoid arthritis. I have personally and sadly witnessed both of these conditions in relation to the anthrax vaccine. After receiving pressure from the Government Accountability Office (the watchdog arm of Congress), the maker of the vaccine, Bioport, was forced to list autoimmune diseases and all the symptoms of Gulf War Illness on the anthrax vaccine product insert.[9]

In May 1999, the USAF Surgeon General's Office arranged a "town hall meeting" at Dover Air Force Base to assuage the concerns of many who had become ill following anthrax vaccination. In attendance were the chief public health officer of Air Mobility Command, Colonel Denise VanHook; the USAF Surgeon General of the USAF Anthrax Vaccine Action, Officer Lieutenant Colonel Jack Davis; the medical head of Dover's anthrax program, Lieutenant Colonel Thomas Fadel-Luna; Wing Commander, Colonel Felix Grieder; and the men, women, and families of the 9th Airlift Squadron. It was during this meeting that Davis let the proverbial "cat out of the bag." While fielding a question from a military member in the audience, Davis stated that the military did indeed possess an experimental anthrax vaccine that contained an illegal adjuvant called squalene, and flippantly stated that no one at Dover received that vaccine. Another assertion came from my former squadron commander, Dr. and Lieutenant Colonel Thomas Fadel-Luna. When addressing the so-called rumor that *there was an increase of people with an autoimmune illness at Dover*, Luna stated the following:

> That is incorrect. Autoimmune disease is something we do not typically see in flyers. If we do see it, it sends up a flag. In my nine-year career, I have seen five or less. We are simply not seeing it. We are looking at other clinics in the Medical Group to see if they are seeing an increase. And, basically they have not seen a problem.

Unfortunately, that was not the case. At that time, we had over fifteen cases of autoimmunity and over thirty cases of illnesses exactly the same as the ones the veterans had reported as associated with Gulf War Illness. In fact, many of our patients were sent for evaluation at Walter Reed's Gulf War Illness clinic.

In response to these town hall meetings, Dover AFB Wing Commander Colonel Felix Grieder temporarily halted anthrax injections on his base because of concerns about safety and after seeing his troops that had become severely ill after receiving the anthrax vaccine. It cost him his career. Greider has since alleged that his troops were used as guinea pigs in undisclosed military medical experiments.[10]

He was right to be concerned. At Dover, the FDA tested anthrax vaccine lots in 2000 and discovered the presence of an experimental adjuvant and immune "turbo booster" called squalene or MF59—an adjuvant with potentially devastating health consequences. I want you to burn these two letters and two numbers into your consciousness so you will remember them because squalene will next be used in civilian vaccines.[11] I strongly urge you to read award-winning journalist Gary Matsumoto's powerful exposé, *Vaccine A*, which is an outstanding example of the dying field of authentic investigative journalism.[12] The book blows open the military's shameful history of secret medical experiments on its soldiers and especially its illicit use of an unlicensed additive in the anthrax vaccine that is proven to cause fatal autoimmune diseases in animals.

I have personally borne witness to the devastating effects of the anthrax vaccine. I will forever have etched upon my memory the vision of a young enlisted woman screaming and crying as she was forcibly held down while the needle delivering the anthrax vaccine was pushed into her body.[13] I will never forget the sad day when my dear friend, Technical Sergeant Clarence Glover, died after anthrax vaccination.[14] My memory holds the stories of those whose skin literally burned off due to anthrax vaccine-induced Stevens-Johnson syndrome[15] and of the infants under my care who were born with severe birth defects after their pregnant mothers were vaccinated with the anthrax vaccine.

Corruption within government for the profit of private industry is nothing new. Go behind the curtain of the anthrax vaccination program and you will see bailouts, lucrative financial contracts,[16] and a revolving door of crony capitalism with strong ties to the Department of Defense and the Department of Health and Human Services. Furthermore, all studies that tout the safety of the anthrax vaccine, reported by so-called independent experts, were anything but independent. The studies were funded by the Department of Defense.

I swore an oath to defend this nation against all enemies, foreign or domestic. However, I most certainly did not swear allegiance to defense contractors and the hard-hitting lobbyists of the pharmaceutical industry. Forensic evidence now exists to launch a full investigation into the use of experimental adjuvants as a controlled experiment on our troops from the first Gulf War to the present. The potential implications of such an investigation are shattering.

We can no longer tolerate the cruel and dishonorable use as guinea pigs of America's sons and daughters who have sacrificed their lives to fight for our freedoms. We must loudly protest when the international legal standard of informed consent does not apply to those who courageously defend our country. We must end the current mandatory anthrax vaccination program and, in its place, launch a full-scale investigation.

I truly mean it each time I say, "God bless America." I have fought to protect our freedoms and rights. I believe that vaccination choice is a fundamental human right for all Americans, including our brave military personnel.

Please read the appendix to this chapter for Captain Rovet's March 25, 2005, letter of complaint to the Department of Defense Inspector General, and excerpts from his report to the House Committee on Oversight and Government Reform.

Editor's Note: The appendix to this chapter contains excerpts from Captain Rovet's report on anthrax vaccine injury to his superiors while serving at Dover Air Force Base.

AFTERWORD BY COLONEL FELIX M. GRIEDER (USAF, RET)

In April 1999, late in my two-year tenure as Wing Commander at Dover Air Force Base (AFB), I became aware that several members of my command were experiencing adverse reactions shortly after receiving the anthrax vaccination. At that time, Dover was one of the first military bases to receive and to administer the anthrax vaccine (a protocol of six shots), a vaccine designed to protect our military against what was perceived to be Saddam Hussein's weapons of mass destruction (WMD), including biological weapons such as anthrax.

At Dover, we took pride in leading the way, whether it be with our strategically important airlift mission or, in this case, protecting our troops from the potential WMD threat. My strong support for the administration of the anthrax vaccine changed when I became aware of the adverse effects that a small, yet significant, number of my airmen were experiencing.

I called for an informational briefing. On May 5, 1999, when the Pentagon's supposed "expert" on the anthrax program stated in response to concerns being articulated in the briefing room, "I don't know and I don't care," I immediately ordered a "time-out" and suspended the vaccination program at Dover AFB.

I was simply doing my job as a commander. All military members take an oath to protect and defend our great country against all enemies, foreign and domestic. In this case, I clearly had the responsibility to protect my troops who were being put in harm's way by internal forces.

Six days later, after Lt. Gen. Charles Roadman, Air Force surgeon general, gave us his word "as an officer" during a public briefing session that squalene, an experimental vaccine adjuvant, was never used in the anthrax vaccine, I reinstated the vaccination program.

Subsequently, in September 2000, the FDA documented that squalene had been found in six lots of anthrax vaccine. Of these, four of them—FAV 08, FAV 030, FAV 043, and FAV 047— had been sent to Dover.[17] Not only did these lots contain squalene, but they contained dosages of squalene which approximately doubled in value (from 10 to 20, to 40, to 80), similar to a dose-range study. Separately, Tulane Medical School confirmed that three additional lots of anthrax vaccine were found to induce anti-squalene antibodies, of which two—FAV 041 and FAV 070—were also sent to Dover.[18] It is my belief that the Department of Defense, in its zeal to establish an effective and more immediate anti-anthrax vaccine, experimented with the untested adjuvant squalene at a few military installations, including Dover AFB.

Those responsible for this decision need to be held accountable. There are parallels in this unfortunate decision to what happened at the prison in Abu Ghraib in Iraq.

A nation and an organization should never forego its core values and standards. The United States stands for freedom, human rights, and the fair and humane treatment of all. When we set aside our values, even temporarily, for what may appear to be a justifiable reason, we lose our way as a nation and as a superpower.

Chapter Seventeen

"GET YOUR AFFAIRS IN ORDER"

Lisa Marks Smith

October 15, 2005, started like any other beautiful fall day. We rose early and went to my son, Nathan's soccer game. While we were there, my mom called to say that CVS Pharmacy was having a flu shot clinic. My dad was having knee-replacement surgery in a few weeks and could be vulnerable to infection. We wanted to be sure that I did not get sick. After the game, I drove to the CVS store and registered for the shot. While I was there, my brother Jeff came in for his shot as well. We received the standard disclaimer form. Nowhere did it say this would be the last normal day I would have for four years.

The truth is, I did not put any thought into whether to get the shot. It was just a quick decision, but one that would change my entire life. If I had taken the time to do a little research, I like to think I would have skipped the shot. I had taken the shot one other time with no adverse reaction when the health department had offered them at my church.

Within days of getting the shot, I knew something was wrong. I called my parents and told them to skip my son's orchestral concert on Tuesday. I wasn't feeling right and didn't want to get my dad sick before his surgery. When I set up at a craft show with my friend Jackie on Friday night, I complained of a tickle in my throat. I woke up terribly sick the next morning. The first appointment I could get with my family practitioner was Monday. By Monday, October 24, I felt like I was going to die.

That morning, my husband, Greg, rushed me to Mercy Hospital in Western Hills, Ohio. I was discharged the same day with a diagnosis of pneumonia. They gave me antibiotics and other medications, but I never improved. Early the following Saturday, Greg and Nathan left on a scheduled Boy Scout campout. When I woke up, my legs were strangely weak and they continued to get weaker throughout the day. By evening, I could no longer stand or walk. I was home by myself and had to wait alone and in pain on the bathroom floor for my older son, Matthew, to come home from a babysitting job and call for help. I had never felt so powerless or been in so much pain.

Matthew finally came home and called my brother, Jeff. Jeff came over and carried me into the family room. Just being carried was painful. Jeff called an ambulance. The paramedics had to hoist me onto the stretcher, as I was unable to use my legs. The pain was tremendous. If anything touched my legs, I cried out in agony. The ambulance took me back to Mercy Hospital. Right away, the nurses started to ask me if I had received a flu shot. I wondered why everyone kept asking. At that point, I had no idea that spreading paralysis, or Guillain-Barré syndrome (GBS), was a known side effect of the flu vaccine. After several hours in the emergency room, they moved me to a regular room. A nurse urged me to go to a different hospital, one with a higher level of care. I wish I had listened, but I was just too sick to act. The nurses felt that I should be moved to intensive care; the paralysis was spreading and they could not provide the level of care I needed. I spent several days in intensive care. A nurse told me to call my family and get my affairs in order. It is very difficult to tell your children good-bye forever. It was the most heartbreaking moment of my life.

Nothing improved over the next week. I had two spinal taps, multiple IVs, and more tests and doctors than I can remember. I was on the strongest pain medicine available; nothing relieved the spasms and pain in my legs. I couldn't even bear the weight of a sheet on my legs. All I remember from this period was the overwhelming pain. At one point, I could only move my head. My feet dropped and my legs were totally limp from the damage done to my nerves. To this day, friends still comment on how my feet lay parallel to my legs.

The Mercy neurologist told me nothing was wrong with me. He said my symptoms were psychogenic; in other words, I was crazy and making up this entire disease. I asked how I could fake symptoms—including shaking and dropped feet—in my sleep. He didn't respond and, instead, forced me to do physical therapy, which further increased my pain. The neurologist ignored my neighbor, a doctor, who told him I was sane. He ruled out GBS since I didn't have the protein in my spinal column that normally is present with GBS. He never considered transverse myelitis or postinfectious myositis, well-known vaccine injuries with similar symptoms.

Greg and Jeff arranged to transfer me to a better hospital. Mercy tried to dissuade me from leaving. The caseworker insisted my insurance would not pay for the transfer or cover the bills. My Cigna representative assured us this was not true. While I was waiting to be transported to the new hospital, a nurse approached me with a syringe. I asked her what it was. The doctor had ordered a pneumonia vaccine! I told her if I could move, I would break her arm. I was furious. I was lying there paralyzed, possibly from the flu vaccine, and the doctor had ordered another vaccine.

When I got to Christ Hospital in Cincinnati, things improved immediately. Within two hours of my arrival, the hospitalist, Dr. Rajan Lakhia, had a diagnosis. He said, if the Mercy neurologist had simply repeated the creatine phosphokinase (CPK) blood test for muscle degradation, which was standard procedure, he would have noticed the protein level in my blood was extremely high. Anything over 100 is a problem. At Mercy, they only checked once and it was 60. When I arrived at Christ, the protein level was at 900 and and climbed to a high of 1,600. The muscles in my legs were breaking down. Despite the bad news, I wanted to cry with relief because they had proof that I was sick. One major hurdle had been cleared. I could focus on getting well.

I felt safer at my new hospital. The nurse put in a peripherally inserted central catheter (PICC) line to deliver Dilaudid so I wouldn't require so many IVs. After five days at Christ, the protein level began to drop. Muscle damage had stopped, and I regained some movement. I could use my arms and sit up. I was able to support myself enough to use a portable toilet—a wonderful improvement after two weeks of lying flat on my back. The nurses told me about another patient who was paralyzed by the flu shot. When I asked how she was doing, they didn't answer. I later learned that she never recovered.

Dr. Lakhia fought with my insurance company to get me intensive therapy. I was scared to go to a long-term care facility; I felt like I would be forgotten there. Instead, Christ transferred me to the acute rehabilitation unit. The therapy was grueling. Propped up, I stumbled two steps. I had "Barbie feet"—my feet pointed straight down and my heels would not rest on the floor. My immune system had destroyed the muscles and nerves in my legs. My official diagnosis was postinfectious myositis due to the flu vaccine.

I spent the next two weeks in the rehab unit learning to walk again. I had four hours of therapy each day. Braces forced my swollen feet into the proper position so I could walk. I missed wearing my regular shoes. They wouldn't let me go home until I could dress myself and navigate without help. I had to learn how to get in and out of a car safely, and how to get off the floor if I fell while home alone.

My occupational therapist wanted to see if I could make a bed. I told her I needed practical skills, for example, how to carry a food tray while using a walker. The therapist put a few balls on a tray and had me practice walking. It helped to laugh. I hadn't laughed much in the last few weeks. My doctor ordered a psychiatric exam to see how I was coping. I told the doctor I wasn't so depressed that I wouldn't do everything in my power to recover. I wanted to go home. She ruled me emotionally healthy and well-adjusted. One day, I nearly fell when trying to get to the bathroom. I cried with frustration that I wasn't getting better fast enough. The nurses told my mom they were glad to see me cry; they thought I had been too stoic.

I sobbed every time Greg and my boys left. I wanted to go home so badly. However, family and friends came to see me constantly. The nurses saw all the flowers and visitors and teased me that I must've been a lot of fun to be around when I was well. They said they would go to my house for a party when I recovered.

I went home on November 22, two days before Thanksgiving. I had been in the hospital for twenty-four days. I had to promise not to drive or go up and down the stairs without someone watching me. Matthew and my neighbors took turns driving me to therapy three times a week for twenty-two weeks. After all the hard work, I am proud to say that I don't limp.

My neurologist said that traditional medicine could do nothing further for me. She confirmed all of my problems stemmed from the flu shot and encouraged me to try alternative medicine.

I filed my own report with the government's Vaccine Adverse Event Reporting System (VAERS). Only one doctor ever mentioned the possibility of filing a report. The doctor who performed my EMGs at Aring Neurology told me that the flu shot had paralyzed another one of his patients. She has a diagnosis of transverse myelitis and is permanently confined to a wheelchair. He showed me her VAERS report. My VAERS ID is 251221.

Although the media has not been interested in my story, people with flu shot injuries continue to bombard me with messages. The official estimate of vaccine injury is one in one million people. My experience tells me otherwise.

My neighbor, Chris Sullivan, sat next to a woman who was on a flight to Cincinnati to see her brother who was just paralyzed from a flu shot.

A friend, Tracy Kroger, developed Guillain-Barré syndrome from her flu shot. Tracy still struggles to use her arms.

I recently reconnected with a high school friend, Marianne Madaris. She has Rasmussen's encephalitis, also known as chronic focal encephalitis. While in nursing school, Marianne didn't have any vaccination records so they required her to repeat her vaccines. The doctors later told her that the encephalitis was caused by the MMR vaccine. Marianne has been sick for twenty-one years. She remains today in the same condition that I was in at the hospital. How rare can these reactions be when I seem to be adding to my list of vaccine victims every week?

In April 2006, I hired the law firm of Douglas and London, P.C. to pursue a vaccine injury claim through the National Vaccine Injury Compensation Program (VICP). The firm told me that I had a strong case since my doctor was willing to state for the record that I had a vaccine injury.

Most people have no idea that vaccine makers are protected from most liability by federal law. They are equally shocked that our government compensates people for vaccine injuries. The federal agency that oversees the VICP added the flu vaccine to the table of compensable injuries in July 2005, just three months before I was injured. To collect, you have to prove that a vaccine caused damage lasting more than six months. My attorney requested affidavits from my family and friends who could describe how my life had changed after the flu shot.

At one point during the case, the special master (judges do not administer the program) told my lawyer that I couldn't prove I ever had the vaccine. Following my paralysis and hospital stay, I couldn't find my flu vaccine paperwork. I pointed out that my VAERS report contained my lot number. Since I hadn't given the CDC or the FDA that information, they must have obtained it from Maxim Health Systems. Maxim never responded to my request for this information. The nurse who administered my vaccine didn't complete the section of the form containing the lot number. My attorney had to file a Freedom of Information Act (FOIA) request to get a copy of the bill from the government. Jeff and Greg filed affidavits that I went to CVS Pharmacy; Jeff confirmed that he witnessed my shot. I was lucky. How many people can provide a witness to their shot? In July 2009, Special Master Christian Moran ruled that a preponderance of evidence supported my claim that I received the flu shot.

It took over three years to receive official acknowledgement of my vaccine injury. In December 2009, I received a phone call from my attorney who said that the government had conceded my case. There would be no trial; I was to receive a settlement. I was elated. The money would help pay for my out-of-pocket treatment expenses. Most importantly, though, I felt vindicated. The government agreed that a flu shot caused my medical problems. If you google my name, Lisa Marks Smith, and HHS, you can read the court papers.

The check arrived on May 27, 2010, nearly five years after the fateful shot. Since then, I have heard about many vaccine-injured people who have not received compensation. Based on my own experience, I know it isn't easy.

Wherever I go, I talk about the dangers of the flu shot. The fall is an especially difficult time for me, when the big push for flu shots is on. A few weeks ago, a Walgreens clerk asked if I wanted a flu shot. I told her it took me four years to recover from my last one. She didn't know that the flu shot could cause paralysis. Visibly shocked, she said that our conversation would make her think twice. Every aisle in Walgreens has merchandise tagged with "get your flu shot" stickers. I see them at other drugstores and supermarkets too. I wonder who pays for all of these promotions. I still find it difficult to accept what happened to me. I am deeply fortunate because I got better.

How many people know that one simple flu shot could change your life forever? It's ironic: Vaccines are supposed to be safe and effective. Vaccines are supposed to keep us healthy. Before my flu shot, I was so healthy. Had I gotten the flu, I probably would have been sick for a week or two, at most. After the shot, I was sick for years. Don't think a vaccine injury can't happen to you. It can. I don't tell people not to vaccinate—but I tell my story.

Part III

THE TOPICS IN DEBATE

THE BUSINESS OF VACCINES, MERCURY AND AUTISM, THE LAW AND OUR CHILDREN, THOUGHTFUL MEDICAL VIEWS, DR. WAKEFIELD AND THE SUPPRESSION OF SCIENCE

Vaccines can cause brain damage... An adverse vaccine reaction that causes brain damage (encephalitis) is the same thing as a complication from an infectious disease.

—Michael Belkin

How is disinterested vaccine safety governance even remotely possible when HHS employees stand as heroes at the head of the parade when a new vaccine is invented within its walls, while agency leaders are leading the cheering section, approving the new product's launch, making the market for the product with its policy recommendations, and then turning around to cash multi-million dollar checks?

—Mark Blaxill, MBA, and Dan Olmsted

How about the children who die or are injured by vaccines? Aren't their injuries or deaths worth preventing? If immunizations prevent that one death but do irreparable harm to thousands or more, is that okay? Of course not. It's insane!

—Julian Whitaker, MD

It appears that every nation has been set on a course to inject every human on the planet with increasingly complex mixtures of chemical and biological agents beginning at birth. The science supporting this bold experiment, however, is subject to government and industry manipulation on a scale that few other areas of science have ever experienced. This, combined with silencing reputable scientists who question vaccine safety, is a prescription for turning the hope science offers for future generations into a global disaster.

—David Lewis, PhD

Chapter Eighteen

THE VACCINE BUBBLE AND THE PHARMACEUTICAL INDUSTRY

Michael Belkin

Vaccines can cause brain damage. Most people are completely unaware of this, but that is exactly how *The Merck Manual*, the largest-selling medical textbook, defines an adverse reaction to a vaccine:

> Encephalitis is inflammation of the brain that occurs when a virus directly infects the brain or when a virus or something else triggers inflammation . . . Encephalitis can occur in the following ways: A virus directly infects the brain. A virus that caused an infection in the past becomes reactivated and directly damages the brain. *A virus or vaccine triggers a reaction that makes the immune system attack brain tissue (an autoimmune reaction)* [emphasis added].[1]

An adverse vaccine reaction that causes brain damage (encephalitis) is the same thing as a complication from an infectious disease. Any pediatrician, doctor, or state or federal public health official who tells you that vaccines are completely safe, that adverse reactions to vaccines don't exist, or that vaccine-induced injuries are so rare that they virtually never occur is either ignorant or is committing scientific fraud. Is it worse to have your child vaccinated by a doctor who does not know the possible adverse reactions, or to be lied to by a doctor or government bureaucrat who does know the terrible damage vaccines can cause?

This is no trivial matter. Every day, uninformed physicians administer vaccines to vast numbers of children and adults with little thought about the possibility of adverse reactions. When an adverse reaction occurs—in the form of brain inflammation, convulsions, or another injury—the typical first step is to blame someone else. Doctors and the government accuse parents of child abuse (i.e., shaken baby syndrome) or bad luck (i.e., defective genes) and accuse teenagers of bad behavior (i.e., using illicit drugs). Medical professionals do not step up and ask whether they hold any responsibility for causing an adverse reaction to a vaccine.

The Merck Manual further defines the symptoms of encephalitis: "Symptoms of encephalitis include fever, headache, personality changes or confusion, seizures, paralysis or numbness, sleepiness that can progress to coma and death." Many tens of thousands of parents whose children were diagnosed with autism spectrum disorder reported that their kids were progressing normally until they received one or many vaccines, after which they had fevers, headaches, seizures, personality changes, and were never the same again. The symptoms reported by parents are the same symptoms of encephalitis that are defined in *The Merck Manual.* Health authorities in charge of defending and expanding universal immunization programs label these same symptoms "a coincidence."[2]

When my five-week-old daughter, Lyla, died hours after receiving her hepatitis B vaccine, the New York medical examiner was more concerned about examining our apartment for evidence of child abuse than about the possibility that a vaccine caused her death. The medical examiner initially told us, our pediatrician, and an investigating pathologist that Lyla's brain was so swollen that it led to her death. After consulting with Merck (the manufacturer of Lyla's hepatitis B vaccine), the medical examiner left me, a devastated father who just lost his precious firstborn child to an avoidable vaccine-induced death, with these parting words: "We've changed our minds; her brain was not swollen. Vaccines do a lot of good things for people, Mr. Belkin." Former *New York Times* journalist Melody Petersen, who covered the pharmaceutical industry in her book *Our Daily Meds,* reports that doctors who fill out death certificates are instructed to call a 'therapeutic misadventure' a natural death.[3]

Vaccine-caused diseases and deaths are an unacknowledged epidemic. The Centers for Disease Control and Prevention (CDC), state public health departments, Bill Gates, and doctors incessantly repeat the mantra calling for more vaccines to eliminate "vaccine-preventable disease." However, my daughter experienced the opposite effect—the prophylactic treatment that was supposed to prevent a disease instead caused severe harm. The medical term for this outcome is iatrogenesis, the "inadvertent and preventable induction of disease or complications by the medical treatment or procedures of a physician or surgeon."[4] To put this in real world terms, you walk in to a doctor's office in perfect health and you walk out with a lifelong neurological disability, or you even die, from prescribed vaccines. That is what happened to my daughter. Through immunization programs, modern medicine is creating the epidemic of neurological damage that it takes credit for preventing.

THE VACCINE HOLOCAUST

If you doubt the existence of adverse reactions to vaccines, you must first examine with your own eyes the U.S. Food and Drug Administration's (FDA) Vaccine Adverse Events Reporting System (VAERS), which is available and searchable online at www.medalerts.org. As of November 17, 2010, VAERS listed 352,650 reports of vaccine adverse events.[5] Former Food and Drug Administration (FDA) commissioner David Kessler wrote in the *Journal of the American Medical Association* that "only about 1% of serious adverse events are reported to the FDA."[6]

Using that FDA commissioner's own estimate, there have been 35.2 million adverse reactions since the inception of VAERS in 1990, the vast majority of which were never acknowledged by doctors or public health officials. Regulators pull drugs such as Vioxx, Rezulin and Lotronex from the market after a few hundred or one thousand adverse reaction reports. Yet vaccines remain on the market, with hundreds of thousands of adverse reaction reports. The double standard is clear.

The following two recent VAERS reports, offered verbatim, are representative of thousands of others.

> VAERS ID: 393346
> Vaccinated: 2010-06-11
>
> On 11 June 2010 the subject received 2nd dose of INFANRIX HEXA (intramuscular, unknown injection site), 2nd dose of ROTARIX (oral), 2nd dose of PREVENAR (intramuscular, unknown injection site). Lot numbers were not provided. On 12 June 2010, 23 hours after vaccination with INFANRIX HEXA, PREVENAR and ROTARIX, the subject experienced seizures. The healthcare professional considered the event was disabling. The subject was treated with EPILIM. At the time of reporting the event was unresolved. One month later, the

subject was still having seizures. He was referred to a neurologist. The healthcare professional considered the event was possibly related to vaccination with INFANRIX HEXA, ROTARIX and PREVENAR.

VAERS ID: 391797
Vaccinated: 2010-06-19

A 6-month-old male subject who was vaccinated with ENGERIX B pediatric (GlaxoSmithKline). On 19 June 2010, the subject received 3rd dose of ENGERIX B pediatric (0,5 ml, intramuscular, unknown injection site). Lot number not provided. On 19 June 2010, within hours of vaccination with ENGERIX B pediatric, the subject experienced high fever, 10 episodes of diarrhea and vomiting. 7 hours after vaccination, the subject was taken to emergency unit with severe dehydration and he experienced convulsion. At 19:35, the baby experienced cardiac arrest. The subject was treated with STESOLID, cardiopulmonary resuscitation and electrolytes. The subject died on 19 June 2010 at 20:18, cause of death was not reported. It was unknown whether an autopsy was performed.[7]

The CDC's Vaccine Safety Datalink dismisses seizures after vaccination without mentioning the difference between simple febrile seizures (70%-75% of seizures, transient, not caused by encephalitis), complex febrile seizures (20%-25% of seizures, prolonged, multiple seizures), and symptomatic febrile seizures (5% of seizures, neurological abnormality or acute illness).[8]

While *The Merck Manual* acknowledges that febrile seizures "occur after certain vaccinations such as measles, mumps, and rubella,"[9] vaccination proponents claim that seizures are not associated with learning disabilities or death. The second VAERS report cited above suggests otherwise. The CDC's Vaccine Safety Datalink mentions cases like those described in the VAERS reports above and says, "Febrile seizures can occur after vaccination." Febrile seizures "often result in a visit to an emergency room and can be very frightening for parents" and "cannot be prevented. . . . [C]hildren who have febrile seizures after receiving an MMR vaccine are no more likely to have epilepsy or learning or developmental problems than children who have febrile seizures that are not associated with a vaccine."[10] This is a roundabout way of saying exactly what *The Merck Manual* declares—adverse reactions to vaccines are the same thing as complications from an infectious disease. Let me paraphrase what it seems the CDC, with its contorted and deceptive syntax, is admitting:

Seizures after vaccination—that may lead to permanent developmental problems—are no different than seizures from a nebulous other source. Clearly, the CDC needs to revise this information on the Vaccine Safety Datalink—it is too self-incriminating. People might decode it and realize the truth—that vaccines can cause (and have caused) permanent neurological damage to children. The CDC is in denial about the holocaust of vaccine adverse reactions being inflicted on America's children and adults. All of the CDC's published material appears to be carefully designed so that you—the prospective parent or patient—have complete faith in vaccines' safety and efficacy. Anyone who carefully studies VAERS cases can see that the CDC's vaccine safety proclamations reek of blatant propaganda.

After Lyla died, I learned about the risks of vaccinations. Perhaps you, like me, are wondering why the government doesn't do something about this.

Starting with the hepatitis B vaccine, I investigated the vaccine business and regulatory process. I followed the hepatitis B vaccine through a maze of FDA and CDC advisory committee meetings. I testified before the Congressional Government Reform Committee about my research and my family's experience,[11] and I continue to challenge the qualifications of so-called vaccine industry experts. I am an investment strategist, and therefore I am qualified to analyze and form an opinion of the statistical expertise of those who present themselves as medical statistics "experts." The following is a summary of my conclusions.

In my opinion, the epidemiologist-statisticians from the CDC and state health departments are pseudoscientists. In my work, I use statistical models to forecast markets, so I am particularly aware of the capabilities and limitations of statistics. In the financial markets, statistical fraud has a cost. If you fake the numbers, you will eventually get caught, fired, and prosecuted (as in the cases of Bernie Madoff, Enron, and others). The medical profession and government agencies like the CDC and the World Health Organization (WHO) seem to operate under a different standard. Estimates of disease prevalence are exaggerated,[12] academic medical studies are ghostwritten by pharmaceutical employees,[13] and the WHO's declared global pandemic of H1N1 influenza fizzled out and turned into nothing more than a marketing campaign for flu vaccine manufacturers[14]—and yet, there is little accountability in the world of medicine for cooking the books or for putting pharmaceutical companies' profits above the public's interest.

When public health officials rely on questionable statistics to make blanket statements about the absolute safety and efficacy of vaccines, I am profoundly disturbed on personal and professional levels. Statistics is a world of probabilities, not absolutes. Statisticians make estimates about causation and the likelihood of future events based on mathematical odds, not certainties. It seems that vaccine epidemiology has been reduced to a pseudoscience, and its practitioners have sold out. They have become a de facto marketing department for the CDC and vaccine manufacturers.

I don't use these words lightly. With the Association of American Physicians and Surgeons, I filed a Freedom of Information Act (FOIA) request for two pieces of information more than ten years ago. We wanted all the safety data the CDC had prior to recommending the hepatitis B vaccine for new babies at birth, and we wanted the statistical model they used to prove its safety. We are still waiting for a response today. Their failure to respond is damning. The implication is that the at-birth hepatitis B vaccine recommendation was made without conducting proper safety studies in babies beforehand. Our babies are their guinea pigs.[15]

The government's FOIA refusal is a sad reflection on the dishonesty and incredibility of health officials. With a Wall Street background, I am particularly sensitive to fraud in the form of financial fiction and conflicts of interest. The government and industry have a cozy relationship and the public health arena offers no exception. There is a name for this: the revolving door. Julie Gerberding, the former head of the CDC, resigned from her government-appointed position in January 2009 and was named the president of Merck Vaccines in December 2009. Gerberding began her new job in January 2010, one year after leaving the CDC, which is the minimum amount of time she was legally required to wait before joining an industry that she previously regulated. It is clear that Dr. Gerberding received a professional reward for expanding universal immunization policies and, in effect, pharmaceutical company profits, for marginalizing the plight of victims of adverse reactions to vaccines.[16] New Jersey's former deputy health commissioner, Eddy Bresnitz, was responsible for mandating the flu vaccine for babies and toddlers in the Garden State in 2008. He now oversees adult vaccines at the same New Jersey–based pharmaceutical firm, Merck & Company, Inc.[17] Gerberding and Bresnitz are not the only ones to jump from governmental to industry positions.

Poul Thorsen—a Danish contributor to two articles[18] that are regularly cited by CDC and Institute of Medicine (IOM) officials, which allegedly prove that vaccines and/or thimerosal do not cause autism—has effectively been disowned by his former employer, the University of Aarhus, in Denmark. Thorsen managed the Danish data supporting those CDC-financed studies. According to a public statement released by the managing director of the University of Arhus on January 10, 2010, "a considerable shortfall in funding at Aarhus University associated with the CDC grant was discovered. . . . [I]n March 2009, Dr. Thorsen resigned his faculty position . . . [but] has continued to act in a manner as to create the impression that he still retains a connection to Aarhus University." The statement concluded, "Aarhus University wishes to confirm that Dr. Poul Thorsen no longer has any connection to Aarhus University and that Aarhus University will not be able to collaborate with Poul Thorsen in the future."[19]

HOW VACCINES ARE LICENSED AND MANDATED

Vaccines affect you and your children through a three-step process:

1. Licensure—the FDA says the vaccine is safe and effective;
2. Recommendation—the CDC's Advisory Committee on Immunization Practices (ACIP) places the vaccine on the U.S. recommended vaccination schedule; and
3. Compulsion—state health departments mandate vaccines for day care and public school admission. Federal funding and the relentless lobbying efforts[20] of drug makers powerfully influence state decisions.

Every step of this process is tainted by financial conflicts of interest, scientific fraud, and statistical bias.

Licensure

The FDA must license a vaccine before it can be administered to the general public. A vaccine manufacturer is required to conduct prelicensure safety and efficacy studies. Most people don't realize several key points about the related vaccine safety science.

First, an industry performs its own studies to justify the licensure and selling of its products. The FDA reviews these studies and often provides important guidance to help facilitate the approval process. The FDA does not conduct independent research on proposed vaccines.

Second, the government uses epidemiological studies to disprove causation between adverse events and vaccines or vaccine ingredients. As Drs. Stott and Wakefield explain in chapter 7, the studies the government uses to report "no association" between vaccines and injuries are flawed, either because "they have been badly designed, or they have not been designed with the right hypothesis in mind in the first place."

Third, our public health officials do not require randomized controlled studies.[21] They say that it is unethical to withhold vaccination from anyone, so they do not uphold a scientific gold standard in order to answer ethically troubling questions, especially regarding vaccine safety. Industries typically use other vaccines as placebos rather than using a truly neutral placebo such as saline solution.[22] What better way to prove that your vaccine is safe than to show that it is precisely as safe as another unsafe vaccine? Remember that adverse neurological reactions to vaccines, such as encephalitis and febrile seizures, are no different than encephalitis and febrile seizures caused by disease.Using another vaccine as a placebo is a fail-safe mechanism used to ensure that a test vaccine will be upheld as safe.[23] No regulations govern placebo composition, which can influence research outcomes and merit reporting.[24] For another stark example, suppose a pharmaceutical company invented a new tranquilizer but discovered that it caused birth defects in the unborn children of pregnant women. Now suppose that company designed an FDA safety study using thalidomide (which is known to cause birth defects) as the placebo, to prove its new product was no less safe than other tranquilizers. Vaccine manufacturers follow this pattern in FDA licensure studies.

Recommendation

After obtaining licensure for an unsafe-as-any-other vaccine, the next step is to get it recommended. This involves getting the stamp of approval from the CDC Advisory Committee on Immunization Practices (ACIP). The ACIP is responsible for the federally recommended childhood vaccination schedule. This group tells doctors how many doses of which shots every child in this country should receive. Adult immunization is a new focus for this group. I attended an ACIP meeting and it was truly a farce. A dozen or so doctors from various medical centers around the country sat around a rectangular table and unanimously approved almost every proposal placed before them. Most of the voting members had conflicts of interest.[25] Members abstained from voting on specific proposals in which they were involved (typically, the proposals were vaccine studies financed by pharmaceutical companies, which the members or their medical centers managed), but they voted to approve every other proposal on the agenda. It was a classic case of "you scratch my back and I'll scratch yours." Vaccine safety was not honestly debated. There were no vaccine safety advocates as voting members. For members of this self-serving committee, the path towards professional advancement is clear: never rock the boat and approve every proposal put in front of you.

Compulsion

With an ACIP recommendation in hand, the next task as a vaccine promoter is to get a mandate enacted into law at the state level. This is a relatively simple task, but it is time consuming because every state is different and some states have pesky vaccination choice advocates that assert varying degrees of opposition. Epidemiologists in a state's public health department can reliably produce school mandates that are enacted by state legislative bodies whose members are not educated about vaccine risks, have never read *The Merck Manual* (or any other medical textbook), and have no idea that adverse reactions to vaccines, such as encephalitis and febrile seizures, exist and are indistinguishable from complications from infectious diseases. It is relatively easy to get the votes. Campaign contributions from Big Pharma grease the skids. There is a customary obeisance to modern vaccine technology as the miracle that has banished the ravages of infectious disease. Legislators pass laws requiring children to receive these vaccines to attend day care and public school. While not required, many private schools follow suit and adopt those mandates as well.

Consider, for a moment, schools' vaccination mandates from the drug manufacturers' perspective. Day care and school systems become free vaccine marketing departments. There is no need to train a sales force or to incur other marketing expenses. If you can force your customers to acquire your product, it eliminates all of the messy uncertainty and hassle of a competitive market. Instead of free market competition, their business strategies are built on compulsion. The public school system is an ATM for pharmaceutical companies—who are laughing all the way to the bank.

PATENT CLIFFS AND GOLDEN PARACHUTES

The clout that drug companies wield belies an emerging truth. In the stock market, pharmaceutical companies are no longer the gold-plated, recession-proof investments they once were. The Big Pharma business model has been simple and profitable: Find a natural substance known to have healing properties, invent a synthetic substitute for its active ingredient, give it a clinical name, add toxic additives and preservatives, press it into small white or colorful pills, and charge a price that is thousands of times the cost of production. Most notably, protect yourself from price competition via patent protection. Using this business model, the pharmaceutical industry has turned its stocks into popular, dividend-paying, high-yielding investments.

However, industry analysts understand what the rest of us are coming to realize: Patent protection does not last forever. Pharmaceutical industries now face what is known in the investment world as a "patent cliff." Specifically, they are dreading the near-term expiration of about $140 billion' worth of patents on many of their key, blockbuster drugs.[26] When drugs lose patent protection and face generic competition, prices plummet and profits vanish. The pressure on research departments to invent viable replacement products has mostly drawn a blank. The industry is now experiencing financial pressure that has taken the bloom off stocks and industry prospects. Making drugs is now a cutthroat business. Blogs such as *Café Pharma*[27] offer an insider's perspective.

PHARMA'S SURVIVAL STRATEGY

The pharmaceutical industry has devised a three-pronged survival strategy:

1. Expand to emerging markets,
2. Increase drug prescription compliance, and
3. Expand the vaccine market.

Expand to Emerging Markets

Drug companies are enthusiastic about growth prospects in countries that have low per capita healthcare spending, such as Brazil, China, and India; however, their ambition to gain big profits from patented blockbuster drugs in emerging markets probably won't work. Copycat knockoffs of high-priced American goods are endemic in China. The best-case scenario for pharmaceutical companies includes incremental generic drug profits, theft of intellectual property, and state-encouraged competition from national champions in China and India.[28] Drug companies seem intent on exporting vaccines. The United States can't manage to manufacture many products that the world wants to buy anymore, which is proven by the persistent U.S. trade deficit. However, the United States may end up successfully exporting its sky-high rates of autism and chronic childhood disorders. Recently, the drug industry–dominated World Health Organization and Bill Gates's Global Alliance for Vaccines and Immunisation (GAVI) have stated their intention to extend a U.S.-style childhood vaccination schedule to other countries.[29] Health officials and the pharmaceutical industry seem to think that selling the U.S. vaccination schedule globally can help balance U.S. terms of trade, through greater pharmaceutical profits. Foreign governments should be very wary.

Increase Drug Prescription Compliance

Vaccination mandates have worked so well that drug companies intend to extend the "strong recommendation" model to prescription drugs. The push for prescription drug compliance seems ironic, given the data provided in the latest CDC National Center for Health Statistics Data Brief Number 42, which details Americans' prescription drug use from 2007 to 2008:

- One of every five children and nine out of ten older Americans reported using at least one prescription drug in the month prior to being surveyed.
- 22.4% of kids up to age 11 used at least one prescription drug.
- 29.9% of young people aged 12–19 used at least one prescription drug.
- 48.3% of people aged between 20 and 59 used at least one prescription drug.
- 88.4% of Americans aged 60 and over used at least one prescription drug, more than 76% used two or more prescription drugs in the past month, and 37% used five or more.[30]

This study suggests that Americans are already overmedicated and prospects are poor for increased prescription compliance. It does not mean, however, that the pharmaceutical industry will not try.

The growth of the drug compliance movement is most evident in the recent "deputization" of pharmacists to market and administer flu shots at big-box stores and chain drugstores. The pharmaceutical industry is now commandeering drugstores in the same manner that it captured the public school system. Drugstores appear delighted to have this potential, new revenue source. They are conscripting pharmacists as foot soldiers in the crusade against infectious diseases. The public now faces a marketing onslaught for flu vaccination every time it walks through a drugstore or a grocery store that has a prescription drug department. *Medical News Today* has reported, "Last year, retail pharmacists were particularly influential in driving vaccinations, administering more than sixteen million vaccines and acting as a source of information for concerned citizens. Cardinal Health offers an array of support services to help retail pharmacy customers earn certification to offer and promote vaccination services to their local community."[31] I expect drug makers to expand the drugstore vaccine promotion with a flood of forthcoming adult vaccines; one example is the herpes zoster (shingles) shot. In the business world, this is called "channel stuffing."

Expand the Vaccine Market

Big Pharma wants you to believe that you are vaccine deficient. There are currently 145 new vaccines in development.[32]

Media-stoked public fears of terrorist biological-weapon attacks (e.g., anthrax) following the attack on the World Trade Center on September 11, 2001, and the subsequent overhyped H1N1 "pandemic" have convinced the drug industry that there are huge profits to be made in vaccines. The global vaccine revenues were $22 billion in 2009—an increase of 16 percent since 2008. Industry analysts predict that vaccine revenues will hit $35 billion by 2015.[33]

Despite tremendous growth, vaccine revenues cannot blunt the entire impact of the $140 billion that will be lost when blockbuster drug patents expire. That won't stop the vaccine development binge, however. The drug industry sees vaccine development as a golden parachute to take them safely over their patent cliff.

As I previously discussed, the drug industry dominates the regulatory process and can jam new vaccines through the approval process by disguising adverse reactions with bogus placebos. The implications for a major increase in neurological damage are staggering. Having dominated the child vaccine market through school mandates, the vaccine industry is champing at the bit to compel adults to comply with their forthcoming campaign for adult immunization.

Moreover, there are severe quality limitations on vaccines for both adults and children. Some vaccines are ineffective or only protect against a small number of strains of a particular disease.[34] Some vaccines lose their supposed protective qualities when disease strains, like influenza, shift.[35] Other vaccine-derived antibodies only persist a few years.[36] These vaccines are economically attractive to vaccine manufacturers, though, because people will need repeated doses, or boosters. The industry's repeat-revenue model favors rapidly obsolescing vaccines.

While far more dangerous due to its health implications, this model is reminiscent of the calculated obsolescence business model used by the U.S. automotive industry in the 1960s, which Ralph Nader criticized. The U.S. auto industry deliberately made junky cars that wore out or went out of fashion—a business strategy that ultimately destroyed the U.S. automotive industry because the Japanese and German manufacturers made better cars. According to this business model, just like American cars made in the 1960s and 1970s, ineffective vaccines that are quickly obsolete are preferable to ones that provide lifelong immunity. From the vaccine manufacturer's perspective, it is essential to keep the customers coming back for more.

Medical sociologist Dr. Donald Light from the School of Public Health at the University of Medicine and Dentistry of New Jersey (UMDNJ)[37] recently presented a paper touching on this subject at the American Sociological Association's annual meeting in Atlanta on August 17, 2010 (session 487), entitled "Pharmaceuticals: A Two-Tiered Market for Producing 'Lemons' and Serious Harm." He argued,

> The pharmaceutical market for "lemons," differs from other markets for lemons in that companies develop and produce the lemons. Evidence in this paper indicates that the production of lemon-drugs with hidden dangers is widespread and results from the systematic exploitation of monopoly rights and the production of partial, biased information about the efficacy and safety of new drugs.

I anticipate that vaccine makers will use their monopoly power and influence over the regulatory process to press for more vaccination mandates. Pharma's business model is compulsion: Take the right of informed consent away from the individual and replace it with the dictatorial requirement to be a captive customer for lemon vaccines and prescription drugs.

VACCINE BUBBLE PSYCHOLOGY

In the investment world, I have lived through several speculative bubbles. Lessons learned from those experiences apply to what I call the "vaccine bubble." Two key forces create bubbles:

1. Promoters, who typically have a financial interest in creating mass public hysteria. In the case of the housing bubble, this would include the banking, mortgaging, and construction industries.
2. Do-gooder policy makers, who want to wave a magic wand and improve the world through well-intended interventions. These do-gooders are oblivious to the unintended consequences of their involvement, otherwise known as "blowback" in the political and investment worlds.

Osama Bin Laden is one example of blowback. CIA do-gooders funded a mujahideen resistance against the Soviet occupation of Afghanistan in the 1980s, providing training and weapons. What seemed like a great idea at the time became the most obvious example of blowback on September 11, 2001, when the World Trade Center was attacked. CIA-equipped terrorists, who had initially joined us in fighting our Soviet enemies, evolved into our enemies. Meanwhile, the Soviet Union had collapsed and its central successor, Russia, was no longer our enemy. Few, if any, policy makers anticipated this chain of events when the CIA originally orchestrated the mujahideen resistance. At the time, nearly everyone thought it was a fine idea.

Another glaring example of blowback from well-intentioned policies is the collapse of the U.S. housing market. The well-intentioned efforts of Congress to expand home ownership fostered a subprime mortgage bubble, allowing low-income renters who probably never would have qualified for a normal home loan (with the required 20 percent down payment and verifiable employment and income) to purchase homes. This process encouraged real estate speculation and mortgage-market insanity, which drove up prices to unrealistic levels in many areas of the country (e.g., Miami, Las Vegas, and Southern California). The blowback from this policy error is now obvious to everyone—homeowners who lost their homes, banks that busted, and other grim economic news.

The blowback from the vaccine bubble is a rising epidemic of neurological damage. Approximately one in ninety-one children (1.1 percent of children) in the United States now receives an autism diagnosis.[38] This is thirteen times more than the average autism rate (0.08 percent) found in a 1993 survey—before the U.S. childhood vaccination schedule expanded dramatically.[39] The latest CDC study shows that 9.5 percent of parents surveyed reported that their child had received a diagnosis of attention deficit hyperactivity disorder (ADHD).[40]

This plague of neurological damage is expanding in direct proportion to the greater number of vaccines imposed on an unsuspecting public. The future economic costs to society of medical care and welfare for a neurologically disabled population are staggering.[41] Policy makers must consider these costs now—and Americans must hold the corporate perpetrators economically and criminally liable.

Every bubble contains an element of crowd psychology, or a collective conception that shepherds unsuspecting individuals into wholeheartedly participating in an ill-fated popular delusion. Examples of this include buying the NASDAQ at the 5,000 level in early 2000, and buying an unaffordable house in 2007. During a bubble, people are dying to participate. The bigger the bubble, the more powerful the promotional messages encouraging participation.

A similar force is at work regarding vaccination. Television, print, and online publications are constantly quoting CDC and public health officials, who tell everyone how dangerous the flu, measles, chickenpox, meningitis, pertussis, and cervical cancer are. Their messages include exhortations to vaccinate against vaccine-preventable diseases. *These health authorities are essentially selling fear and disease.*[42] It appears that their intention is to whip up mass hysteria and persuade people to wait in line for the latest miracle vaccine against a deadly disease. These unsuspecting people remain unaware that adverse reactions to vaccines exist and are defined in medical textbooks as the same thing as infectious disease complications. We cluck at the bad judgment of those who invested their life savings in overpriced technology stocks when stockbrokers trumpeted a Dow advance to 24,000 (its peak was 14,164), and at others who bought Miami condominiums and Las Vegas ranch houses when mortgage brokers urged them to avoid missing the housing boom. Today, the pharmaceutical industry has convinced the government and the medical profession to do its bidding. Another group of jive-talking promoters, who have a vested personal and financial interest in fanning the flames of mass hysteria, is exploiting the public.

Incidentally, institutional portfolio managers (my clients) who resisted the NASDAQ and real estate bubbles lost customers who only wanted portfolio managers who purveyed the popular investment flavor of the month. In other words, there was tremendous institutional pressure to play the game, even though they did not believe in it and knew it would end in tears. There is undoubtedly similar pressure in the medical world for doctors to play along with CDC vaccination recommendations and to make patients into pincushions for pharmaceutical companies, even if they suspect that "something is rotten in Denmark," as Shakespeare wrote in *Hamlet.* The overall result is that enormous psychological pressure is placed on people by medical professionals to receive vaccines, either in normal primary-care settings or during emergency-room visits for unrelated injuries or illnesses.[43]

The aftermath of deviant psychology bubbles is typically a backlash against the bubble promoters. In the old days, the perpetrators would be tarred and feathered, or dispatched. After the NASDAQ bubble, investors' sentiments toward stockbrokers and bubble cheerleaders soured. As we now know, through endless stories of people losing their homes through foreclosure, the public's sentiment toward estate brokers, mortgage brokers, politicians, and overpaid "banksters" is vindictive. Victims want retribution.

When the vaccine bubble finally bursts, public sentiment toward government health officials, pharmaceutical companies, and doctors is likely to turn negative in the same way that it did after the collapse of the stock market and real estate bubbles. Today, the vaccine regulatory system is primed to approve and advance all of the vaccines Big Pharma can dream up. The sales pitch is germophobia and protection against diseases you didn't know you had or could get, such as genital and anal warts—Merck's latest justification for giving *male* sixth graders the female-oriented human papillomavirus (HPV) cervical cancer vaccine. In all likelihood, this will end badly (as if VAERS isn't already disastrous enough). The greedy fearmongers will overstep their boundaries and create a catastrophically dangerous vaccine, like the 1976 swine flu vaccine that caused an estimated five hundred cases of Guillain-Barré-syndrome-induced paralysis and at least twenty-five deaths.[44] A similar disaster would ignite today's vaccine safety controversy and rapidly turn the tide of public ideology against vaccines and vaccination mandates. Fraud always becomes a public issue after a bubble bursts. All of the evidence is currently here before everyone's eyes, including fudged safety studies, the revolving door of health officials and pharmaceutical executives, denial about the basic scientific definition of neurological damage, and more. The public simply needs to connect the dots.

THE REFUSERS

In the meantime, we must preserve the right to opt out of compulsory vaccination requirements for our children and ourselves. Infectious disease specialists use the term "refuser" pejoratively, to cast aspersions on those who are intelligent, informed, and strong willed enough to "refuse" the corporate and governmental assault on our families' bodies with their ineffective, unsafe, and toxic vaccines.[45] I admire their supposedly pejorative term "refuser" so much that I trademarked it and created a musical band around the concept—The Refusers.[46] This is my attempt to raise public awareness about vaccination risks that the pharmaceutical industry and the CDC are foisting on us. I fully support the right of all parents to make informed vaccination choices for their children, whatever their decisions may be. Indeed, the CDC claims that over ninety percent of American children younger than three years of age are now fully vaccinated.[47] So be it. After losing a daughter and investigating the so-called science behind our current vaccines, *I am a refuser.* Medical authorities and uninformed parents want to turn that informed decision around and paint people like me as a menace to society. The VAERS reports show that *adverse reactions to vaccines are the real menace to society.*

In my opinion, health officials and drug companies intend to abolish that right of refusal; meanwhile, they are causing an epidemic of neurological damage from vaccines and denying responsibility for it. They take credit for preventing neurological damage from disease epidemics. Nevertheless, *The Merck Manual* and VAERS tell us the truth. Public health officials have betrayed the public they pledged to protect. Postbubble forces will unleash a massive reform and bring the perpetrators of this vaccine bubble and the epidemic of neurological damage to justice. Until then, it is essential that we honor an individual's choice to refuse a vaccine as a human right.

Chapter Nineteen

A LICENSE TO KILL?

Mark Blaxill, MBA and Dan Olmsted

This article identifies a pattern of conflicts of interest at the U.S. Department of Health and Human Services—and ultimately throughout the federal government—involving Merck's controversial Gardasil vaccine against HPV, or human papillomavirus.

- Researchers at the National Cancer Institute at the National Institutes of Health (NIH) invented critical technology for the vaccine.
- Another NIH office filed for patents on the technology, licensed those patent rights to vaccine manufacturers, and eventually received royalties from Merck, Gardasil's manufacturer.
- The Food and Drug Administration (FDA) supervised the clinical trials and granted Merck a license application for the first HPV vaccine.
- Just three weeks later, an advisory committee at the Centers for Disease Control (CDC) recommended universal HPV vaccination for women from nine to twenty-six years of age, guaranteeing Gardasil would reach block-buster status for Merck: annual revenues of well over $1 billion.
- Subsequently, agencies within FDA and CDC have been responsible for monitoring Gardasil's safety in the field as officials within the U.S. Department of Health and Human Services (HHS), Health Resources and Services Administration (HRSA) brace themselves to sit in judgment over a new wave of vaccine injury claims.

Meanwhile, key officials involved in the decisions rotate through a revolving door into private industry. These conflicts are both extraordinary in scope and poorly understood by the general public, but they are central to understanding why unsafe and unnecessary vaccines are approved and recommended—why we have a vaccine epidemic.

1. HOW A PUBLIC-PRIVATE PARTNERSHIP MADE THE GOVERNMENT MERCK'S GARDASIL PARTNER

"Perhaps no other recent product on the market demonstrates successful health care technology transfer better than the HPV vaccine, Gardasil, produced by Merck & Co. and approved by the FDA in June 2006," proclaimed a recent NIH newsletter. In a February 23, 2007, article entitled "From Lab to Market: The HPV Vaccine," the *NIH Record* celebrated the pivotal role of government researchers in developing Merck's Gardasil product. "Based largely on technology developed at NIH," the newsletter reported, "the vaccine works to prevent four types of the sexually transmitted HPV that together cause 70 percent of all cervical cancer and 90 percent of genital warts."[1]

The occasion motivating this celebratory article was the "Philip S. Chen, Jr. Distinguished Lecture on Innovation and Technology Transfer" given by Douglas T. Lowy, one of the NIH scientists involved in developing the HPV vaccine. In a ceremony pictured in the article, Lowy received accolades from the head of NIH at the time, Elias Zerhouni, MD, who showered praise on Lowy's team and viewed its work as a model for future efforts. "It's a 'heroic' story about the effort to fight cervical cancer, the second most deadly cancer for women worldwide, said NIH director Dr. Elias Zerhouni," in the *NIH Record*'s account. "He noted that he has talked about the vaccine's creation to Congress and with the President on his recent visit to NIH. How researchers took the technology 'from the lab to the marketplace is a journey we can learn from,' Zerhouni said."[2]

While Zerhouni was bragging to anyone in Washington, DC, who would listen about the NIH team's role in this historic accomplishment, the vaccine's developers were actively spreading the news of their achievement in scientific circles. It's hard to blame them because at the time Lowy and his colleague John T. Schiller, leaders of the team that had invented the technology for the "virus-like particles" (or VLPs) that made Gardasil possible, were in some pretty heady company. In 2008, Harald zur Hausen, the scientist who discovered the role of HPV in cervical cancer during the 1980s, received one half of the Nobel Prize in

Medicine; the two researchers at the Pasteur Institute who had discovered the human immunodeficiency virus (HIV) had to share the other half.

Perhaps campaigning for their own place in the pantheon of medical heroes, Lowy and Schiller described their VLP technology in several review articles on the history and development of the Merck vaccine. These treatments were studiously scientific in tone and at points openly critical of their commercial partner, as the authors commented with disapproval on the high price Merck was charging for Gardasil. But in one May 2006 review in *The Journal of Clinical Investigation,* the pair also made the following disclosure about their own commercial interests:

> Conflict of interest: The authors, as employees of the National Cancer Institute, NIH, are inventors of the HPV VLP vaccine technology described in this Review. The technology has been licensed by the NIH to the 2 companies, Merck and GlaxoSmithKline, that are developing the commercial HPV vaccines described herein.[3]

Attached to an otherwise heroic narrative of the triumph of technology over cancer, this disclosure struck a discordant note. Conflict of interest? Inventors? Vaccine technology? Licenses? Pharmaceutical companies? Commercial vaccines? This isn't scientific language but rather the language of money and commerce. What was this unusual concession doing there in the fine print?

This is not an idle question, for Lowy and Schiller's conflict disclosure forms the basis for an alternative to Zerhouni's triumphalist narrative, one that spotlights the unusually self-contained set of Department of Health and Human Services (HHS) activities that surrounded HPV vaccine development. This alternative account is more of a business story than a scientific one, a tale in which commercial interests were inextricably linked to matters of life and death. In this revised narrative, Gardasil is perhaps the leading example of a new form of unconstrained government self-dealing, in arrangements whereby HHS can transfer technology to pharmaceutical partners and simultaneously both approve and protect their partners' technology licenses while also taking a cut of the profits. Literally and figuratively, HHS has the authority in such situations to allow its business partners to get away with murder for the greater good, effectively granting its private business partners a license to kill.

HHS officials have their own language for such arrangements. They call them *public-private partnerships*, and HHS agencies have become progressively more aggressive about pursuing them. NIH, for example, launched its own

"Program on Public-Private Partnerships" in 2005, shortly before Gardasil's launch. On the website describing this program, the NIH program managers concede that the kind of technology transfer involved with Gardasil carries unavoidable ethical risks and acknowledge that "the potential for conflict of interest exists any time the NIH and NIH staff engage with non-Federal entities to achieve mutual goals." They provide little more than a *pro forma* solution for such conflicts; however, any concerned NIH staffers are encouraged to "contact their Deputy Ethics Counselor." [4]

It's important to shed light on this alternative narrative as a counterpoint to the heroic story promoted by Gardasil's many sponsors. An uninformed observer might assume that the responsible agencies of HHS care not at all about commercial opportunities and exclusively attend in a disinterested but methodical manner to the issues of health and safety that would naturally concern any consumer of vaccine products.

As appealing as that assumption might be, it would be incorrect. By taking a commercial perspective on Gardasil's development and regulation, one is forced to confront a new and disturbing question. How is disinterested vaccine safety governance even remotely possible when HHS employees stand as heroes at the head of the parade when a new vaccine is invented within its walls, while agency leaders are leading the cheering section, approving the new product's launch, making the market for the product with its policy recommendations, and then turning around to cash multimillion-dollar checks? In order to better understand the real lessons of Gardasil under the harsh light of the business interests at work, let's take a closer look at how the Merck-NIH partnership on Gardasil was forged.

Conflicts of Interest in Vaccine Development and Regulation

As the world's largest single sponsor of biological research, NIH frequently funds research with commercially valuable outcomes. When that R&D generates potentially valuable inventions, NIH submits patent applications to the U.S. Patent and Trademark Office (USPTO) and actively pursues the approval of those patents, which when granted become valuable commercial property for HHS, the patents' owner. Since NIH has neither the authority nor the capability to pursue product commercialization efforts, in order to encourage private companies to invest in conducting the necessary clinical trials, NIH's Office of

Technology Transfer (OTT) was created to grant commercial licenses for such HHS patents to commercial partners, including vaccine manufacturers. When new products invented at NIH clear the requisite regulatory hurdles at the FDA and reach the market, OTT then shares in the profits. It also distributes the rewards back to the scientific teams whose products have succeeded in reaching the commercial stage; when license fees flow into OTT's coffers, the federal employees who invented the technology are entitled by NIH policy to a share of the royalties.

From a technology development standpoint, such commercial arrangements are the result of intentional public policy; in fact, they resulted from an act of Congress. The Bayh-Dole Act of 1980 was written with the express purpose of making it easier for federally funded academic research to receive patent protection that would allow the ready licensing of the fruits of commercially valuable R&D to private businesses. At the time, Congress was concerned that federally funded inventions too often languished within the academy because businesses had insufficient incentive to invest in clinical trials since these inventions were often unsupported by the powerful competitive protection afforded by an exclusive patent license.

The policy worked. Within the research universities that receive the vast majority of federal funding, Bayh-Dole has had the desired effect. It has enabled university technology transfer offices all over the world to generate billions of dollars of licensing revenue in the last few decades—especially in the life sciences—by licensing patents from federally funded university research to corporate partners. Bayh-Dole has effectively turned research into big business for many universities and transformed technology transfer offices into important profit centers at academic institutions all over the world.

But when technology licensing takes place *within* federal agencies, Bayh-Dole creates an entirely different problem: an unprecedented web of conflict, one in which the same departments that are tasked with regulating the health and safety of medical products are also profiting from them. As Lowy and Schiller conceded in their review article disclosure, this conflict of interest came into play directly on Gardasil: both men are named inventors on the technology that makes Gardasil possible; NIH filed for and received patents on their invention of the VLP technology; HHS is the owner of the patent family that protects the commercial rights to the invention; in order to bring the product to market, OTT licensed the vaccine technology to Merck; and as Merck has generated billions in highly profitable Gardasil revenue, OTT has received millions in Gardasil profits as well.

But HHS is also responsible for regulating Gardasil in numerous ways. Its agency, the FDA, reviewed the clinical trials in which Gardasil was tested in human populations and passed judgment on Gardasil's safety. An Advisory Committee on Immunization Practices (ACIP) of another of its agencies, the CDC, decided whether or not to recommend Gardasil for young women and children. Together, the FDA and CDC now conduct the post-licensure surveillance to decide whether or not Gardasil is proving safe in larger populations. And as some families are now beginning to seek compensation based on claims that Gardasil caused injury in some of their children, the division of the HRSA, which represents HHS in the National Vaccine Injury Compensation Program (VICP), will soon play a large role in deciding whether Gardasil victims will receive any compensation and, if so, how much.

As you can see in the chart below, all of this activity is supervised in a single department by one cabinet official, the Secretary of Health and Human Services. The sole nongovernmental agency involved in this commercial enterprise is Merck's vaccine division. In effect, the Merck-HHS partnership leaves the business side to Merck while HHS is solely responsible for:

1. Creating the market for Gardasil by funding commercial research, supervising the conduct of clinical trials, judging the outcome of those trials, and promoting a policy of universal vaccination;
2. Collecting the license fees that result from Gardasil revenues from Merck and other vaccine manufacturers and then distributing these financial benefits to federal employees; and
3. Deciding whether or not to protect the policy decisions and profit streams of its sister HHS agencies through post-licensure safety monitoring and vaccine injury compensation rulings.

Is this good government at work or an example of the medical-industrial complex run amok? To answer that question, we need to look at the NIH patent portfolio and the associated license fees that have been flowing into NIH coffers since 2006.

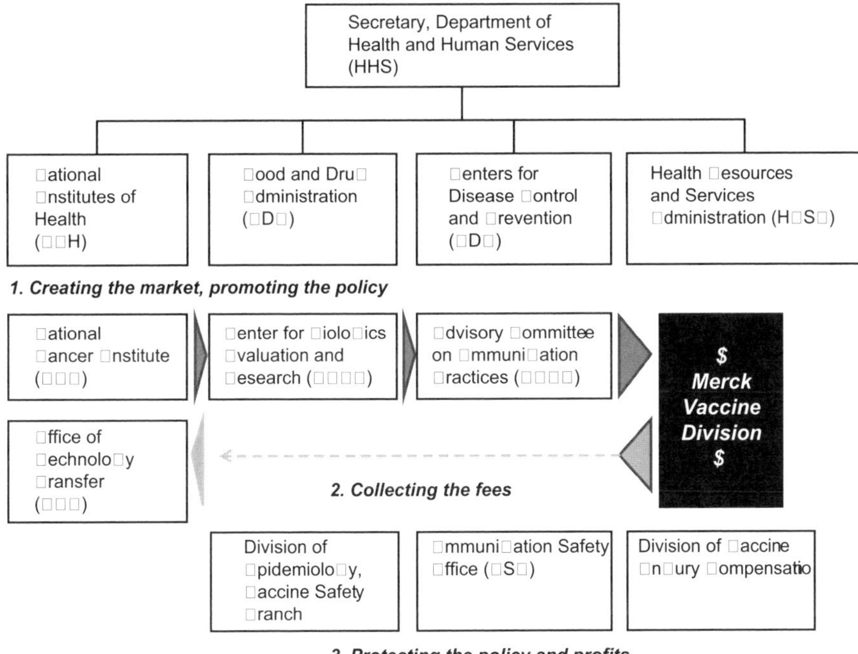

Celebrating the Invention of a New Market

Lowy and Schiller are both employed by the National Cancer Institute (NCI). One of the largest of the NIH institutes, NCI was established in 1937 by President Franklin Delano Roosevelt. For many decades, NCI has been the agency at the forefront of the so-called "war on cancer." Perhaps the earliest inspiration for the both the cancer war and the Gardasil program began during the 1960s, when NCI researchers first began looking in earnest at viruses as a potential cause for cancer. In 1961, NCI leaders created the Laboratory of Viral Oncology to begin the search for cancer-causing viruses; in 1962 the Human Cancer Virus Task Force was first convened, and by the end of the decade, enthusiasm over this research was part of the scientific momentum that persuaded President Richard Nixon to launch the war on cancer in 1971. Unfortunately for Nixon's legacy, and for most subsequent cancer victims, the war on cancer has famously failed to find a cure for cancer or to validate theories of viral causation in the vast majority of human cancers.

But starting in the 1980s, the two exceptions to this litany of failure—hepatitis B virus and HPV—led to the launch of two blockbuster new vaccine products. The infant hepatitis B vaccine was developed in the 1980s and launched in 1991 with an ACIP recommendation that all American infants be vaccinated on the first day of life. And after 1984, when Harald zur Hausen first pinpointed the role of certain strains of HPV in cervical cancer, the work on another anticancer vaccine could begin. By the early 1990s, laboratories all over the world were racing to develop the first HPV vaccine.

Lowy and Schiller's NCI team was among the four most active research teams in this race, all of whom were aggressively filing patents on their HPV inventions. Along with a third NCI colleague, Reinhard Kirnbauer, Lowy and Schiller filed their first application for a patent entitled "Self-assembling recombinant papillomavirus capsid proteins" on September 3, 1992.[5] Since then—and after splitting the original application into twenty-nine "children" in the form of numerous "divisionals," "continuations," and "continuations-in-part"—nine patents from that family have been granted, as well as four more from a branch of the family tree entitled "chimeric papillomavirus-like particles." The ability of the novel "L1 proteins" described in their patent to "self-assemble" into virus-like structures, which when deployed in a vaccine solution could stimulate a protective immune response against HPV, formed the essence of their invention. Although OTT doesn't specify the royalty-bearing patents, the commercially valuable technology that Merck has licensed likely comes from this group of nine "self-assembling recombinant papillomavirus capsid proteins" patents.

The NCI team was among the leaders in HPV technology, but the race to make a commercially viable HPV vaccine involved several other research teams. Most notable among these were the University of Queensland in Australia, Georgetown University, and the University of Rochester. In addition to NCI's filings, each of these university-based research teams filed its own patents; eventually, Merck and GlaxoSmithKline (GSK) got into the act as well. Like many promising areas of technology, the HPV patent landscape became large and crowded in a short period of time.

Amid this blizzard of activity, the USPTO's Board of Patent Appeals and Interferences (BPAI) had to step in to sort out whether these competing patent applications interfered with each other. BPAI distributed the credit, making a series of hotly contested decisions that were ultimately appealed to the U.S. Court of Appeals for the Federal Circuit (CAFC), the nation's most powerful patent court. By 2007, all the BPAI and CAFC rulings had come in and the respective contributions of all four groups were conclusively allocated for commercial purposes. The team led by Ian Frazer at the University of Queensland received credit for being the first to propose the idea of using VLP technology for a vaccine since its application was filed on July 20, 1992, just six weeks earlier than the NCI team's. But thanks to its unique technology of "self-assembly," most of the invention claims of the NCI patent family remained intact as well;[6] Lowy and Schiller's invention has since been generally accepted as a critical advance in the wave of new technology that made Gardasil possible. In terms of the distribution of financial reward, both Rochester and Queensland have reported receiving royalty income for their HPV inventions (in undisclosed amounts) in addition to the revenues reported by OTT.

As the technology transfer officials at OTT were paving the way for the financial benefits from Gardasil to flow back to NIH, Lowy and Schiller were benefiting in other ways as well, especially when it came to scientific credit. Throughout much of 2006 and 2007, they received awards from many quarters for their role in developing Gardasil's "virus-like particles." Their joint awards included the Dorothy P. Landon-AACR Prize for Translational Cancer Research in April 2007 and the 2007 Novartis Prize for Clinical Immunology. In addition, Lowy also shared the Daniel Nathans Memorial Award in September 2007 and received the American Cancer Society's Medal of Honor for Basic Research in October 2007.

In addition to these awards, on September 19, 2007, Lowy and Schiller received what was perhaps their crowning honor. That's when the Partnership for Public Service awarded the pair the Federal Employees of the Year Service to America Medal. According to its sponsors, "The Service to America Medals have earned a reputation as one of the most prestigious awards dedicated to celebrating America's civil servants. Often referred to as the 'Oscars' of government service," they are more commonly known in government circles as the "Sammies." Upon receiving his crowning honor, Lowy was interviewed for the *NIH Record* and professed the requisite modesty in its October 2007 edition, saying "We are simply symbols of the many people who have made critical contributions to understanding the relationship between papillomavirus infection and cervical cancer."[7]

If Lowy was modest, the top brass at NIH could barely conceal their pride over their employees' accomplishments. According to the Partnership for Public Service, "Lowy and Schiller's 20-year partnership has been a boon to the nation's health and for the advancement of scientific discovery."[8]

Collecting the Licensing Fees

Alongside the science and policy celebrations, the business side of the Merck-NIH partnership proceeded with a bit less fanfare and with a different kind of currency. Once the patent was approved, OTT could then turn to extracting its share of the benefits from its commercial partners' new products, which in the case of the HPV vaccine included sales from Merck's Gardasil product and later from GlaxoSmithKline's Cervarix. Merck reached the market first in 2006, and GSK followed in 2007. As each company began collecting revenue from its new vaccine, OTT began collecting royalties. The table below shows our analysis of how Merck and GSK's revenues may have flowed into OTT's coffers.

	Gardasil Revenue ($M)	Cervarix revenue ($M)	NIH Top 20 Revenues ($M)	HPV Rank in NIH Top 20	HPV Vax Revenue: estimated at 1% license fee ($M)
2006	235	--	NA	NR	
2007	1,481	20	71 (est.)	#4	15
2008	1,403	229	77.4	#2	16
2009	1,118	292	75.7	#1	14
2010	989	375	75.1	#2	14
2011	1,209	815	82.5	#2	20

Both Merck and GSK itemize revenue for Gardasil and Cervarix in their quarterly and annual earnings statements. Their annual results are summarized in the first two columns of the table. For Merck, Gardasil has been a blockbuster success, yielding a cumulative total of more than $6 billion in revenue through year end 2010. By contrast, GSK's revenues have been smaller and have not yet reached a cumulative total of $2 billion.

For its part, OTT does not itemize its HPV license revenues. However, it does report its total royalty revenue as well as the cumulative revenue from its "top 20" technology licenses since 2007. These top 20 licenses have been worth more than $70 million annually in profits for NIH in the last five years, and HPV licenses have soared to the top of those rankings quickly. Over the last four years, HPV licensing was either the #1 or #2 revenue generator.[9] OTT doesn't disclose exactly how much the Gardasil and Cervarix royalties contribute to NIH, but if we make the assumption that its patent licenses entitle the agency to 1 percent of the HPV vaccine revenues of its partners (an assumption that appears reasonable based on the available data) then we can safely estimate that OTT has been collecting in the range of $15–$20 million per year from Lowy and Schiller's invention.

In addition to their numerous scientific awards for their discoveries, Lowy and Schiller have received cash distributions from NIH based on their patents. As federal employees, they are each eligible to receive a share of patent royalties up to $150,000 per year, and Gardasil's success has guaranteed that they would receive the maximum reward. That means that since FDA's approval in 2006, each man has earned close to $1 million in royalty revenue.

* * *

This is the HHS vision of public-private partnerships at work. Contrary to the rhetoric, these partnerships aren't simply a high-minded collaboration of scientific visionaries, but rather a large commercial enterprise with extraordinary profits at stake: an enterprise from which NIH receives credit and money and upon which its corporate partners build multibillion dollar businesses.

How does such a partnership affect the incentives of regulators whose job it is to make sure the products are safe? It's not obvious that they do. Just because HHS has a financial stake in Gardasil, that doesn't necessarily mean that every subsequent decision its employees make is a corrupt part of a nefarious conspiracy to kill young women for money. Indeed, HPV royalty revenues of $15 million represent just a small fraction of a HHS budget that rose to well over $700 billion in 2009. In the larger scheme of things, HHS revenues on Gardasil are just a small drop in a very large bucket.

Far more likely to play a role, however, in public-private partnerships like the Gardasil vaccine are the insidious cultural pressures that emerge in a supremely political organization like HHS. Can we really expect the secretary of HHS to take the FDA director to task for implementing lax standards on vaccine approval when the director of NIH is simultaneously praising the "heroic" researchers who invented the product in the first place? Will the CDC apply extra caution in its vaccine policy recommendations when its sister agency is involved, or will it be more likely to activate the "fast track" in the process of making recommendations for Gardasil? What we have observed so far merely suggests the *potential* for bias in the regulation of products in which HHS holds a direct stake. The second question is whether or not there have been *actual* patterns of bias in the ways in which regulators at FDA and CDC have conducted their duties with respect to Gardasil.

2. WHO GUARDS GARDASIL'S GUARDIANS?

In a related process at the FDA, officials in the Center for Biologics Evaluation and Research (CBER) supervised the clinical trials and granted Merck the first "Biologics License Application" (BLA) for a HPV vaccine.[10] Three weeks later, the ACIP of the CDC recommended universal HPV vaccination for women from nine to twenty-six years of age. In one series of votes, ACIP guaranteed that Gardasil would reach blockbuster status for Merck: annual revenues of well over $1 billion.

How Stringent was the FDA's Safety Review for Gardasil?

When the FDA issued its approval of Merck's BLA for Gardasil on June 8, 2006, its decision was based on a review of Merck's data from five separate clinical trials, each of which included efficacy and safety assessments for Gardasil. Four of the five trials approached their efficacy and safety studies in a similar fashion, comparing Gardasil against a "placebo" that contained an active ingredient, with one trial comparing Gardasil against what the CBER reviewers described as a "saline placebo." All together, these five trials examined a total of close to 12,000 subjects who received at least one dose of Gardasil and compared their outcomes to roughly 10,000 subjects who received up to three injections of what Merck and CBER officials agreed to describe as a "placebo."[11]

But what is a placebo, really? One definition describes a placebo as "an innocuous or inert medication; given as a pacifier or to the control group in experiments on the efficacy of a drug."[12] The operative term here is the word *inert.* But in four of the five trials, Gardasil placebos contained a vaccine ingredient called an adjuvant, "a substance which enhances the body's immune response to an antigen." According to one of the trial publications, most of the Gardasil trial placebos actually contained an "amorphous aluminium hydroxyphosphate sulfate adjuvant . . . and was visually indistinguishable from vaccine."[13] Although the majority of the placebo treatments in the Gardasil trials did not include Gardasil VLPs, they were by no means inert. In control populations representing nearly 95 percent of all "placebo" recipients, the study subjects received a formulation that actually included an immunologically active (and potentially harmful) aluminum adjuvant.

One of the five trials, however, was different. In this trial, the only one that examined a younger population of nine- to fifteen-year-olds, the placebo recipients did not receive an aluminum adjuvant. By contrast, and according to most of the FDA documentation, the nearly 600 control subjects in this trial received a formulation most commonly described as either a "non-alum placebo" or a "saline placebo." The safety results of this trial deserve special notice since it's the only trial that compared Gardasil to a solution that could reasonably be described as "inert."

But even that assumption would overstate the case. Although the "saline placebo" did contain water and sodium chloride (ordinary table salt), the FDA was incorrect to suggest that there were no other active ingredients. According to the published description of this trial's methods, "The placebo used in this study contained identical components to those in the vaccine, with the exception of HPV L1 VLPs and aluminum adjuvant, in a total carrier volume of 0.5 mL."[14] Formulations like this, which are comprised of everything in the vaccine except its immunologically active components, are sometimes called a "carrier solution." The correct description of the placebo as a "carrier solution" rather than a "saline placebo" was provided only once in the CBER review, buried in a table on page 301. Nowhere in either the CBER review or the published account of the trial can one find any description of this placebo's ingredients.[15]

It is possible, however, to infer the composition of the carrier solution from Merck's Gardasil package insert, which lists the vaccine's immunologically inactive ingredients. These include: "yeast protein, sodium chloride [table salt], L-histidine [an amino acid], polysorbate 80 [an emulsifier], sodium borate, and water for injection."[16] At least one of these chemicals, sodium borate, is a chemically reactive toxin, one that has many industrial uses as an active ingredient. These include applications as a replacement for mercury in gold mining, an insecticide and fungicide, and a food additive that is now banned in the United States.

Is there any defense for the FDA to allow this approach to placebo selection in the Gardasil trials? From an efficacy standpoint, one can reasonably argue that yes, using an adjuvant in a placebo makes sense since it will provide the most rigorous test of the value of the active ingredient under review, in this case the

VLPs invented at NCI. And, in fact, the returns from all five clinical trials provided convincing evidence that when the VLPs were added to a vaccine formulation containing the aluminum adjuvant, a strong immune response resulted. CBER therefore drew the reasonable conclusion that Gardasil works, at least against the end points it was able to measure.

But is it safe? When it comes to the accurate measurement of adverse effects of Gardasil, there is little justification for reliance on a placebo with ingredients that are not inert. There is some limited value, perhaps, in comparing adverse events that are introduced solely by the addition of VLPs to the vaccine solution. But a truly rigorous safety assessment would investigate the full safety profile of the VLPs *in combination with* the aluminum adjuvant and compare that profile to the profile of an inert solution. After all, the adjuvant is present precisely because it is not inert.

If the FDA trial standards were truly to enforce a high standard of safety, they would require the comparison of Gardasil's safety profile to a true saline placebo. But Merck performed no such analysis and CBER permitted the company to apply a lesser safety standard of safety analysis. As a result, CBER issued its BLA approval without any idea whatsoever of the true risks of Gardasil. Not surprisingly, most of the comparisons between adverse outcomes for those receiving doses of Gardasil and those exposed to an aluminum adjuvant "placebo" showed little evidence of injury risk from Gardasil.

Unfortunately, the conclusion that Gardasil was therefore safe was horribly wrong.

A Different View of the Gardasil Trial Data

Based on the data provided in CBER's review of the Gardasil trials, it is possible to piece together an alternative view of Gardasil's adverse event profile by examining three separate populations: 1) the subjects who received actual doses of Gardasil (over 96 percent got all three doses); 2) the subjects who received a "placebo" containing an aluminum adjuvant (over 98 percent got 225 micrograms of amorphous aluminum hydroxide sulfate) formulated in a carrier solution that made it visually indistinguishable from the full vaccine; and 3) the subjects who received only doses of the carrier solution. For the Gardasil and aluminum adjuvant groups, safety results were described in two ways: a narrow set of reported outcomes was measured for the entire trial group (the "general safety population"), while for a smaller group (the "detailed safety population"), including the entire carrier solution group, a more extensive set of outcome data was collected. The respective sizes of these three safety assessment groups are shown below.

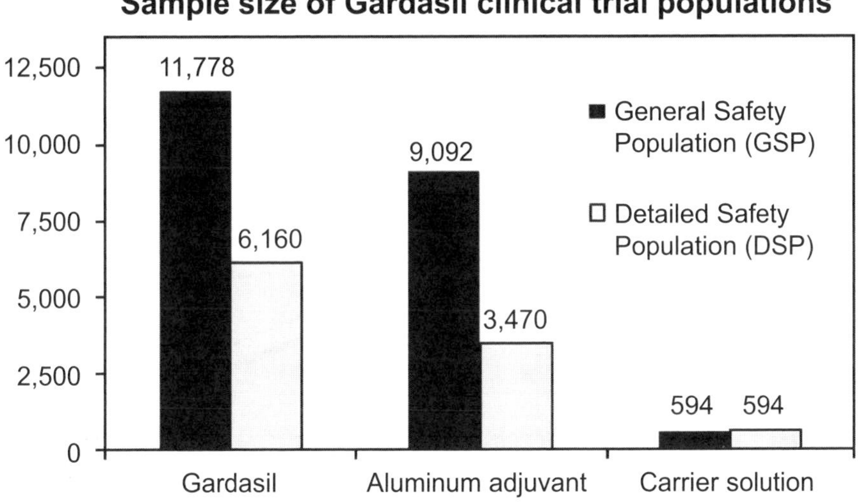

Unfortunately, the small relative size of the carrier solution group reduces the statistical power of comparisons between the three groups. In addition, the vaccine recipients in the only group that could be defended as a true placebo group were a bit different from the larger two groups: the age profile in the carrier solution group was younger (nine to fifteen years of age) than vaccine recipients in the other four trials (ten to twenty-six years of age, with the bulk falling between sixteen and twenty-three years old), while a smaller proportion of the carrier solution group was female (54 percent) than the Gardasil recipients (more than 90 percent female) and the aluminum adjuvant recipients (100 percent female). Nevertheless, the results of this three-way comparison are the closest thing we have to a valid safety analysis, and they show striking differences in safety profiles, none of which can be attributed to sample bias.

There are several ways in which the CBER trial review permits a comparative safety analysis across all three groups. The first is by comparing immediate adverse events at the injection site: events such as pain, swelling, "erythema" (redness of the skin), hemorrhage, and pruritis (itching). These events are highly specific and show up in the first few days; they can, however, vary quite a bit in terms of severity. The Gardasil trials reported their results for these injection site adverse events in the "detailed study population" within five days after any vaccination visit. The comparison of these outcomes is shown below (using a scale that keeps the ratios between the rates of the adverse events constant).

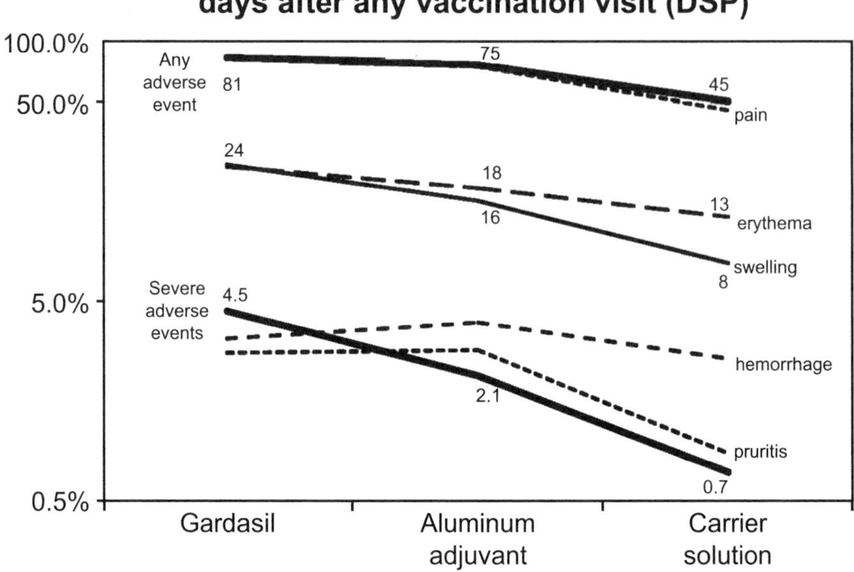

As the chart shows, the vast majority of the Gardasil (81 percent) and aluminum adjuvant (75 percent) groups reported some kind of adverse event, most of which involved some kind of pain. By contrast, less than half of the carrier solution group (45 percent) reported an adverse event. This pattern continues in almost all of the individual categories, with the Gardasil group showing the largest rate of local reactions, followed closely by the aluminum adjuvant group, and then with a clear drop off in the frequency of adverse events in the carrier solution group. On a retrospective basis, all but one of the reduced risks for the carrier solution group were statistically significant.

The most striking difference between the three groups is in the area of "serious adverse events." Although less frequent than minor instances of pain or swelling at the injection site, these serious events were disturbingly common in the groups exposed to active substances. Nearly 5 percent of the Gardasil recipients had a serious adverse event, well over six times the rate of the carrier solution group. And more than 2 percent of the aluminum "placebo" recipients had severe reactions, more than three times the rate of adverse events in the carrier solution group. Based on this finding alone, it's hard to defend the choice to classify Merck's adjuvant as an "inert" placebo.

A second approach to comparative safety analysis involves examining the adverse events that caused the participants to withdraw from the trial in a two-week period after any vaccine visit. These withdrawals included a range of adverse reactions, only a small fraction of which the investigators designated as "severe." But sudden deaths (which need not be specific to the vaccine) were also included. The comparison of the discontinuation rates in the three groups is shown below.

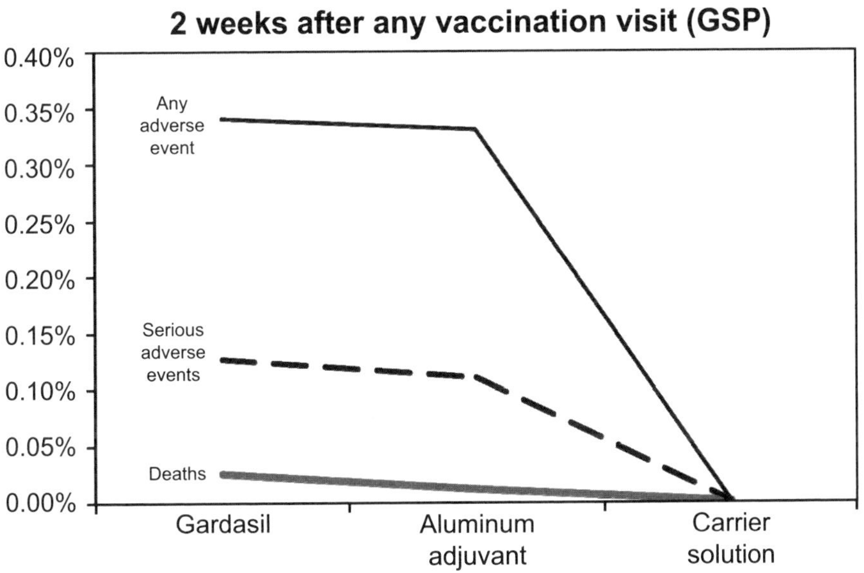

Overall, the rate of discontinuation was low at less than half a percent. But in the carrier solution group, not a single recipient chose to drop out of the trial. In addition, there were three discontinuations after two weeks due to deaths in the Gardasil group and one such death in the aluminum adjuvant groups whereas there were zero deaths at any point in the carrier solution group. Seven discontinuations (four in the Gardasil group and three in the aluminum adjuvant group) were due to other severe adverse events. These are obviously small numbers, and the deaths were dismissed by the reviewers as unrelated to vaccination. And, in fact, the rate of discontinuation in the Gardasil and aluminum adjuvant groups was nearly identical. As a result of this similarity in outcomes, the CBER reviewers dismissed any effect of vaccination on withdrawal decisions, in all likelihood because the vast majority of the officially designated "placebo" group was exposed to the aluminum adjuvant.

A third approach to a comparative safety analysis takes a longer view of adverse events, using data for serious adverse events over a twelve-month period after the beginning of the trial. The FDA review includes voluminous data on these events, but one of the easiest to measure is simply the overall rate of serious adverse events. The trial data show rates for such serious events that were similar between the Gardasil and placebo group. Indeed, the rate of serious adverse events in the Gardasil group (1 percent) was actually lower than the placebo group as a whole (1.1 percent). Not surprisingly, however, this result was driven entirely by a high rate of serious adverse events in the aluminum adjuvant group. When one examines the rate of serious adverse events in two distinct placebo groups, the rate of serious adverse events in the aluminum adjuvant group rises even higher, to 1.27 percent, while the rate in the carrier solution group comes out at zero. This comparison is shown below.

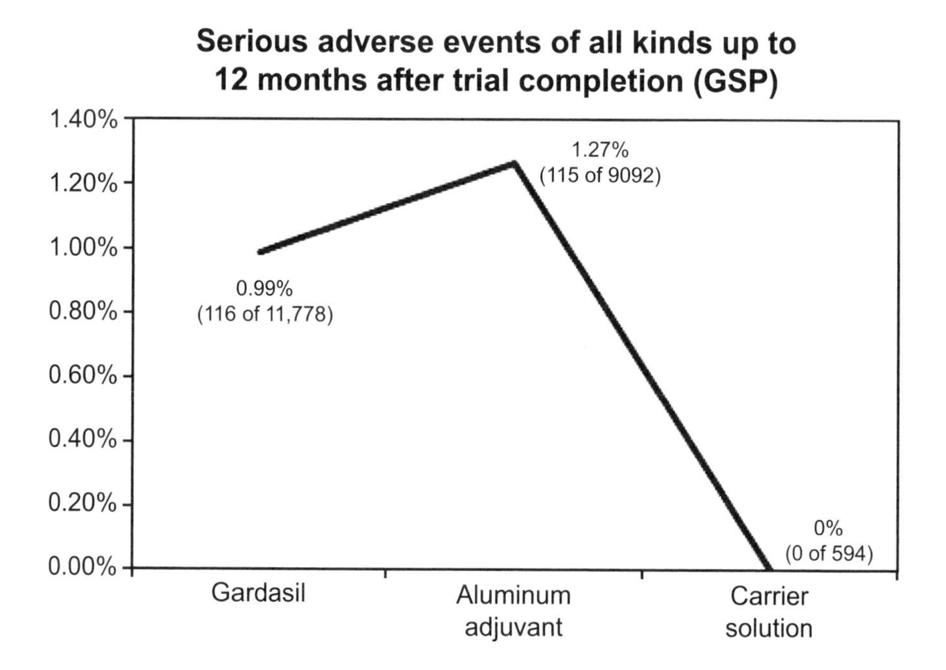

A final approach to safety assessment takes the extensive twelve-month data on the medical conditions in all trial subjects and examines the longer-term adverse events in specific categories of interest. Several such categories show disturbing patterns. Autoimmune conditions like arthritis, lupus, and thyroiditis were sharply higher in the Gardasil group when compared to the overall "placebo" group and were even noted by the FDA reviewer as a source of concern. These occurred at a rate of more than 1 in 1,000 in the Gardasil group; there were, however, zero reported cases of autoimmune disorders in the carrier solution group. As in the two-week analysis, death rates over twelve months were higher in the Gardasil and aluminum groups. By contrast, the carrier solution group had no deaths in the longer period. The chart below shows the results for the twelve-month analysis.

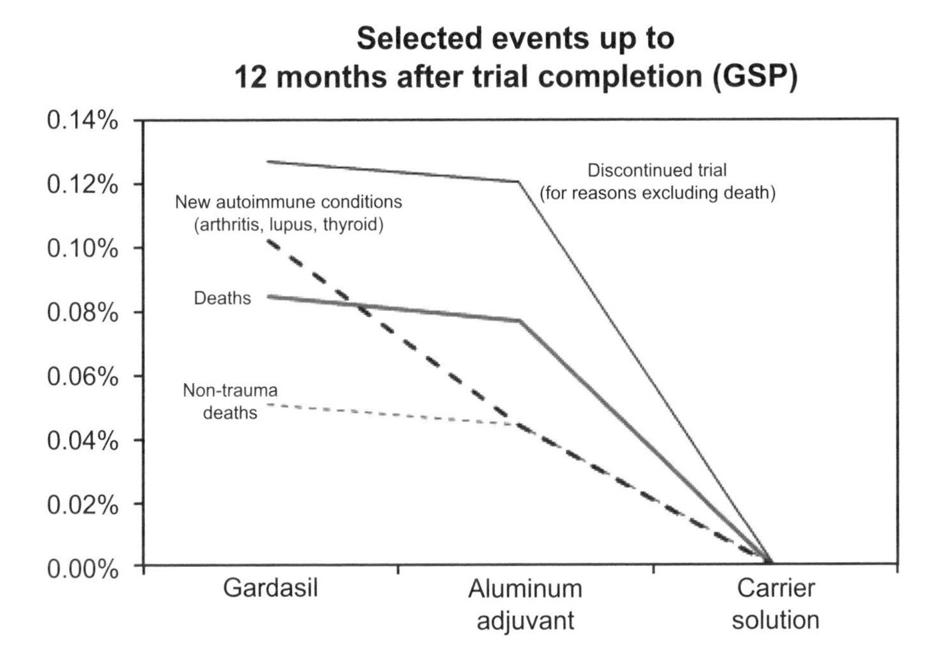

How much of the low rate of adverse events in the carrier solution group (officially designated the "018 Protocol") was due not to real differences in outcome but rather to sample bias, the fact that the population for the 018 Protocol was younger and had a lower percentage of females than the other four trial populations? The short answer is not very much. There are several ways to test the effect of sample bias. These include: comparing the adverse event rate in the 018 Gardasil group to the Gardasil groups in the other four protocols (higher adverse events show the 018 population was more vulnerable); comparing differences in adverse event rates between boys and girls in the 018 Gardasil group (higher adverse events in boys also show the 018 population was more vulnerable); and comparing differences in rates between the nine- to twelve-year-olds and the thirteen- to fifteen-year-olds in the 018 Gardasil groups (higher adverse events in younger subjects show the 018 group was more vulnerable). If anything, most of these comparisons suggest the use of the carrier solution group understates the adverse event rate for Gardasil. For example, the younger subjects in both 018 groups had a higher rate of injection site adverse events and the 018 Gardasil group also had a higher rate of severe adverse events than the other groups. Only the findings on deaths and discontinuations (which were most frequent in the Protocol 018 Gardasil boys and thirteen- to fifteen-year-olds) might have been influenced by sample bias.

The FDA Downplayed Deaths During the Clinical Trial

When it came to the most serious adverse event of all, death, the FDA review effectively gave Gardasil a free pass. It failed to mention, of course, that the deaths in the "placebo" group actually received the entirety of the vaccine's contents excepting the VLPs. Nevertheless, the review did report briefly on each individual case of death. In cases of death due to traumatic events like motor vehicle accidents, however, no details were reported (could a seizure or heart attack while driving have caused some traumatic events?). In most of the biologically related deaths, the FDA found reasons not to make any connection to Gardasil or to blame the victims' behavior ("they were on birth control pills") or family history ("the family had a history of arrhythmia"). Here is the FDA reviewer's summary of the deaths in the trial:

> There were 10 deaths in the Gardasil recipients (0.8%), and 7 deaths in the placebo group (0.7%). The majority of the deaths were due to trauma in both groups. These deaths did not appear related to vaccine administration.
>
> In each treatment group, there was a death related to a deep vein thrombosis and/or pulmonary embolism, and both subjects were on hormonal contraceptives. The Gardasil recipient with this event had symptoms of leg pain prior to the first vaccination. The other Gardasil recipients who died included one subject with pancreatic cancer 578 days after dose 3, and one young male who died of arrhythmia 27 days after dose 1. This latter subject had a strong family history for arrhythmia. These events did not appear related to administration of the vaccine.[17]

Even if all of these deaths could be explained away one way or another, this certainly sounds like a lot of deaths for such a young and overwhelmingly female group (sixteen of the seventeen deaths were in females; one fifteen-year-old male Gardasil recipient died of a heart attack). What kind of death rate is normal for young women? The trials provide no such reference rate, but such statistics are readily available. Carnegie Mellon has a website called Death Risk Rankings[18] that provides an interactive tool for calculating death rates within a wide range of demographic categories. For American females in the age range of the Gardasil trials (nine to twenty-six years of age), the rates are as follows: 2.75 per 10,000 in ten- to nineteen-year-olds and 5.03 per 10,000 in twenty- to twenty-nine-year-olds. [Note: the majority of trial subjects were from the U.S. and Europe. European deaths rates from young women are 30 percent lower than American death rates, making this a conservative comparison].

Out of 11,778 Gardasil recipients, more than 90 percent of them young women between the ages of nine and twenty-three, one would expect an annual death rate to be a mix of the rates for the two reference groups, or less than 4 per 10,000 in an entire year. But in the trials, there were three "sudden deaths", i.e., deaths that occurred within just the two weeks of the Gardasil injections, in a review period of less than forty-five days. That's a death rate close to ten times higher than would be expected in such a short period. And the overall Gardasil death rate of 8.5 per 10,000 (10 deaths out of 11,778) for the twelve-month period of the trial is more than twice what one would expect. The FDA review evinced little concern over this high death rate, preferring instead to compare

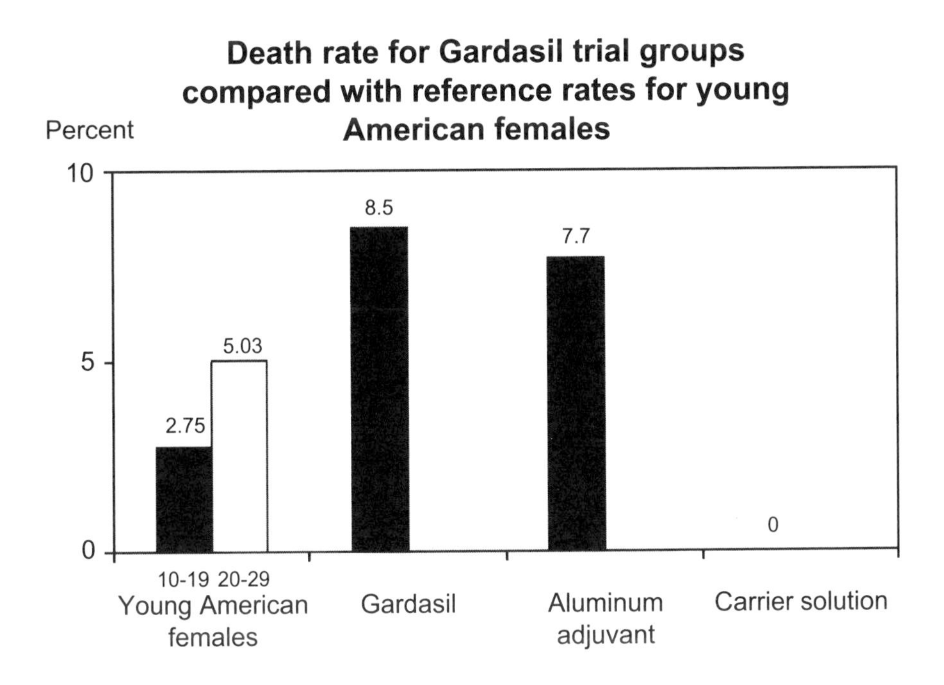

the deaths in the Gardasil group to that of the "placebo" group. But as one can see from the chart on page 195, the death rate in the aluminum adjuvant group was higher than the reference groups as well.

In short, the CBER review of Gardasil condoned the use of an immunologically active placebo and not an inert solution. So instead of adopting an increased measure of diligence in light of the potential for bias due to the HHS conflict of interest in Gardasil, it appears that FDA permitted Merck to use a lower standard of safety. Only by using the single set of trial data in which the placebo solution was relatively (although not entirely) inert can one assess the impact of this relaxed standard. Based on an analysis of this data, one is drawn to an inescapable conclusion: Gardasil was not safe.

ACIP's Recommendation for Universal Vaccination of Young Females

If the FDA was less than diligent in its review of the safety profile of the anti-cancer vaccine invented at its sister agency at NIH, how did its counterparts at CDC compare in terms of their own decision processes? Faced with a choice either to take a deliberate course, allowing a period of observation to follow Gardasil's BLA approval, or to rush Gardasil into widespread use, CDC's approach provides another standard of comparison for HHS's conduct. Would their key decision-making group, the ACIP, choose the deliberate or the hasty path?

There is very little ambiguity in this answer. ACIP wasted absolutely no time in recommending Gardasil for universal use among young women. Indeed, it would have been hard for them to have moved any faster. In June 2006, almost immediately after FDA approval, ACIP recommended the HPV vaccination. An account in the March 23, 2007, edition of the CDC publication *Morbidity and Mortality Weekly Report* (MMWR) showed that ACIP was preparing for near instantaneous approval even before the FDA's final reviews were completed.[19]

The date of the BLA approval for Gardasil was June 6, 2006. In the June 29 ACIP meeting, just twenty-three days after the FDA's decision, ACIP gave Gardasil its formal support. The vote was unanimous, with two of the fifteen members abstaining due to their financial involvements with Merck.

But not only was the vote unanimous, the mood in the meeting was celebratory. Numerous vaccine safety advocates attended the June meeting due to their concerns over the influenza vaccines. One attendee recounted the reaction to the Gardasil decision: "After the vote the place erupted in applause. There was handshaking and back-slapping. It seemed kind of odd and inappropriate to us." Asked why it seemed inappropriate, the observer explained that the concern arose "because they were so clearly cheering the recommendation. It was clear and absolutely a celebratory reaction."[20]

The momentum for an aggressive rollout was strong. Representatives from nine different organizations gave formal statements in support of Gardasil during the public comment period. In the meantime, and with a strong push from Merck, some state officials stood in line to move even faster than ACIP. In February of 2007, Republican Texas Governor Rick Perry bypassed the legislature and mandated Gardasil for all eleven- and twelve-year-old girls in the state.[21]

The commercial results were powerful and immediate. Merck reported its first revenues from Gardasil in the second quarter of 2006 (presumably from sales after the June BLA approval), and its revenues began to climb rapidly: $70 million in the third quarter and $155 million in the fourth quarter, all leading to a blockbuster year in 2007 during which Gardasil recorded revenues of $1.5 billion. The financial bonanza had begun in earnest.

A few years later, Merck would attract criticism for its aggressive marketing of Gardasil during this period. Professors Sheila and David Rothman wrote a sharp critique in *The Journal of the American Medical Association* (hardly a radical hotbed of vaccine consumerism) in which they neglected the conflicts described here but noted the extreme measures that Merck adopted:

> The marketing of this vaccine broke with traditional practices. Heretofore, vaccines had been identified by the disease they were preventing (measles, mumps) or by their creators (Salk or Sabin). This HPV vaccine followed a different model. It was identified by a trade name, Gardasil, and promoted primarily to "guard" not against HPV viruses or sexually transmitted diseases but against cervical cancer. The marketing campaign that followed, according to Merck's chief executive officer, proceeded "flawlessly." In 2006, Gardasil was named the pharmaceutical "brand of the year" for building "a market out of thin air."[22]

But the Rothmans' critique would do little to delay or disrupt the launch. At the time of the HHS rush to market, there were few dissenting voices and none of them was heard at ACIP.

One lone voice stood out. In March 2007, just as the vaccine was reaching its peak revenue numbers, one of the doctors who had guided the clinical trials voiced an objection. On March 14, 2007, an article in a small newspaper in Fort Wayne, Indiana, reported on an interview with one of the scientists involved in the clinical trials. The scientist was named Diane Harper, and she expressed dismay at the ACIP recommendation:

> "Giving it to 11-year-olds is a great big public health experiment," said Diane M. Harper, who is a scientist, physician, professor and the director of the Gynecologic Cancer Prevention Research Group at the Norris Cotton Cancer Center at Dartmouth Medical School in New Hampshire."It is silly to mandate vaccination of 11- to 12-year-old girls. There also is not enough evidence gathered on side effects to know that safety is not an issue."[23]

Harper didn't have much to gain from the commercial success of Gardasil, but she also was taking considerable risks by breaking ranks with her colleagues. One can only imagine how things would have been different if she had been in charge of the review process at FDA rather than running one branch of the clinical trial. In the midst of such a widespread degradation in regulatory ethics and standards, it's interesting to consider why she made that choice.

The Fort Wayne reporter, Cindy Bevington, was frank as to why Harper was telling her story to them rather than a larger media outlet. "For months, Harper said, she's been trying to convince major television and print media to listen to her and tell the facts about the usefulness and effectiveness of this vaccine." Why was an inside critique of the Gardasil promotion campaign not already big news? "No one will print it," Harper said.[24]

Over the ensuing months, the assessment of adverse effects of Gardasil would be transferred to a different group within HHS, from CBER and ACIP to the "post-licensure safety surveillance" groups within FDA and CDC. At this point, the deaths and serious adverse events would leave the realm of closely held statistics within a vaccine manufacturer's actively monitored trial sample and into the realm of passive surveillance in the general population; soon watchdog and vaccine safety groups like the National Vaccine Information Center[25] and

Judicial Watch[26] issued critical analyses, and even establishment voices like *The New England Journal of Medicine*[27] and *The New York Times*[28] joined the critical chorus. And Harper's concerns over the inadequate safety data would prove prophetic. Why had Harper broken ranks so early? When Bevington asked Harper why she was speaking out despite the momentum to the contrary, her answer was refreshingly simple: "I want to be able to sleep with myself when I go to bed at night."[29]

3. AFTER GARDASIL'S LAUNCH, MORE VICTIMS, MORE BAD SAFETY ANALYSIS, AND A REVOLVING DOOR CULTURE

On July 20, 2008, the *New York Post* reported the vivid account of a mother who claimed her daughter was killed by Gardasil. In a story titled "My Girl Died As 'Guinea Pig' For Gardasil," Lisa Ericzon's description of her daughter's tragic death was detailed and disturbing. As told by the *Post*'s reporter, the story began like this:

> She loved SpaghettiO's, pepperoni, lilies, listening to her iPod, and making her pals laugh. In her senior yearbook, she wrote, "The best things in life aren't things, they're friends." Now that's the quote chiseled into her gravestone.
>
> Jessica Ericzon, 17, was "an all-American teenager," as described by one of her upstate LaFargeville teachers. Last February, she was working on her softball pitches, getting ready for a class trip to Universal Studios in Florida, and hitting the slopes to snowboard with her older brother. Then one day, the blond, blue-eyed honors student collapsed dead in her bathroom. It started with a pain in the back of her head.
>
> On the advice of her family doctor, Jessie had taken a series of three Gardasil shots.[30]

Sadly, Jessica Ericzon's death was not an isolated incident. Since Gardasil's launch in late 2006, a rising number of parents have stepped forward to report the deaths of their daughters at the hands of the vaccine. Gardasil has now become a global product, so these reports have come from around the world; but the

United States is by far Merck's largest market, so most of the reported fatalities have come from closer to home. Jessica Ericzon lived in upstate New York, about a mile south of the Canadian border, and her parents were among the first to go public about Gardasil. But they haven't been the last. There are at least ten public reports of young women allegedly killed by Gardasil in the months since FDA approved Merck's BLA on June 8, 2006. Many more have been reported privately to the CDC. More than one hundred deaths have been reported to the Vaccine Adverse Events Reporting System (VAERS).

In contrast to Jessica's sudden death and just a few weeks before the *New York Post* headline and article, another family went public with their daughter's vaccine injury plight in a blog named "Jenny's Journey." In their introductory blog post, Jenny Tetlock's parents relayed an urgent request. Their daughter wasn't dead, but she was dying from what Jenny's doctors had theorized was a rare degenerative neurological condition, an unusual form of early onset amyotrophic lateral sclerosis (or ALS, popularly known as Lou Gehrig's disease). The bloggers, Barbara Mellers, Philip Tetlock, and Barbara Shapiro, were in no sense activists and weren't eager to join what they later termed the "anti-Gardasil movement." What they wanted most of all was to find a way to save Jenny's life. "One of the major things that would help her doctors figure out what to do," they wrote on June 6, 2008, "is to find other people like Jenny (called "comparables")—people that share her medical condition and perhaps have had luck with certain treatments."[31]

The list of Gardasil victims who have gone public—parents of young women like Jessica Ericzon and Jenny Tetlock—provides only a fragmentary view of the death toll associated with Gardasil. Many more deaths have been reported to the VAERS in cases where the family has chosen not to go public with their tragic loss. Among the short list of publicized cases, most simply dropped dead like Jessica Ericzon within days of receiving a dose of the vaccine; these cases most closely resembled the three cases of sudden death from Gardasil reported during the clinical trials. Cases of clear "comparables" to Jenny Tetlock, young women who could satisfy Jenny's parents' quest, were less common. Nevertheless, there were a number of these publicly reported cases in which a Gardasil shot seemed to trigger a downward spiral of ill health encompassing a diverse range of symptoms—many of which came on suddenly as well—that would culminate in death.

The table on the next page summarizes the connection between Gardasil and ten deaths that have been publicly associated with the vaccine. All of these stories have been reported elsewhere, most of them assembled in a website called

	Name	Date of First Vaccine	Date of Death	Reported Cause, Complications	Timing of Death
1	Santana Valdez	Dec. 2006?	Aug. 31, 2007	Sudden death with airway papillomatosis	Less than 4 months after 3rd dose
2	Jenny Tetlock	Mar. 2007	Mar. 15, 2009	Juvenile amyotrophic lateral sclerosis	Degenerative condition with onset after 1st dose
3	Brooke Petkevicius	Mar. 2007	Mar. 26, 2007	Sudden death with seizure and pulmonary embolism	2 weeks after 1st dose
4	Jessica Ericzon	July 2007	Feb. 12, 2008	Sudden death, after neurological symptoms	1 day after 3rd dose
5	Jasmin Soriat	Sep. 2007	Oct. 12, 2007	Sudden death with "respiratory paralysis"	Less than 1 month after 1st dose
6	Amber Kaufman	Mar. 2008	Apr. 7, 2008	Sudden death with seizure, "cardiac disturbance of undetermined etiology"	1 week after 2nd dose
7	Christina Tarsell	June 2008	June 23, 2008	Sudden death after symptoms of dizziness and fatigue	2 weeks after 3rd dose
8	Moshella Roberts	Apr. 2008	Apr. 5, 2008	Sudden death	4 days after 1st dose
9	Megan Hild	May/Jun. 2008?	Nov. 15, 2008	Sudden death after "severe headaches" and "severe stomach pain"	2 months after 2nd dose
10	Jasmine Renata	Sept. 2008	Sept. 21, 2009	Sudden death after increasing cardiac and neurological symptoms	6 months after 3rd dose

"The Truth About Gardasil." You can go to this website to read more about the stories of many of these young women.[32]

The Revolving Door Culture and a Lack of Diligence in Post-Licensure Safety Surveillance

Responsibility for what public health officials call "postlicensure safety surveillance" falls to a small set of HHS departments. Two of these are the FDA's Vaccine Safety Branch (VSB) and the CDC's Immunization Safety Office (ISO). After CBER approves a vaccine and ACIP recommends it, the baton within HHS passes next to VSB and ISO. In the passing of this baton, as stipulated previously, the presence of a conflict of interest does not mean that regulatory activity will necessarily reflect bias, negligence, or lack of diligence on the part of the next group of regulators: Each department's work deserves to be judged on its own merits. But in light of what appears to be a clear pattern of bias in the prelicensure activities of HHS, it's reasonable to approach an assessment of post-licensure activities with some skepticism.

HHS has a clear conflict of interest with respect to Gardasil at the institutional level since it shares directly in Gardasil's profits. We've also seen now that this conflict of interest is echoed by (and possibly sustains) a pervasive pattern of regulatory bias in favor of Gardasil during multiple stages of the decision process. But as the Gardasil body count rises, one natural question to ask is why, at a personal level, more HHS officials haven't taken the principled stand of Diane Harper, who spoke up against the ACIP Gardasil recommendation.

In the case of several senior officials involved in overseeing key HHS decisions during the Gardasil era, some portion of that answer is provided by their subsequent career moves. Indeed, these moves reveal a cultural problem that is in many ways more troubling than the direct Gardasil financial conflicts: a widespread pattern of senior officials cashing in on their careers in public service in order to obtain lucrative corporate and consulting jobs. The career moves of these senior officials show that a virtual revolving door between regulators and the pharmaceutical, vaccine, and biologics companies they are supposed to regulate erodes any meaningful sense in which these officials truly serve consumer interests, especially when it comes to product safety. This revolving door provides the cultural foundation that undergirds some of the more egregious institutional conflicts.

A short account of the recent careers of just a few of the officials involved in regulating Gardasil shows this revolving door in action.

- Mike Leavitt was named on December 13, 2004, as secretary of HHS, where he subsequently was responsible for most of the critical regulatory decisions involving Gardasil. In January 2009, he left HHS and formed Leavitt Partners,[33] a Washington, DC, consulting firm that helps its clients "enter new markets, enhance the value of their products and navigate dynamic regulatory and reimbursement systems." In his consulting work, Leavitt could certainly teach his clients about Gardasil and how they could follow Merck's example in forging a model "public-private partnership."
- Julie Gerberding was named director of CDC on July 3, 2002, and served in that role until she resigned on January 29, 2009. Gerberding watched over Gardasil policy at CDC during the period of FDA review in which ACIP put Gardasil on a fast track for approval in June 2006. She also was in charge of the oversight for CDC's postlicensure safety activities during much of the period when portions of the VAERS analysis were reviewed with ACIP. Less than a year after leaving public service, on December 21, 2009, Merck announced Gerberding's appointment as president of the Merck vaccine division, effective January 25, 2010, the minimum interval allowed for a federal official to assume a position at a company they used to regulate. Gerberding, who once regulated Gardasil, is now directly responsible for its growth and profitability.
- Karen Goldenthal was the director of the Division of Vaccines and Related Products Applications within CBER, the FDA division responsible for approving Gardasil's BLA in June 2006. In 2007, shortly after Gardasil's approval, Goldenthal left CBER to become executive director of PharmaNet Consulting. PharmaNet is "a global, drug development services company, provides a comprehensive range of services to the pharmaceutical, biotechnology, generic drug, and medical device industries." Other FDA executives have taken leadership positions at PharmaNet,[34] including William Egan, former head of Vaccine Research and Review at FDA, now a vice president in PharmaNet's consulting practice.

These departures provide just a few small examples of a pervasive exodus of FDA officials, many of whom leave the FDA in order to provide advice to pharmaceutical companies on how to make their way successfully through the pre- and postlicensure processes. And when it comes to vaccines, there is a specific market for former CBER officials to coach vaccine manufacturers on how to get their

BLAs approved and make their launches more profitable. Like PharmaNet, a consulting company called the Biologics Consulting Group (BCG) shows how active the revolving door between FDA and industry has become. Here's how BCG describes itself:

> Biologics Consulting Group, Inc. (BCG) is a team of consultants who provide national and international regulatory and product development advice on the development and commercial production of biological, drug and device products. Our staff consists of experts in regulatory affairs, product manufacturing and testing, pharmacology/toxicology, facility inspections, statistics, program management, and clinical trial design and evaluation. **Many of our consultants are former CBER, CDER, and CDRH reviewers** [emphasis added].[35]

In an environment so steeped in both direct and indirect conflicts of interest, is it any wonder that Gardasil regulators have leaned so steeply in favor of industry while overlooking serious safety concerns?

The Early Victims of Gardasil Look for Justice

The final phases of regulatory activities surrounding Gardasil have yet to play themselves out. These involve the process of adjudicating claims of injury and death due to the vaccine. The HHS agency responsible for this work is the HRSA, which houses the Division of Vaccine Injury Compensation, the HHS group responsible for representing HHS in the VICP, more commonly known as "vaccine court" (although it bears little resemblance to a real court).[36] In light of the pattern of bias we've observed in the approach other HHS agencies have taken to Gardasil, one might reasonably question the prospects for fair treatment of Gardasil victims in the VICP. Can we really expect the director of HRSA to encourage a fair and generous compensation policy on Gardasil when her colleagues over at NIH are profiting from the patent license, the CDC is actively promoting its use, and her former colleagues at FDA are providing consulting services to companies to help them avoid regulatory pitfalls and keep their profits intact?

As in the case of its sister agencies, the presence of conflicts of interest does not necessarily mean that HRSA will demonstrate bias, negligence, or failures of diligence in its approach to Gardasil. In advance of any record of decisions, however, it's simply too early to tell how HRSA will respond. In other controversial areas such as the Omnibus Autism Proceeding, petitioners have been deeply disappointed in their treatment at the hands of the VICP. And it seems likely that Gardasil families are destined for their own day there. HRSA provides "table injuries" that provide compensation for a short list of adverse outcomes on a few vaccines, but there are no table injuries yet specified for Gardasil. (HRSA commissioned the Institute of Medicine to develop such a list for a number of vaccines, including HPV vaccines. A list of proposed injuries can be found in the proposed causality from the Committee to Review Adverse Events of Vaccines; only anaphylaxis is listed for HPV).[37] So as Gardasil petitioners find their way into the VICP process, HRSA officials will be setting Gardasil injury compensation policy for the first time.

And the Gardasil girls are coming to seek justice. In an early action, on March 13, 2010, the parents of Jenny Tetlock filed the following petition with the VICP:

> The above captioned Petitioners request compensation under the National Vaccine Injury Compensation Program, 42 U.S.C. 300aa-10 et seq. (Supp. 1996), for the death of minor, Jennifer Tetlock who received the third series of the Gardasil vaccination on March 1, 2007 from James Cuthbertson, M.D. in Berkeley, California and thereafter suffered from atypical amyotrophic lateral sclerosis (ALS)-like lower motor neuron disease which was caused in fact by the above stated vaccination.[38]

Will Jenny Tetlock receive justice? It's hard to imagine how any agency so inextricably linked to the Gardasil program—from invention to approval and recommendation to protection and profit—can possibly be trusted to be a fair arbiter.

But one can always hope. After all, the guardians of vaccine safety in HHS have children themselves. And like Diane Harper, they also need to sleep with themselves when they go to bed at night.

Chapter Twenty

THE ROLE OF GOVERNMENT AND MEDIA

Ginger Taylor, MS

REVELATION OF HIGH MERCURY LEVELS IN THE CHILDHOOD VACCINATION SCHEDULE

The most recent chapter in the long and controversial history of vaccine safety began in the United States in July 1999. That is when the American Academy of Pediatrics (AAP) and the United States Public Health Service (USPHS) issued a joint statement through the Department of Health and Human Services (HHS) on mercury and vaccines. They stated that in the U.S. vaccine program at the time, "some children could be exposed to a cumulative level of mercury over the first six months of life that exceeds one of the federal guidelines."[1]

The truth was that the amount of mercury in the childhood vaccination schedule grossly exceeded the Environmental Protection Agency's (EPA) maximum daily adult exposure for methylmercury, the form of mercury most closely related to thimerosal for which the government had established a guideline. The EPA sets the daily limit at 0.1 micrograms per kilogram of weight.[2] Based on that guideline, a baby weighing approximately five kilograms (eleven pounds) at two months of age should not receive more than 0.5 micrograms of mercury on the day of a doctor's visit. At the time the AAP and USPHS joint statement was issued, infants at their two-month visit routinely received 62.5 micrograms of mercury, or 125 times the EPA's limit. Studies have suggested that, for thimerosal (ethylmercury), "the accepted reference dose should be lowered to between 0.025 and 0.06 micrograms per kilogram per day," meaning that the exposure at the two-month visit could be as high as 500—rather than 125—times the safe level.[3]

In November 2002, Dr. Neal Halsey, director at the Institute for Vaccine Safety at Johns Hopkins Bloomberg School of Public Health, told the *New York Times*:

> My first reaction was simply disbelief . . . if the labels had had the mercury content in micrograms, this would have been uncovered years ago. But the fact is, no one did the calculation.[4]

At the time, USPHS claimed that, "there [are] no data or evidence of any harm caused by the level of exposure that some children may have encountered in following the existing immunization schedule."[5]

However, the government made this safety claim *before* it had begun to look for evidence of harm. In November 1999, the Centers for Disease Control and Prevention (CDC) initiated a study to evaluate whether children receiving the highest amounts of thimerosal had suffered any ill effects. Thomas Verstraeten,[6] the study's lead epidemiologist, did not begin the study until four months after the government's "no evidence of harm" claim. The CDC did not publish the results until 2003.[7] The first phase of the Verstraeten study found an association between higher doses of thimerosal and neurodevelopmental disorders. In the second phase of his study, Verstraeten described his findings as "neutral."[8]

HHS further asserted in July 1999:

> Given that the risks of not vaccinating children *far outweigh the unknown and much smaller risk, if any, of exposure to thimerosal-containing vaccines* over the first six months of life, clinicians and parents are encouraged to immunize all infants even if the choice of individual vaccine products is limited for any reason.[9] (emphasis added)

With this single statement, the government took the position that the risk posed to children from exposure to thimerosal was both "unknown" and a "smaller risk" than exposure to childhood diseases. This suggested that public health officials could perform a risk-benefit analysis with no risk information for half of the equation.

HHS further asserted, "[i]nfants and children who have received thimerosal-containing vaccines do not need to be tested for mercury exposure."[10] On what basis could HHS make this statement? It had not done (and still has not done) the underlying research to show that these children were not at risk and should not be screened for mercury toxicity. Without hard evidence, the government nonetheless seemed eager to reassure parents that "no evidence of harm" meant "no harm"—even as it failed to look for evidence.

The mainstream media did not investigate HHS's claims or recommendations, nor did it investigate those of vaccine safety advocates. None of the problems with the joint statement, the investigation, or the CDC's handling of the thimerosal question came to light until 2005, when investigative journalist and author David Kirby's book *Evidence of Harm* was released. The searing and detailed account exposed the questionable behavior and judgments of the CDC and HHS.[11] Likely sensing the potential for public outrage, the CDC quickly took action and posted a notice on its website explaining that it would review the book and respond. By the end of 2005, however, the CDC had taken the notice down without responding. To this day, the government has offered no response to the book, and the media has done nothing to hold public health authorities accountable.

THIMEROSAL-CONTAINING VACCINES ARE LEGALLY CLASSIFIED AS HAZARDOUS MATERIALS

As potential hazards from vaccination began to be explored by the autism community, the EPA's rules on mercury disposal and their application to thimerosal-containing vaccines became a focus. The realization that mercury levels in vaccines were so high that they were legally classified as hazardous materials, and that the law required they be disposed of according to hazmat protocols, led many to further doubt the government's safety claims.

For example, Wisconsin's hazardous waste disposal guidelines state that:

> Some vaccines are preserved with 1:10,000 or 0.01 percent Thimerosal (see the vaccines in the table titled "Thimerosal Content in Some U.S. Licensed Vaccines" at www.vaccinesafety.edu/thi-table.htm that have .01% in the Thimerosal Concentration column). Thimerosal contains about 50 percent mercury by weight. Vaccines with 1:10,000 or 0.01 percent Thimerosal have about 50 mg/L mercury, which exceeds the 0.2 mg/L hazardous waste toxicity characteristic regulatory level for mercury. According to state and federal hazardous waste management

> requirements, discarded Thimerosal-preserved vaccines may need to be managed as hazardous waste, using the waste code D009 (mercury).... It is illegal to manage Thimerosal-preserved vaccines as infectious waste or regular trash.[12]

The mercury concentration in the previously used thimerosal-containing vaccines and the current flu vaccines is exponentially larger than what is considered hazmat material, or that is even allowed in safe drinking water, as outlined in *Pediatrics*:

> 0.5 parts per billion (ppb) of mercury kills human neuroblastoma cells.
>
> 2 ppb of mercury is the US EPA's limit for mercury in drinking water.
>
> 20 ppb of mercury destroys neurite membrane structure.[13]
>
> 200 ppb of mercury is the level in liquid that the EPA classifies as hazardous waste.
>
> 25,000 ppb of mercury is the concentration in the hepatitis B vaccine administered at birth in the United States from 1990 to 2001.
>
> 50,000 ppb of mercury is the concentration in multidose DTaP and Hib vaccines, administered four times each in the 1990s to children at two, four, six, twelve, and eighteen months of age; it is also the current "preservative" level in multidose flu vaccines (in 94 percent of the supply), meningococcal vaccines, and tetanus vaccines (for children age seven and older). This can be confirmed by analyzing the multidose vials.[14]

Parents who reviewed this information understood the inherent problem with the CDC's recommendation that was in place until approximately 2003, which stated that hazardous materials should be injected into infants beginning days after birth, and with its current recommendation to inject hazardous materials into all Americans beginning at six months of age, via the current yearly flu vaccination campaign. They did not overlook that the vaccine being injected into children at flu shot clinics, if spilled onto the floor, would trigger laws requiring immediate evacuation of the building and clean up by hazmat teams. This information was, however, seemingly lost on the media, who neither carried the information to the public nor demanded that the federal government explain this dramatic contradiction.

The media's failure to bring this to light continued through the conclusion of the 2009–2010 flu vaccine campaign, even as toxic-waste management companies promoted their services to dispose of the more than eighty million unused doses of the H1N1 vaccine in publications like *Occupational Health and Safety Magazine*:

> Clean Harbors, based in Norwell, Massachusetts, is offering the service to healthcare providers because multiple doses of the vaccine contain enough mercury-based Thimerosal to be treated as a hazardous waste.... Clean Harbors of Norwell, Massachusetts, now offers H1N1 Vaccination Incineration Services that will profile, collect, and dispose of unused 2009 H1N1 vaccine for healthcare customers nationwide. Multiple doses of the vaccine contain enough mercury-based Thimerosal to be treated by EPA as a hazardous waste and will be incinerated.[15]

VACCINE ENCEPHALOPATHY AND AUTISM

The Health Resources and Services Administration's (HRSA) Vaccine Injury Compensation Program has a table of known vaccine-induced injuries for which the government offers compensation. The table lists "encephalopathy" as an outcome for the combination MMR (or any of the various individual measles, mumps, or rubella vaccines) and for the DTaP (or any pertussis-containing vaccines). The symptoms of this encephalopathy (a medical term meaning brain disorder, brain damage, or a change in brain functioning) in a child who is eighteen months or older include a "significantly decreased level of consciousness" which HRSA describes as follows:

> (1) Decreased or absent response to environment (responds, if at all, only to loud voice or painful stimuli);
> (2) Decreased or absent eye contact (does not fix gaze upon family members or other individuals); or
> (3) Inconsistent or absent responses to external stimuli (does not recognize familiar people or things).[16]

Many parents have reported these symptoms in their previously typically functioning children after neurological regression following measles, mumps, rubella (MMR) and pertussis-containing (DPT or DTaP) vaccines. These parents, however, reported that those symptoms were not used to diagnose their children with a vaccine-induced encephalopathy but rather to diagnose them with autism. In addition, one of the signs of encephalopathy is seizure activity. Estimates suggest that one-quarter to one-third of those with an autism diagnosis also suffer from seizures.[17]

Were "vaccine-induced encephalopathy" and "autism" merely the same phenomenon, described from the vantage point of two different disciplines, medicine and mental health? Were cases of autism merely misdiagnosed vaccine-induced encephalopathy, due to the lack of physician training regarding the recognition of vaccine injury? These questions never surfaced when the media ran stories regarding parental concerns about vaccine-induced autism—that is, until 2008, when the Vaccine Injury Compensation Program (VICP) became national news.

THE HANNAH POLING CASE

In early 2008, Jon and Terry Poling announced to the press that HHS had conceded their daughter's case of vaccine-induced autism. Ten-year-old Hannah Poling had regressed into autism after receiving nine vaccines in five shots during one office visit. The Polings argued that their daughter had a preexisting, asymptomatic, and undiagnosed mitochondrial dysfunction and sustained a neurological regression into autism from receiving vaccines at eighteen months of age. Jon Poling is a well-respected neurologist who was at Johns Hopkins at the time, and his wife Terry is a registered nurse and an attorney. The Polings' medical testing following their daughter's regression was so thorough and their case so strong that HRSA conceded the case and elected to pay compensation without a hearing before the VICP. The government acknowledged, albeit in very evasive language, that vaccines were the culprit that led to Hannah Poling's autism.

While the media had yet to rigorously scrutinize the vaccine-autism story, national and local consumer-safety and autism-awareness groups were organizing to share information and advocate for change. When CNN broadcast the Polings' press conference live, the event poured gasoline on the already fiery vaccine safety debate. Federal public health officials were forced to comment on how vaccines cannot cause autism, even though they seemed to have done just that in little Hannah Poling. The government's position on the Polings' case and on vaccine-induced autism were completely at odds with one another, and the government's clumsy and conflicting answers raised even more questions about vaccine safety:

- Did vaccines cause Hannah's autism?
- Is mitochondrial dysfunction rare?
- Did the government deliberately mislead the public about Hannah's injury?
- Did the media pursue this news story appropriately?

Our government would not say that Hannah had autism, which she indeed does have.[18] The concession document[19] said that Hannah has "a regressive encephalopathy with features of autism spectrum disorder."[20] By definition, a person diagnosed with a disorder will have features of that disorder. Government attorneys had full access to Hannah Poling's extensive medical files, which disclose that she has DSM-diagnosed, full-syndrome autism. Yet, they referred to her neurological disorder using terms that sounded ambiguous, as if she has something like autism, but not autism. Hannah's parents repeatedly clarified to the media that their daughter has full-syndrome autism. A scientific journal article[21] further confirmed her diagnosis.

Where was the mainstream media? It failed in two respects. First, it continued to repeat the government's euphemistic words, "autism-like symptoms," thereby attempting to dodge the burning question—is the dramatic increase in the number of childhood vaccines causing the dramatic increase in autism incidence? Second, the media gave extensive airtime to vaccine-program defenders who seemed to turn the case on its head, blaming the victim for her own injury. In a twist of logic, they inferred that it wasn't really the vaccines' fault that Hannah was permanently injured; on the contrary, Hannah was merely a poor receptacle for lifesaving vaccines.

An article in the *New Scientist* declared, "Significantly, the government's decision says nothing about whether vaccines cause autism. Instead, government lawyers concluded only that vaccines aggravated a preexisting cellular disorder in the child, causing brain damage that included features of autism."[22] This vague government pronouncement prompted the tongue-in-cheek response from a commenter, "Do cigarettes only aggravate preexisting genetic factors, causing lung damage including features of cancer?"[23]

In late 2010, reporter Sharyl Attkisson of CBS News summed up HHS's position, "In acknowledging Hannah's injuries, the government said vaccines aggravated an unknown mitochondrial disorder Hannah had which didn't 'cause' her autism, but 'resulted' in it."[24]

CDC DIRECTOR DR. GERBERDING APPEARS ON CNN

A few days after the announcement, CDC Director Dr. Julie Gerberding appeared on CNN with Dr. Sanjay Gupta to explain the government's position on the Poling case and vaccine-autism causation.

Gupta began the interview by noting that a child with regressive autism had been compensated and that the government had conceded that vaccines had caused her "autism-like symptoms." He zeroed in on a key question. Gupta asked whether Hannah had "autism" or "autism-like symptoms."

Gerberding never answered.

She instead claimed that she had not read the Poling case file. Gupta failed to challenge this extraordinary and implausible statement. Gerberding was at the helm of the government agency responsible for the U.S. vaccine program and reported directly to Congress. A government agency conceded that vaccines caused Hannah Poling's autism-like symptoms and Gerberding had not read her case file before appearing on national television?

In another extraordinary statement, Gerberding proceeded to explain a way in which vaccines can cause autism:

> My understanding is that the child has what we think is a rare mitochondrial disorder and when children have this disease, anything that stresses them creates a situation where their cells just can't make enough energy to keep their brains functioning normally. Now we all know that vaccines can occasionally cause fevers in kids, so if the child is immunized, got a fever or other complications from the vaccine then, if you are predisposed with a mitochondrial disorder, it can certainly set off some damage, some of the symptoms can be symptoms that have characteristics of autism.[24]

Gerberding had just said that vaccines can cause autism in children with mitochondrial disorders.

Gupta passed right by this statement as well. Seeming not to have heard her, he instead asked, "As it stands, are we ready to say that vaccines do not cause autism?" Off the hook of the vaccine-autism causation question, Gerberding quickly responded,

> What we can say absolutely, for sure, is that we don't really understand the causes of autism. We've got a long way to go before we get to the bottom of this, but there have been at least 15 very good scientific studies, and the Institute of Medicine which has searched this out and they have concluded that there really is no association between vaccines and autism.[25]

Dr. Julie Gerberding, director of the CDC, had just explained an association between vaccines and autism on national news. She then said there is no association between vaccines and autism.

Gupta did not challenge her on her conflicting statements. Instead, he allowed Gerberding to reinforce the government's position that there is no link between vaccines and autism.

Two weeks earlier, the CDC had held a conference call with concerned physicians and insurance companies to discuss the Poling case.[26] During the call, experts presented information that Hannah's preexisting mitochondrial dysfunction may not be so rare. An unpublished study of thirty children with regressive autism revealed that they all shared Hannah's same biomarkers.[27] On the call, it was estimated that up to one in fifty children, or 2 percent of the general population, may have a genetic mutation that places them at risk for mitochondrial dysfunction.[28] This information had been in the press for three days when Gerberding gave the CNN interview and made the claim that Hannah's condition was "rare," but Gupta didn't challenge her claim.

In *The Washington Post*, Gerberding offered additional, unsubstantiated words of reassurance to a concerned public:

> While we recognize, and have recognized, mitochondrial disorders are associated with . . . autism-like syndrome, there is nothing about this situation that should be generalized to the risks of vaccines for normal children.[29]

Gerberding failed to explain the seemingly simple phrase "normal children." Hannah seemed "normal" before her shots, as did tens of thousands of children who regressed into autism after their shots. In fact, Hannah was above average socially and so highly verbal that, at the age of sixteen months, she had been chosen to be a "typical peer" to model appropriate social skills to developmentally disabled children in an early intervention program. Millions of concerned parents wonder about vaccine safety and which of their "normal" children might be at risk of developing autism after vaccination. How could they know?

Gerberding's *Washington Post* statement raised several troubling questions:

- By definition, regressive autism means that the children were, by all appearances, neurologically "normal" before their diagnosis. In the absence of criteria to identify susceptibility, aren't all children "normal" before they regress into autism after vaccination?
- How many other children with regressive autism following vaccination have asymptomatic, undiagnosed mitochondrial dysfunction like Hannah Poling? Was Hannah diagnosed only because her father is a neurologist?
- In the Hannah Poling scenario, a seemingly healthy child suffered a vaccine regression that gave her autism. Autism affects one percent of all U.S. children. Why aren't we screening children before vaccination to make sure they are not susceptible, just like Hannah was?

Dr. Anne Schuchat, the assistant surgeon general and director of the National Center for Immunization and Respiratory Diseases at the CDC at the time, answered the last question in an interview in *The Atlanta Journal Constitution*:

> Some have suggested that infants and children be screened for mitochondrial disorders before getting recommended vaccinations. Unfortunately, mitochondrial diseases are very difficult to diagnose and it is usually not possible to identify children with such disorders until there are signs of developmental decline. A definitive diagnosis often requires multiple blood tests and may also require a muscle or brain biopsy (removal of a portion for testing, usually under anesthesia). Therefore, providing routine screening tests on children who have no symptoms would bring other medical risks and raise many ethical questions.[30]

Schuchat failed to mention that a simple blood test to screen for "soft biomarkers" of mitochondrial dysfunction is available and reasonably predictive.[31] She further failed to mention the medical risks and ethical questions raised by blindly vaccinating nearly all children when we know that some will have mitochondrial dysfunction that puts them at risk for neurological injury.

The following year, Gerberding resigned from the CDC and joined Merck & Co., Inc., the pharmaceutical giant, as head of its vaccine division. Merck manufactures several childhood vaccines including the MMR. Notably, the MMR is the vaccine HRSA has admitted causes an encephalopathy that progresses into autism, and was among the vaccines that resulted in Hannah Poling's regression into autism. While the autism advocacy community vigorously discussed and debated the Poling concession, Gerberding's public statements on vaccine encephalopathy and autism, and her new employment, mainstream media once again remained mute.

AMERICAN ACADEMY OF PEDIATRICS PRESIDENT DR. TAYLOE APPEARS ON THE *TODAY SHOW*

Pediatricians are on the front lines of vaccine administration and sales. Their professional association, the American Academy of Pediatrics (AAP), has gone even further than the government in asserting vaccine safety. AAP President Dr. David Tayloe, Jr. appeared on the *Today Show* immediately following the Poling concession announcement and claimed that vaccines don't cause any permanent injuries. When asked if vaccines should be used on every child, he responded,

> Yes. I think any of the vaccines we have today have been tested and proven to be safe, and the credible studies don't show any relationship between vaccines and permanent injury. So we favor this and we know that unless we have vaccination rates that are in the 90 to 95 percent range we are not going to prevent epidemics from coming into this country of measles, of polio, from countries where these diseases are still endemic. So it's very important that we vaccinate all our children.[32]

Tayloe was incorrect. Not only are vaccines legally classified as "unavoidably safe," vaccine package inserts list dozens of permanent injuries that occur following vaccination, and HHS itself oversees a table of known, compensable vaccine-induced injuries that include brain damage, paralysis, anaphylactic shock, seizure disorders, and death.[33]

Despite the inaccuracy of Tayloe's statement that there is no relationship between vaccines and permanent injury, the AAP never offered a correction. Neither the CDC nor HRSA nor the FDA nor HHS offered clarification to the public. And the media, including *Today Show* interviewer Hoda Kotb, never challenged Tayloe's statements, even though the show invited Tayloe on the air to discuss Hannah Poling, who had received compensation because the government conceded that vaccines caused her autism.

It is important to note that Tayloe's father, physician David Tayloe, Sr., was the defendant in a well-known vaccine injury case.[34] The child sustained severe, lifelong brain damage when the elder Tayloe inappropriately administered a DTP vaccine after a previous adverse reaction to that same vaccine. The judgment was the largest jury verdict in North Carolina for a medical malpractice case at that time, amounting to $3.5 million.[35] The reaction from doctors and pharmaceutical companies to this case and similar ones was swift and unequivocal. Led by the elder Tayloe, who was a local AAP official at the time, they demanded tort liability protection from injuries caused by vaccines. In 1986, Congress passed the National Childhood Vaccine Injury Act, which includes provisions to shield doctors and manufacturers from liability and to compensate vaccine injury from taxpayer funds.[36]

HRSA'S PRESS STATEMENT ON VACCINE-AUTISM CAUSATION

During reporter David Kirby's investigation of the Poling case, he requested clarification of the government's position on whether or not vaccines could cause autism in light of the VICP decision. HRSA's Office of Communications responded,

From: Bowman, David (HRSA) [mailto:DBowman@hrsa.gov]
Sent: Friday, February 20, 2009 5:22 PM
To: 'dkirby@nyc.rr.com'
Subject: HRSA Statement
David,
In response to your most recent inquiry, HRSA has the following statement:
The government has never compensated, nor has it ever been ordered to compensate, any case based on a determination that autism was actually caused by vaccines. We have compensated cases in which children exhibited an encephalopathy, or general brain disease. Encephalopathy may be accompanied by a medical progression of an array of symptoms including autistic behavior, autism, or seizures.
Some children who have been compensated for vaccine injuries may have shown signs of autism before the decision to compensate, or may ultimately end up with autism or autistic symptoms, but we do not track cases on this basis.
Regards,
David Bowman
Office of Communications
Health Resources and Services Administration
301-443-3376[37]

Bowman asserts that vaccines don't cause autism, but that they do cause brain damage that can result in autism. However, HRSA doesn't track that. Kirby and Robert F. Kennedy, Jr., published this email, but the mainstream media again failed to report it to the public.

ASSISTANT SURGEON GENERAL SCHUCHAT'S 2009 APPEARANCE ON *THE DOCTORS*

In the fall of 2009, Dr. Schuchat appeared on the television program *The Doctors*[38] and told the public, "You can't get seriously ill from a regular flu vaccine." Schuchat is incorrect. Flu vaccine injury is well documented. The VICP has compensated cases of serious injury from the traditional flu vaccine, as we have described in Lisa Marks Smith's case. Schuchat goes on to say,

> Now the other questions people have—and, I get this all the time—is about mercury. It's about the thimerosal preservative. I want to say there have been a lot of studies about that. There's no scientific link between the thimerosal preservative and any kind of long-term problem.[39]

Again, Schuchat is incorrect. In scientific and medical literature, there is extensive research linking mercury and the mercury-based vaccine preservative thimerosal to serious health problems. A sampling of this research is available in the appendix to this chapter. Again, the media failed to fulfill its duty to the public. The media made no challenge, inquiry, or even follow-up investigation after *The Doctors.*

MORE QUESTIONS RAISED

The response of the autism community to the Poling decision was to ask the logical question: "How many other cases of vaccine induced autism have been paid from the VICP that were never announced to the public?"

In April 2011, a paper in the *Pace Environmental Law Review*[40] (see chapter 5) revealed that eighty-three children who have received compensation in the VICP for vaccine-induced brain damage also have autism or an autism spectrum disorder. These findings once again raised concern about a causal relationship between vaccines and autism, but the government did nothing to investigate. Then, in April 2012, the CDC announced that the autism rate had increased from 1 in 110 to 1 in 88 in just two years.[41] There was no explanation from the government for this dramatic increase. Following a New York City press conference by several autism groups calling for new hearings on the autism epidemic and the VICP, as well as the firings of several federal public health officials tasked with overseeing the autism research and response efforts,[42] Indiana Congressman Dan Burton joined the demand. In an article in *The Hill,* "It is time to re-engage on the autism epidemic," Burton, referring to previous oversight hearings under his leadership, wrote, "Our investigations found that over the years the system had broken; and what was supposed to be quick and fair became slow and contentious. There has been no Congressional oversight of VICP in the last decade, and the system has not improved; if anything it has gotten worse. It is time for Congress to revisit this issue and consider substantially reforming this program."[43] Autism parents are still waiting.

PUBLIC HEALTH STATEMENTS ON VACCINE SAFETY TODAY

Thimerosal

Despite the continuing, alarming rise in autism prevalence and significant concerns, on its website discussing thimerosal, the CDC states,

> There is no convincing evidence of harm caused by the low doses of thimerosal in vaccines, except for minor reactions like redness and swelling at the injection site.[40]

In carefully crafted language, the CDC no longer claims "no evidence of harm" as it did in 1999 but rather that there is "no *convincing* evidence of harm," implicitly recognizing that there is some evidence of harm but the CDC has decided not to be "convinced" by it.

Vaccines and Autism

The current CDC website offers this response to the question, "Do vaccines cause autism spectrum disorders?"

> A: [There are] many studies that have looked at whether there is a relationship between vaccines and autism spectrum disorders (ASDs). To date, the studies continue to show that vaccines are *not* [*sic*] associated with ASDs.[41]

This statement does not accurately depict the state of vaccine safety science today. While some studies do not find evidence of an association between vaccines, heavy metal components such as thimerosal, and autism, many do. The peer-reviewed meta-analysis released by DeSoto and Hitlan addressed in Chapter 7 ("An Urgent Call For More Research"), found that 74 percent of the relevant studies support an association between autism and heavy metals such as thimerosal.

There are now approximately sixty studies that support the autism-vaccine causation theory."[42] However parents, attempting to make fully informed decisions about vaccinating their children who Google "Do vaccines cause autism spectrum disorders?" will see the page containing the CDC's inaccurate claim as the first search result that is returned to them. The trust those parents have that the government is giving them accurate information is misplaced.

In early 2012, in preparation for the paperback version of Vaccine Epidemic, the CDC was contacted directly to ascertain its current stance on vaccine-autism causation. Thomas W. Skinner public affairs officer from the Office of the Associate Director for Communication responded:

> "Subject: Re: MI-Normal-Book author-Autism/Vaccine
> Date: Sat, 28 Apr 2012 20:32:40 +0000
> From: Skinner, Thomas W. (CDC/OD/OADC) <tws3@cdc.gov>
> To: 'ginger@adventuresinautism.com'
>
> Autism presents difficult challenges for thousands of families across the United States. Scientists do not know what causes autism. However, very thorough studies conducted by some of the world's brightest scientists simply do not point to an association between vaccines and autism. Hopefully additional research will someday provide answers as to what is the cause or causes of autism."

Because this statement is inconsistent with current research, I sent Mr. Skinner a follow-up email, in which I brought to his attention a list of sixty studies (listed in appendix starting on page 389) that point to an association between vaccines and autism. I requested three pieces of information: (1) the list of studies that "do not point to an association between vaccines and autism;" (2) the reasons for the CDC's failure to mention any of the studies that point to an association between vaccines and autism; and (3) the person or panel responsible for approving his statement.

No reply was forthcoming from the CDC as of the publication of this edition.

HHS Secretary Sebelius's "Don't Give Equal Weight" Request to the Media

This chapter presents just a few of many examples of incomplete or misleading government statements on vaccine safety. Why does the press allow them to continue? One answer may be because the government has asked the press to do so.

In March 2010, while discussing the H1N1 flu, *Readers Digest* asked HHS Secretary Kathleen Sebelius, "What can be done about public mistrust of vaccines?" Sebelius replied,

> There are groups out there that insist that vaccines are responsible for a variety of problems despite all scientific evidence to the contrary. We have reached out to

media outlets to try to get them to not give the views of these people equal weight in their reporting to what science has shown and continues to show about the safety of vaccines.[44]

Neither the Obama Administration nor *Readers Digest* clarified this remarkable disclosure, so it remains unclear which press outlets HHS contacted, what HHS asked the press not to report, or who complied with the request. But again, the mainstream media failed to pursue a story on potential vaccine injury, as well as a compelling story about the government's request for censorship.

A Modern-Day Semmelweis Reflex

Just as the government hails vaccines as a cornerstone of public health, the medical community upholds vaccination as a miracle of modern medicine. If it seems almost impossible that public denial of vaccine injury could exist on such a huge scale, it should be recognized that there is an established precedent for such a phenomenon.

In the mid-1800s in Vienna, Austria, mothers were dying shortly after childbirth from a now-extinct illness known as puerperal fever, or "childbed fever." A woman entering the hospital to give birth had roughly a 16 percent chance that she would die before taking her baby home. The mortality rate of mothers giving birth in the midwife centers, however, was lower.

In 1847, Dr. Ignaz Semmelweis, professor of obstetrics at the University of Pest, performed an autopsy on a colleague who had died from the fever and then fell ill with it himself. He postulated that small particles of the disease may have been left on his hands, and surmised that the maternal death rate from childbed fever was so high because doctors and medical students at the teaching hospital were not properly washing their hands after exams and autopsies of fever patients, before delivering newborns.

He instituted new sterilization guidelines and the death rate in the obstetrics and gynecology ward fell to 1.27 percent.[45]

When Semmelweis's colleague published this information, rather than finding it cause for celebration, the medical community lashed out against Semmelweis. He was mocked, attacked, and run out of the profession. He subsequently suffered a nervous breakdown. Semmelweis was invited by a colleague to visit an asylum for the mentally ill under the pretense of offering his professional opinion, but was instead locked inside, where he died two weeks later. Conflicting stories report that he died after being physically assaulted by the staff, or alternatively, that he died from puerperal fever. Two decades would pass after Semmelweis's discovery before the work of Louis Pasteur and Joseph Lister helped to usher in the modern era of sanitation and hygiene, including medical sterilization.

In serving the public, Semmelweis delivered the unwelcome news to doctors that they were largely responsible for the deaths of new mothers. It was bad news that they were not prepared to hear. This phenomenon has come to be known as the Semmelweis reflex—the reflexlike rejection, often in the medical community, of new scientific information without proper investigation.[46]

Today's vaccine injury denialism is a modern-day Semmelweis reflex. Pediatricians who care passionately about the welfare of children understandably find repulsive the idea that autism is largely iatrogenic. Statements offered by government agencies (i.e., HHS and the CDC) and medical professionals (i.e., the AAP) offer plausible deniability to those who do not want to know or admit that the vaccines they are administering are capable of causing serious damage to a population, let alone to the individual children in their own practices.

The Government and Media's Lack of Accountability

It is the responsibility of the government to pursue public health and to base its decisions on sound science.[47] It is the responsibility of the media to challenge governmental assertions and investigate their claims. The media's unwillingness to follow the story on vaccines and autism and to ask hard-hitting questions has resulted in a larger failure: an absence of accountability.

The public's trust in the government and the press continue to erode as stories of corruption, bad faith, and abuse of authority continue to surface. Despite government and media attempts to manage the vaccine safety narrative, parents, the main stakeholders of this debate, are catching on. In its October 2010 survey,[48] the University of Michigan revealed that vaccine safety is the top research priority for 89 percent of parents. As always, the American people are their own last line of defense. Today, the American people must bring reason, responsibility, and prudence to the national childhood vaccine program.

Chapter Twenty-One

MERCURY TOXICITY AND VACCINE INJURY

Boyd Haley, PhD

I firmly believe that the Centers for Disease Control and Prevention's (CDC) vaccination program, recommended at the federal level but strongly influencing state mandates, has led to the dramatic rise in autism starting in or around 1988. That is a strong statement; however, it is supported by sound science, including the well-described toxicity of thimerosal- and aluminum-containing vaccines.

There is no doubt that thimerosal is toxic. For example, in 1977, a report[1] was published regarding the deaths of ten of thirteen infants treated for infected umbilical cords. These infants died from mercury toxicity due to the application of a thimerosal-containing antiseptic[2] externally to their skin. Thimerosal releases ethylmercury, which is insoluble in water and penetrates the skin and all organs of the body with great ease. Due to their lack of mature liver and kidney functions, infants could not effectively excrete this mercury, so it built up to toxic and ultimately lethal levels.

It is this same thimerosal that the CDC regards as safe to inject into children on the day of birth, at levels the Environmental Protection Agency (EPA) asserts are safe to be exposed to in your diet—if you are an adult weighing 275 pounds. Infants are not adults; they are dramatically weaker in their ability to fight off toxic environmental effects, and they are, of course, much smaller. Moreover, the EPA calculation for an exposure limit (from eating seafood) does not incorporate other factors that are known to synergistically increase the toxicity of thimerosal. Many of these cofactors are found in thimerosal-containing vaccines, with aluminum being the one of most concern. Not only is aluminum a neurotoxin on its own, but aluminum dramatically increases the neurotoxicity of thimerosal many times over.[3] Given this knowledge, I find it incredible anyone could claim that a vaccine containing these two ingredients is safe.

How certain is it that vaccine-delivered thimerosal is a culprit in the current autism epidemic? After spending millions of dollars looking for "autism genes," we can certainly say that no significant finding has occurred. Autism does not appear to be a solely genetic disorder. However, studies do indicate that certain individuals may have a genetic susceptibility to a particular environmental toxin.

How can we be sure that thimerosal is such a toxin? One way is by process of elimination. The toxin that caused the autism epidemic had to increase in all fifty states around 1988 to 1990. This excludes many factory-produced toxins. Children had to be exposed to the toxin before they were two years old. This excludes many other toxins used for discretionary reasons. The toxin has to have a gender bias, in that it is more likely to affect boys than girls. Only mercury-containing compounds can explain such a profound difference; they have a stronger negative effect on human males than on females, in the form of neurological impairment and lethality.[4] This makes thimerosal, delivered via the universal childhood vaccination program, a very likely suspect. In fact, it is the only item that I know of that fits these requirements and observations.

The toxin causing autism has a corresponding symptom: It also effectively and rapidly induces oxidative stress.[5] Nothing is more effective than mercury at inducing oxidative stress, which is identified by exceptionally high mitochondrial dysfunction. Over 85 percent of autistic children who have been tested have been diagnosed with such a dysfunction.[6] Biochemically, the exceptionally tight binding of mercury to the electron transporting system of mitochondria has been proven to cause high production of toxic chemical intermediates, leading to oxidative stress and to many of the biochemical abnormalities found most often in children with autism.[7]

Unlike the genetic studies performed by many scientists, biochemical studies have identified numerous biological abnormalities that are found in people with autism more often than in age-matched controls, and these studies have repeatedly shown that some of the same biochemical abnormalities can be induced under experimental conditions in test animals by the application of thimerosal.[8] We have the list of biochemical abnormalities associated with autism and we have the published scientific observations that mercury can cause these same abnormalities. Yet we also have the CDC and the American Academy of Pediatrics denying the existence of any studies showing the role of thimerosal in autism spectrum disorder.

In addition to oxidative stress, thimerosal and mercury will 1) suppress the immune system;[9] 2) decrease methyl-B12 production and, among other things, the proper chemical modification of DNA and RNA;[10] 3) cause membrane-tight junctions to become leaky, leading to intestinal dysbiosis;[11] 4) decrease molybdenum levels, which leads to 5) higher sulfite and lower sulfate levels,[12] which further leads to 6) abnormal myelination of the nervous system and corresponding neurological damage.[13]

I could continue listing many problems that are found in children with autism that are also known to be found in infants exposed to mercury. I know of nothing else that could do this so effectively and be so readily identified as a known exposure. I, and other involved researchers, would gladly debate these issues in public with CDC and other experts—if they had the courage to show up. They don't.

They would rather set up a kangaroo court system where mothers of children with autism are drawn into debate with self-identified experts, and where the input of knowledgeable researchers is excluded. And they would rather participate in biased media pieces where the input of anyone who offers dissenting opinions of the vaccine establishment orthodoxy ends up on the cutting-room floor, as recently happened in the PBS "Vaccine War" presentation on *Frontline*.[14] This represents documentary manipulation at its worst, and is very common in our national media today. It is symptomatic of the media's pandering to the pharmaceutical and medical industries for advertising dollars.

Finally, of all countries, the United States has by far the earliest and highest level of vaccinations given to infants. Yet stunningly, we have nearly the worst record for infant mortality[15] and autism incidence when compared to other developed countries of the world. How can this be? Recent research from academic medical groups in North America has shown that the hepatitis B vaccine given in the United States on the day of birth leads to increased rates of developmental disability.[16] It is fair to demand that our medical authorities and Congress tell us why our infants are dying at a shocking rate, relative to our peer nations. This nationally demonstrated crisis, using the government's own data on infant mortality, defies any claim of vaccine safety by the CDC. To the contrary, this exceptionally high death rate supports the concept that the CDC vaccination program is causing severe damage to our infants who live.

I am greatly in favor of a safe vaccination policy, but today's vaccination policy is not proven safe and appears to me to do more damage than good. Vaccination policy and safety in this country are biased. Those with decision-making authority are deeply attached—either directly or indirectly—to the pharmaceutical industry that produces the vaccines. These decisionmakers need to stay in the good graces of the vaccine manufacturers and thus rarely question what the industry proposes. It is obvious to me that we need a fundamental change in who determines vaccine safety if we are ever to accomplish a safe and effective national vaccine program. Until then, there can be no doubt that parents must be able to make their own free and informed vaccination choices for their children.

Chapter Twenty-Two

FLU VACCINE MANDATES FOR U.S. HEALTH CARE WORKERS: POLICY WITHOUT REASON

Toni Bark, MD, MHEM, LEED AP

Each year the federal government encourages Americans of all ages, including infants, pregnant women, and the elderly, to get the flu vaccine. The U.S. Food and Drug Administration (FDA) asserts that "vaccination remains the cornerstone of preventing influenza."[1] The Centers for Disease Control and Prevention (CDC) states: "The single best way to protect against the flu is to get vaccinated each year" and recommends that everyone six months and older should get vaccinated.[2]

Based on these strong federal recommendations, you would think that health care workers (HCWs) would be among the most eager to get flu vaccines. But you would be wrong. According to the U.S. Department of Health and Human Services (HHS), the rate of vaccination against the seasonal flu among HCWs in 2008 was 45 percent.[3] Twenty-one studies from nine countries of HCWs who refuse flu vaccination confirm that fear of adverse reaction is a primary reason for their decision.[4] To combat this HCW vaccine refusal, the federal National Vaccine Advisory Committee (NVAC) recommended in February 2012 that health care employers mandate that HCWs take the flu vaccine or potentially lose their jobs.[5] HHS tasked NVAC in 2010 to provide it with guidance on how to reach HHS's ten-year "Healthy People 2020" objectives. Among these long-range goals is that at least 90 percent of HCWs receive annual flu vaccination.[6]

NATIONAL VACCINE ADVISORY COMMITTEE'S FLU VACCINE RECOMMENDATIONS FOR HCW

Created by statute in 1987, NVAC advises HHS on vaccine policy and is composed primarily of physicians, federal health officials, and vaccine researchers with ties to the pharmaceutical industry.[7] While it includes a handful of union representatives and members of the public, they are a small minority. In 2010, HHS assigned NVAC to recommend strategies to achieve HHS's "Healthy People 2020" objective to obtain a rate of compliance among HCWs of 90 percent for annual flu vaccination.[8] While continuing to recommend education programs for HCWs and further research, NVAC now urges HCW mandates as terms of employment. The NVAC recommendations are intentionally vague about specifics, leaving details to individual institutions about possible exemptions and sanctions for noncompliance. But the overall message could not be clearer: HCWs must start getting annual flu vaccines or risk losing their jobs.

The NVAC recommends that all "health care personnel" receive annual flu vaccination. The NVAC report notes that such HCWs include doctors, nurses, therapists, technicians, pharmacists, students, trainees, volunteers, and others not involved in direct patient care.[9] The definition is sweeping, and NVAC notes that it is not exclusive. NVAC similarly provides an extraordinarily broad definition of "health care employers" to which it makes these recommendations. These employers include hospitals, nursing homes, outpatient clinics, physicians' offices, rehabilitation centers, and home health care agencies. Again, NVAC suggests this definition should not be exclusive.[10] There are no limiting factors in these definitions; in other words, it is easy to imagine how the mandates could eventually include all visitors to hospitals, all patients in outpatient clinics, and all residents in nursing homes.

One of NVAC's "overarching themes" in their 64-page report on the mandate is that "immunization is the most effective way to protect patients and HCP [health care personnel] from influenza infections."[11] The scientific references for this assertion are footnotes 5 and 22:(5) Rasmussen, S., D. Jamieson, and J. Bresee, *Pandemic influenza and pregnant women*, and (22), Cox, N. and K. Subbarao, *Influenza*.[12] Neither of these published studies even remotely relates to the assertion that immunization is the most effective means to protect patients. The conclusion of the Rasmussen et al. article is the need for new assessments and treatments for pregnant women in pandemics, and the conclusion of Cox et al. is: "The availability of new diagnostic tests, new antiviral drugs, and new vaccines will undoubtedly alter our approaches to influenza control and have an impact on clinical practice." The only other scientific citation to buttress this assertion is a statement by the CDC's Advisory Committee on Immunization Practices, which is not derived from scientific data.[13]

In its conclusion, NVAC acknowledges that "there are individuals and groups that may be opposed" to its recommendations for reasons including "the quality of the evidence in the literature regarding the impact of HCP [health care personnel] vaccination on patient risk of health care-associated influenza and the issue of workers' rights."[14] NVAC overcame these reservations, however, to make its bold recommendations. It is worthwhile that we examine the issues of efficacy, safety, workers' autonomy, and the scientific evidence.

THE GOVERNMENT'S DEARTH OF SCIENTIFIC EVIDENCE

The stated rationale behind the HCW mandate is to protect the sick, the weak, and the elderly in hospitals and nursing homes. But is the rationale justified? An examination of the science behind influenza vaccines casts a long shadow of doubt. The CDC devotes substantial attention on its website to endorse vaccination of HCWs.[15] The CDC claims the scientific evidence supports the recommendation that HCWs who get vaccinated help reduce transmission of influenza, staff illness and absenteeism, and influenza-related illness and death.[16]

The idea to protect the sick from a potentially life threatening illness by using a safe and effective prophylaxis is valid. The problem is that the vaccine is neither proven safe nor effective in carrying out this mission. Considering the cost and effort to vaccinate millions of HCWs in this country, one would assume that extensive research supports such a broad mandate. Yet the federal government has failed to identify unbiased, reliable studies supporting its recommendations, instead relying on studies conducted or funded by the pharmaceutical industry itself.[17] Importantly, according to the National Vaccine Information Center, adult flu vaccine injuries are the leading cause of injury claims in the National Vaccine Injury Compensation Program, the federal program to compensate individuals for injuries caused by federally recommended vaccines.[18]

The evidence the CDC uses simply does not support its position. In one randomized controlled trial (RCT) using uncorrected death rates, the influenza infection rates were similar between hospitals with vastly different HCW vaccination rates.[19] In other words, vaccination of HCWs did not make a significant difference in patient infection rates. Moreover, there was built-in bias in the study protocol because the sampling included almost twice the number of individuals who died in the low-vaccinated hospitals as in the more-highly vaccinated hospitals. The study also tested the serum of the deceased even if influenza was not listed as a cause of death. By sampling serum from all deceased regardless of flu status, the study created an invalid sample, hence the use of the unusual designation "uncorrected death rates."[20]

The CDC relies on another RCT that similarly provides almost no support for its recommendations. The study concluded that influenza immunization failed to reduce episodes of respiratory infections (1.8 episodes/study period among vaccinees vs. 2.0 among controls) and failed to affect the total number of days those vaccinated suffered from respiratory infections (13.5 days vs. 14.6 days, respectively).[21]

There have been no large-scale randomized placebo controlled trials of flu vaccines. Thomas Jefferson, MD, of The Cochrane Collaboration has called for this type of study to be done, not only for effectiveness but also for assessing adverse reactions.[22] Defenders of the lack of trials claim it would be "unethical" to conduct such trials because half of the subjects would have the flu vaccine withheld. But Dr. Jefferson argues that this rationale is backwards—placebo controlled trials are exactly what should be done when there is uncertainty and especially when public policy is based on scant scientific evidence.[23]

Other CDC references similarly cite articles that merely endorse the influenza vaccine but give no science to support that position. Even more egregious, some of the articles have nothing to do with reduction of infection; they discuss how to increase HCW vaccination uptake. The federal government's cited studies are far from the last word. In fact, to date, there is not even accurate data on influenza infection rates among HCWs.[24]

INDEPENDENT RESEARCH RAISES SEVERE DOUBTS ON FLU VACCINE EFFICACY

In sharp contrast to government sources, The Cochrane Collaboration, a highly respected, independent research institute, has carefully analyzed influenza vaccine studies. Cochrane does not accept funding from the pharmaceutical industry.[25] The Cochrane Collaboration review on flu vaccines states:

> We conclude that there is no evidence that only vaccinating healthcare workers prevents laboratory-proven influenza, pneumonia, and death from pneumonia in elderly residents in long-term care facilities. Other interventions such as hand washing, masks, early detection of influenza with nasal swabs, antivirals, quarantine, restricting visitors and asking healthcare workers with an influenza-like illness not to attend work might protect individuals over 60 in long-term care facilities and high quality randomized controlled trials testing combinations of these interventions are needed.[26]

Sumit Majumdar, MD, MPH, with colleagues at University of Alberta, Canada, conducted an RCT of the flu vaccine among the elderly to see if the widely held belief that flu shots reduce morbidity and mortality survived close scrutiny.[27] It did not. The study included 700 patients ages sixty-five and older matched for various demographics. Half of the patients received the flu shot and half did not. Once mortality risk controlled for confounding variables such as age-adjusted mortality risk, the relative risk of death was statistically insignificant. The study concluded that "previous observational studies may have overestimated the mortality benefits of influenza vaccination" by, among other things, not controlling for appropriate variables.[28]

Flu vaccine science has not adequately explored the impact of confounding variables on infection rates. This is particularly critical because the method by which the influenza virus transmits itself is poorly understood. In addition to the dearth of studies on the efficacy of vaccination in preventing transmission, there have been no studies examining the effects of HCWs' hand washing or taking time off when ill with an upper respiratory infection on transmission. Studies with volunteers subjecting themselves to direct exposure to flu droplets have not shown the virus to be highly transmissible.[29] This is contrary to conventional wisdom.

What if the assumptions underpinning annual flu vaccines are wrong? We can identify many treatment protocols that were assumed to work but were later shown to actually harm or kill patients. A few examples include: radiation for tonsillitis, which actually increased risk for thyroid and other cancers years later; high-dose chemotherapy followed by a bone marrow transplant for breast cancer patients, which proved to have higher death rates than alternative options; and lidocaine as prophylaxis during myocardial infarction, which later proved to increase mortality once an RCT was performed.[30]

VACCINE EFFICACY IN THE REAL WORLD

One of the key questions in the real world is what percentage of influenza-like illnesses (ILI) are caused by the flu virus? There are more than 200 viruses that can cause flu or flu-like illnesses, all of which produce identical or very similar symptoms. Only laboratory testing of nasal or nasal-pharyngeal swabs can indicate if an ILI is actually flu. Although people often casually report most ILIs as the "flu," testing is not routine in most cases.[31]

According to the CDC, between 5 and 20 percent of the population develops the flu during the winter months. This estimate, however, represents the percentage of the population that develops ILIs. According to limited serological data, only about 10 percent of those with ILIs actually have the flu. Based on 274 studies of influenza vaccines and twenty-eight epidemiological studies from 1966 to 2007, representing more than three million participants, seven out of one hundred people will develop ILIs in the winter, but only one of those cases will be influenza.

In addition, predicting the correct strain or strains of influenza against which to vaccinate presents an enormous challenge for public health authorities and vaccine manufacturers. Flu vaccines aim to be effective against influenza A and B, but these strains comprise only about 10 percent of all circulating viral strains that cause ILI. Furthermore, the flu virus mutates from year to year, so predicting the exact serotype can be elusive. The actual risk that any ILI is actually the flu is low.

The Cochrane Collaboration's meta-analysis on the flu vaccine concluded that the influenza vaccine's efficacy is 30 percent in most years and 59 to 70 percent in a good year.[32] Vaccine manufacturers make effectiveness claims based on antibody response alone; they do not address whether the vaccine in question actually prevents the disease or not. The FDA measures only antibody production, which may or may not be associated with protection against disease. Thus, a significant number of those vaccinated never mount an antibody response or have immunity against a disease. Immunity and immune response might involve a few immune pathways and might involve other cofactors. In addition, the duration of vaccine-induced protection is uncertain as well.

Recent research in the journal *Immunity* may lead to a new understanding of the best way to help protect people against potentially lethal viruses. Ulrich H. Von Andrian, MD, PhD, of Harvard Medical School, showed that mice that survived intentional, potentially fatal viral infections did not always mount high levels of antibodies and that the innate non-memory immune system can effectively fight certain viruses without the presence of any antibodies. This turns conventional wisdom about documented immunity and vaccination science on its head.[33] As researchers at the Wellcome Trust Sanger Institute in England have recently confirmed, antibody response does not necessarily confer immunity to all types of viral illnesses.[34]

The immune system responds to many other stimuli. For example, vitamin D levels strongly influence the innate immune system. Flu is seasonal and more prevalent during the months when there are fewer hours of solar radiation. Thus, flu season occurs during the winter in the northern and southern hemispheres but during the rainy season in the tropics. The amount of solar radiation is directly related to vitamin D levels and inversely related to flu and ILI. As vitamin D levels drop, the risks of ILIs increase, as do hospitalizations from pneumonia and other upper respiratory illnesses. Numerous published studies conclude the same thing: vitamin D deficiency increases morbidity and mortality from upper respiratory infections regardless of etiology.[35] The federal government would do well to spend tax dollars researching vitamin D levels and ILI prevention.

FLU VACCINE SAFETY

Flu vaccines pose many safety issues. Concerns include manufacturing practices and oversight, bacterial contamination, vaccine adjuvants, and the accuracy of vaccine efficacy claims.

Flu Vaccine Recalls

While the federal government goes to great lengths to characterize flu vaccines as safe and effective, all vaccines, including the flu vaccine, are legally classified as "unavoidably unsafe."[36] During the past three years, the vaccine maker Baxter has had two recalls of its flu vaccines. In 2009, the batch of Baxter's seasonal influenza vaccine that it sent to the Czech Republic was contaminated with the deadly H5N1 avian flu virus. The company recalled the vaccines after all laboratory animals died from vaccine testing. This episode should have triggered an investigation into Baxter's manufacturing practices, but it did not; Baxter simply stated that it could not explain what happened without divulging proprietary information, and that ended the inquiry.[37] In 2011, an inordinate number of side effects followed vaccination with Baxter's new flu vaccine Preflucel, resulting in the immediate recall of all 300,000 doses. Whether or not there was follow-up for those who suffered from the side effects of headaches, fatigue, and muscle pain is unknown.[38] The effects of these vaccines could have been catastrophic absent these recalls.

Flu Vaccine Contamination

Until recently, influenza virus for vaccines was cultured in chicken embryonic material that required added antibiotics to prevent bacterial contamination. One of the major concerns for bacterial contamination is *Campylobacter jejuni* (*C. jejuni*), a serious infection. In humans, infection with *C. jejuni* is usually the result of eating undercooked chicken and often precedes Guillain–Barré syndrome, an autoimmune attack on the nerve ganglia rendering an individual partially or fully paralyzed.[39] One of the theories behind the risk of Guillain–Barré after flu vaccination is *C. jejuni*-contamination from vaccines cultured in chicken embryonic material.[40]

Contamination risks and egg allergies have prompted manufacturers to develop flu vaccines in other cell cultures.[41] For example, Baxter is currently producing vaccines using Vero cell lines, which come from African green monkey kidney epithelial cells and originated in Japan in 1962. Vero cells are immortalized cells (such as the HeLa cells of the famed book on Henrietta Lacks) and are considered neoplastic or cancerous cells. They are almost always grown in bovine serum, which also runs the risk of contamination with certain viral species or prions. Laboratories now recognize the need to end the standard use of bovine serum due to these risks.[42]

There has long been concern regarding Vero cells and the potential for tumors or carcinogenicity. To grow new antigenic material for vaccine production, the Vero cells are put through a procedure called "passage," also known as subculture or splitting cells, which involves transferring a small number of cells into a new vessel.[43] Certain Vero cell lines are known to cause cancer and others are only oncogenic at specific levels of passage.[44] There are many uncertainties about the risks of bacterial contamination and the potential for carcinogenicity in flu vaccines.

Flu Vaccine Adjuvants

Adjuvants are another flu vaccine safety concern. An adjuvant is any substance that increases the immune response so that less vaccine antigen is needed. Antigenic material is an expensive component of the vaccine. In June 2011,

Fluad, a Novartis flu vaccine containing the adjuvant squalene, was approved for use in those sixty-five years of age and older.[45] Dozens of articles have been published on the autoimmune effects of squalene.[46] By contrast, there is only one published study, funded by Novartis, Fluad's manufacturer, supporting its safety and countering the other independent research.[47]

Many people in the military and elsewhere are seriously concerned about the use of squalene as an adjuvant in flu vaccines, as Captain Richard Rovet's chapter 16 details. Many scientists have associated squalene with Gulf War Syndrome (GWS).[48] GWS was only present in soldiers from countries that used a specific anthrax vaccine containing the adjuvant squalene. Physicians at Tulane Medical Center identified antibodies to squalene in soldiers suffering from GWS.[49] Several soldiers died and many more have suffered life-long debilitating diseases such as multiple sclerosis, rheumatoid arthritis, cerebellar atrophy, and other disabling autoimmune diseases from GWS.[50]

The United States Army initially denied that squalene was in the anthrax vaccine, but eventually the FDA admitted its presence. Once contradicted by the FDA, the military claimed that squalene was "naturally occurring from the egg it was cultured in"—an impossible explanation as the vaccine is not produced in an egg medium. In the end, the army and the FDA defended the use of experimental squalene in anthrax vaccines based on a scientific study supporting the safe use of topical squalene in cosmetics.[51] No one should be reassured by the safety science on squalene as a vaccine adjuvant.

Past Flu Scares

Many recipients of the flu vaccine have suffered serious adverse events, as Lisa Marks Smith describes in chapter 17. In 1976, the U.S. public was first alerted to the risk of Guillain–Barré and death after the CDC's massive swine flu vaccine campaign.[52] Of the 40 million Americans who received the vaccine, at least 4,000 recipients of that vaccine suffered serious adverse events, including Guillain–Barré. Many vaccine recipients were hospitalized, some were permanently handicapped, and as many as 300 died. The cost to the taxpayers included $3.5 billion in vaccine injury damages.

We now know that the swine flu "epidemic" simply never happened and that the American public was left with a huge bill, along with approximately 4,000 citizens left handicapped or dead. Sadly, it seems that the government did not learn the lessons of 1976; trouble resurfaced in 2009 when the H1N1 swine flu scare commenced.

In April 2009, the World Health Organization (WHO) announced an H1N1 pandemic, altering its definition of a pandemic in response to just more than 600 cases of swine flu worldwide, including seventeen deaths. Sixteen of the seventeen deaths were in Mexico, where the first H1N1 case occurred.[53] When urged to do so by the European Scientific Working group on Influenza (EWSI), a group purporting to be independent but receiving industry funding, the WHO removed criteria relating to the severity of the disease as a precondition for using its highest alert level.[54]

The WHO began an active, visible swine flu scare campaign, which distorted public health priorities across Europe and wasted vast amounts of public funds to stockpile vaccines and antivirals.[55] The declaration of the pandemic also led to accelerated vaccine approvals. Scientists have blamed a resulting inadequately researched vaccine, Pandemrix, for permanent cases of narcolepsy in children aged five to fifteen in Iceland and Ireland.[56]

After the 2009 nonpandemic, the overstock of flu vaccines and doses of the antiviral Tamiflu in national stockpiles prompted the Parliamentary Assembly of the Council of Europe (PACE) to undertake an investigation into the undisclosed financial ties between the WHO, ESWI, GlaxoSmithKline, and Roche. PACE's report highlighted the extent to which industry interests had hijacked public health.[57] A few independent scientists, including Dr. Thomas Jefferson of The Cochrane Collaboration and Dr. Wolfgang Wodarg of PACE, condemned the behavior of those promoting vaccines for financial gain.[58] They accused the WHO of fueling excessive fear and pushing many European governments to stockpile unnecessary vaccines and antivirals.[59]

PREVIOUS FLU VACCINE MANDATES FOR HCWS

The concept of flu mandates is not new, but previous efforts to impose them have been mixed. The most notable example of a failed statewide mandate for HCWs was New York's effort in 2009. By contrast, employers have been better able to impose private mandates.

New York State's Failed 2009 H1N1 and Flu Mandates

On July 17, 2009, New York State Health Commissioner Richard Daines issued a regulation to require all state HCWs to receive both the H1N1 and seasonal flu vaccines by November 30, 2009, or face termination of employment.[60] The proposed regulation applied to New York State's 525,000 HCWs in hospitals,

The Texas Immunization Manual teaches medical professionals how to administer up to eight injections (twelve vaccines) to an infant at one time.

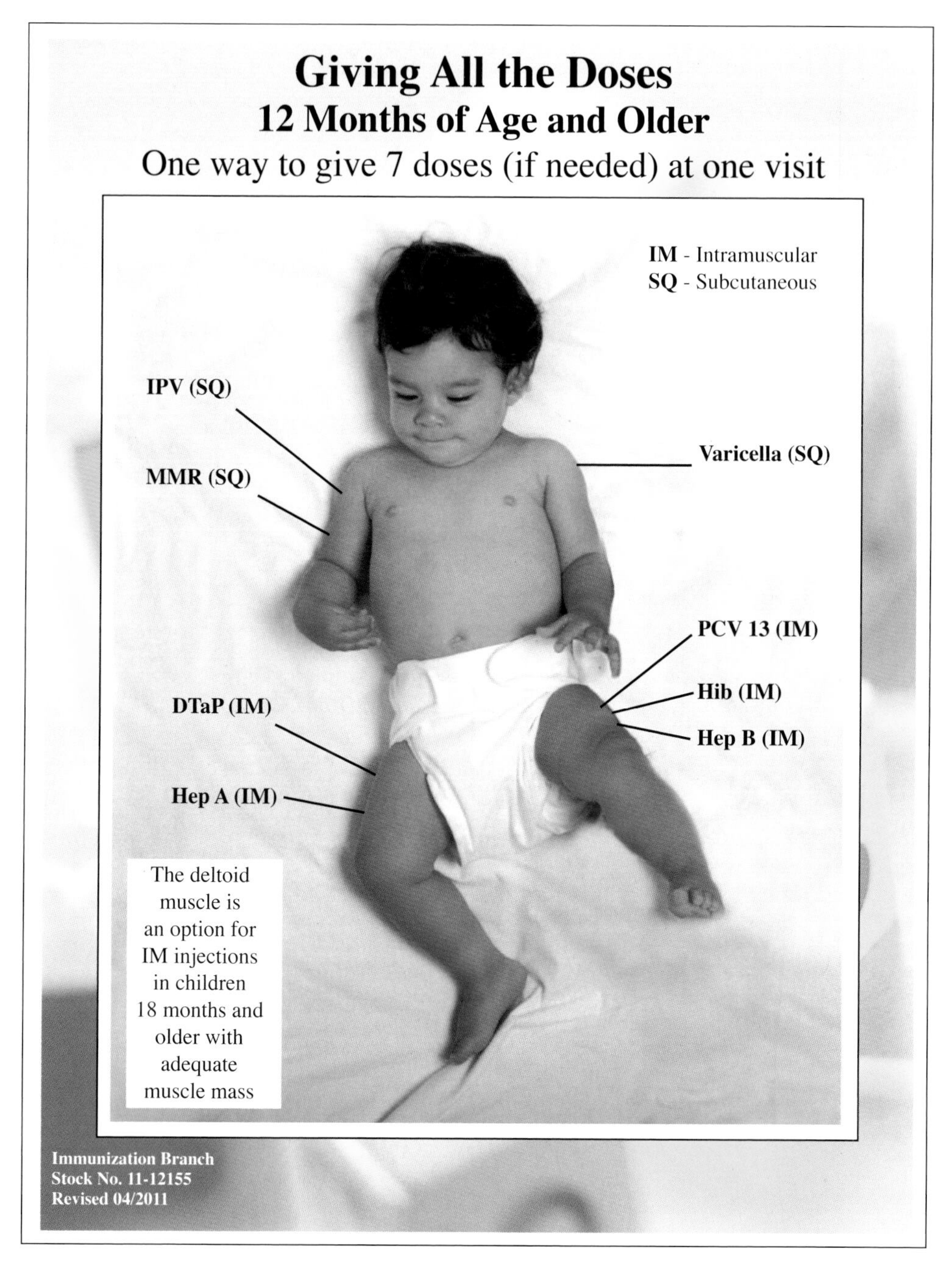

Source: www.dshs.state.tx.us/immunize/docs/11-12155.pdf

Note: Texas manual states seven injections in the title but shows eight

Advisory Committee on Immunization Practices (ACIP) 2012 Recommended Childhood Vaccination Schedules

FIGURE 1: Recommended immunization schedule for persons aged 0 through 6 years—**United States, 2012** (for those who fall behind or start late, see the catch-up schedule [Figure 3])

Vaccine ▼ Age ▶	Birth	1 month	2 months	4 months	6 months	9 months	12 months	15 months	18 months	19–23 months	2–3 years	4–6 years
Hepatitis B[1]	Hep B	HepB			HepB							
Rotavirus[2]			RV	RV	RV[2]							
Diphtheria, tetanus, pertussis[3]			DTaP	DTaP	DTaP		see footnote[3]	DTaP				DTaP
Haemophilus influenzae type b[4]			Hib	Hib	Hib[4]		Hib					
Pneumococcal[5]			PCV	PCV	PCV		PCV				PPSV	
Inactivated poliovirus[6]			IPV	IPV	IPV							IPV
Influenza[7]					Influenza (Yearly)							
Measles, mumps, rubella[8]							MMR			see footnote[8]		MMR
Varicella[9]							Varicella			see footnote[9]		Varicella
Hepatitis A[10]							Dose 1[10]				HepA Series	
Meningococcal[11]						MCV4 — see footnote[11]						

Range of recommended ages for all children

Range of recommended ages for certain high-risk groups

Range of recommended ages for all children and certain high-risk groups

50 doses of 14 vaccines by age six

FIGURE 2: Recommended immunization schedule for persons aged 7 through 18 years—**United States, 2012** (for those who fall behind or start late, see the schedule below and the catch-up schedule [Figure 3])

Vaccine ▼ Age ▶	7–10 years	11–12 years	13–18 years
Tetanus, diphtheria, pertussis[1]	1 dose (if indicated)	1 dose	1 dose (if indicated)
Human papillomavirus[2]	*see footnote*[2]	3 doses	Complete 3-dose series
Meningococcal[3]	See footnote[3]	Dose 1	Booster at 16 years old
Influenza[4]	Influenza (yearly)		
Pneumococcal[5]	See footnote[5]		
Hepatitis A[6]	Complete 2-dose series		
Hepatitis B[7]	Complete 3-dose series		
Inactivated poliovirus[8]	Complete 3-dose series		
Measles, mumps, rubella[9]	Complete 2-dose series		
Varicella[10]	Complete 2-dose series		

Range of recommended ages for all children

Range of recommended ages for catch-up immunization

Range of recommended ages for certain high-risk groups

70 doses of 16 vaccines by age eighteen

Source: www.cdc.gov/vaccines/schedules/index.html

Ingredients and Substances Used in Manufacturing Process
Vaccine Excipient & Media Summary

Excerpted from http://www.cdc.gov/vaccines/pubs/pinkbook/downloads/appendices/b/excipient-table-2.pdf

This table includes not only vaccine ingredients (e.g., adjuvants and preservatives), but also substances used during the manufacturing process, including vaccine-production media, that are removed from the final product and present only in trace quantities. In addition to the substances listed, most vaccines contain sodium chloride (table salt).

Last Updated February 2012

Vaccine	Contains	Mfrs P.I.
DTaP Infanrix	formaldehyde, glutaraldehyde, aluminum hydroxide, polysorbate 80, Fenton medium (containing bovine extract), modified Latham medium (derived from bovine casein), modified Stainer-Scholte liquid medium	Nov 2011
DTaP-IPV/Hib Pentacel	aluminum phosphate, polysorbate 80, formaldehyde, gutaraldehyde, bovine serum albumin, 2-phenoxethanol, neomycin, polymyxin B sulfate, Mueller's Growth Medium, Mueller-Miller casamino acid medium (without beef heart infusion), Stainer-Scholte medium (modified by the addition of casamino acids and dimethyl-beta-cyclodextrin), MRC-5 (human diploid) cells, CMRL 1969 medium (supplemented with calf serum).	Jul 2011
Hib ActHIB	ammonium sulfate, formalin, sucrose, Modified Mueller and Miller medium	May 2009
Hep B Recombivax	yeast protein, soy peptone, dextrose, amino acids, mineral salts, potassium aluminum sulfate, amorphous aluminum hydroxyphosphate sulfate, formaldehyde.	Jul 2011
Human Papillomavirus Gardasil	yeast protein, vitamins, amino acids, mineral salts, carbohydrates, amorphous aluminum hydroxyphosphate sulfate, L-histidine, polysorbate 80, sodium borate.	Mar 2011
Influenza Afluria	beta-propiolactone, thimerosol (multi-dose vials only), monobasic sodium phosphate, dibasic sodium phosphate, monobasic potassium phosphate, potassium chloride, calcium chloride, sodium taurodeoxycholate, neomycin sulfate, polymyxin B, egg protein	Nov 2011
Meningococcal Menactra	formaldehyde, phosphate buffers, Mueller Hinton agar, Watson Scherp media, Modified Mueller and Miller medium	Nov 2011
MMR MMR-II	vitamins, amino acids, fetal bovine serum, sucrose, sodium phosphate, glutamate, recombinant human albumin, neomycin, sorbitol, hydrolyzed gelatin, chick embryo cell culture, WI-38 human diploid lung fibroblasts	Dec 2010
MMRV ProQuad	sucrose, hydrolyzed gelatin, sorbitol, monosodium L-glutamate, sodium phosphate dibasic, human albumin, sodium bicarbonate, potassium phosphate monobasic, potassium chloride, potassium phosphate dibasic, neomycin, bovine calf serum, chick embryo cell culture, WI-38 human diploid lung fibroblasts, MRC-5 cells	Aug 2011
Pneumococcal Prevnar 13	casamino acids, yeast, ammonium sulfate, Polysorbate 80, succinate buffer, aluminum phosphate	Jan 2012
Polio IPV – Ipol	2-phenoxyethanol, formaldehyde, neomycin, streptomycin, polymyxin B, monkey kidney cells, Eagle MEM modified medium, calf serum protein	Dec 2005
Rotavirus RotaTeq	sucrose, sodium citrate, sodium phosphate monobasic monohydrate, sodium hydroxide, polysorbate 80, cell culture media, fetal bovine serum, vero cells, DNA from porcine circoviruses (PCV1, PCV2) detected.	Sept 2011
Tdap Adacel	aluminum phosphate, formaldehyde, glutaraldehyde, 2-phenoxyethanol, ammonium sulfate, Mueller's growth medium, Mueller-Miller casamino acid medium (without beef heart infusion).	Dec 2010
Varicella Varivax	sucrose, phosphate, glutamate, gelatin, monosodium L-glutamate, sodium phosphate dibasic, potassium phosphate monobasic, potassium chloride, sodium phosphate monobasic, EDTA, residual components of MRC-5 cells including DNA and protein, neomycin, fetal bovine serum, human diploid cell cultures	Aug 2011

U.S. Government's Vaccine Injury Compensation Table

Vaccine	Illness, disability, injury or condition covered	Time period*
I. Vaccines containing tetanus toxoid (e.g., DTaP, DTP, DT, Td, TT)	A. Anaphylaxis or anaphylactic shock	4 hours
	B. Brachial neuritis	2-28 days
	C. Any acute complication or sequela (including death) of an illness, disability, injury, or condition referred to above which illness, disability, injury, or condition arose within the time period prescribed.	Not applicable
II. Vaccines containing whole cell pertussis bacteria, extracted or partial cell pertussis bacteria, or specific pertussis antigen(s) (e.g., DTP, DTaP, P, DTP-Hib).	A. Anaphylaxis or anaphylactic shock	4 hours
	B. Encephalopathy (or encephalitis)	72 hours
	C. Any acute complication or sequela (including death) of an illness, disability, injury, or condition referred to above which illness, disability, injury, or condition arose within the time period prescribed.	Not applicable
III. Measles, mumps and rubella vaccine or any of its components (e.g., MMR, MR, M, R)	A. Anaphylaxis or anaphylactic shock	4 hours
	B. Encephalopathy (or encephalitis)	5-15 days (not less than 5 days and not more than 15 days)
	C. Any acute complication or sequela (including death) of an illness, disability, injury, or condition referred to above which illness, disability, injury, or condition arose within the time period prescribed.	Not applicable
IV. Vaccines containing rubella virus (e.g., MMR, MR, R)	A. Chronic arthritis	7-42 days
	B Any acute complication or sequela (including death) of an illness, disability, injury, or condition referred to above which illness, disability, injury, or condition arose within the time period prescribed.	Not applicable
V. Vaccines containing measles virus (e.g., MMR, MR, M)	A Thrombocytopenic purpura	7-30 days
	B. Vaccine-Strain Measles Viral Infection in an immunodeficient recipient	6 months
	C Any acute complication or sequela (including death) of an illness, disability, injury, or condition referred to above which illness, disability, injury, or condition arose within the time period prescribed.	Not applicable

*Time period for first symptom or manifestation of onset or of significant aggravation after vaccine administration vaccine administration

Vaccine	Illness, disability, injury or condition covered	Time period*
VI. Vaccines containing polio live virus (OPV)	A. Paralytic polio	
	--- in a non-immunodeficient recipient	30 days
	--- in an immunodeficient recipient	6 months
	--- in a vaccine assoc. community case	Not applicable
	B. Vaccine-strain polio viral infection	
	--- in a non-immunodeficient recipient	30 days
	--- in an immunodeficient recipient	6 months
	--- in a vaccine assoc. community case	Not applicable
	C. Any acute complication or sequela (including death) of an illness, disability, injury, or condition referred to above which illness, disability, injury, or condition arose within the time period prescribed.	Not applicable
VII. Vaccines containing polio inactivated (e.g., IPV)	A Anaphylaxis or anaphylactic shock	4 hours
	B. Any acute complication or sequela (including death) of an illness, disability, injury, or condition referred to above which illness, disability, injury, or condition arose within the time period prescribed.	Not applicable
VIII. Hepatitis B. vaccines	A. Anaphylaxis or anaphylactic shock	4 hours
	B. Any acute complication or sequela (including death) of an illness, disability, injury, or condition referred to above which illness, disability, injury, or condition arose within the time period prescribed.	Not applicable
IX. Hemophilus influenzae type b polysaccharide conjugate vaccines	No Condition Specified	Not applicable
X. Varicella vaccine	No Condition Specified	Not applicable
XI. Rotavirus vaccine	No Condition Specified	Not applicable
XII. Pneumococcal conjugate vaccines	No Condition Specified	Not applicable
XIII. Hepatitis A vaccines	No Condition Specified	Not applicable
XIV. Trivalent influenza vaccines	No Condition Specified	Not applicable
XV. Meningococcal vaccines	No Condition Specified	Not applicable
XVI. Human papillomavirus (HPV) vaccines	No Condition Specified	Not applicable
XII. Any new vaccine recommended by the Centers for Disease Control and Prevention for routine administration to children, after publication by Secretary, HHS of a notice of coverage[b,c,d]	No Condition Specified	Not applicable

Source: www.hrsa.gov/vaccinecompensation/vaccineinjurytable.pdf

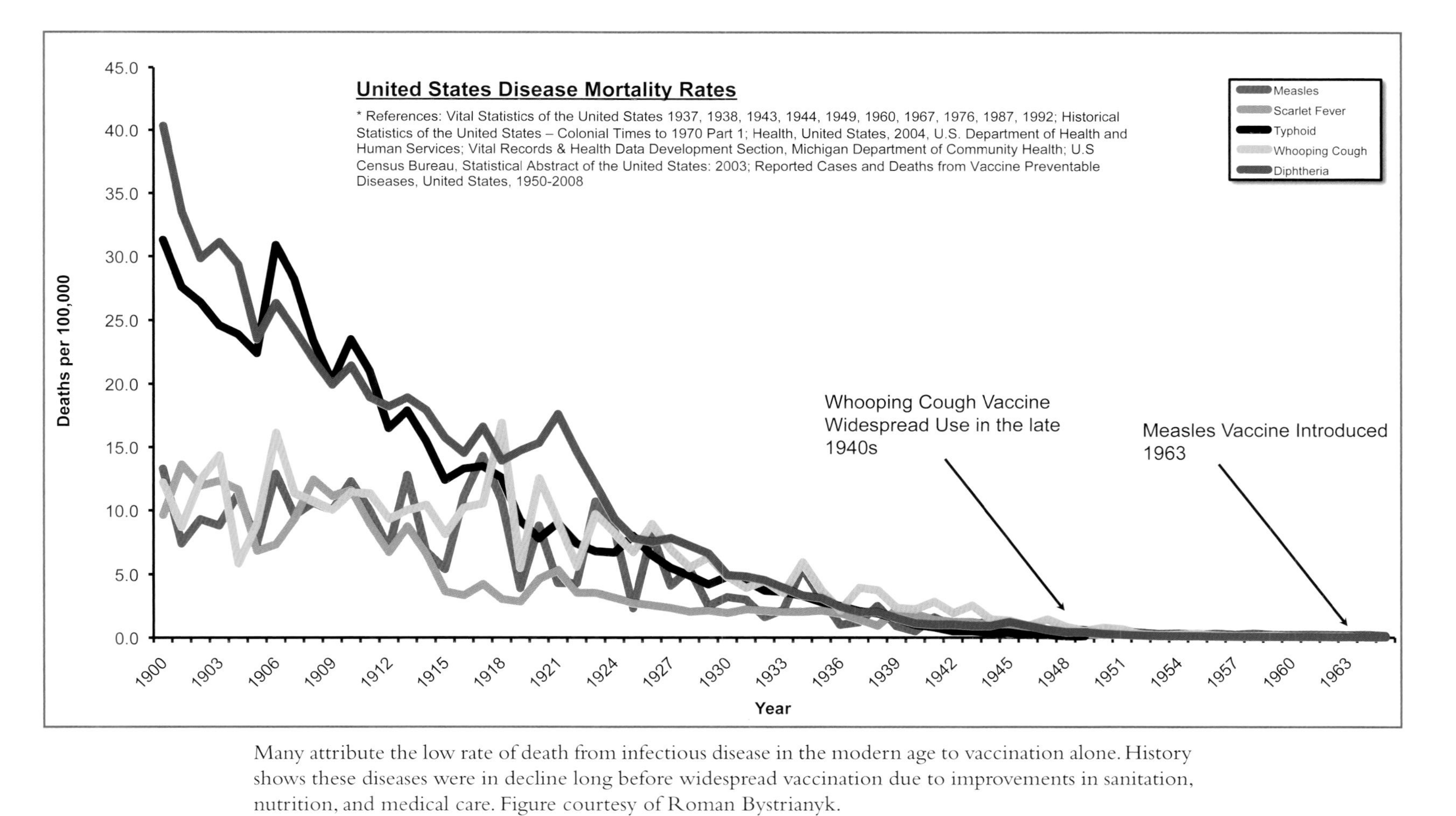

Many attribute the low rate of death from infectious disease in the modern age to vaccination alone. History shows these diseases were in decline long before widespread vaccination due to improvements in sanitation, nutrition, and medical care. Figure courtesy of Roman Bystrianyk.

CDC's Vaccine Information Sheet (VIS) on the HPV Vaccine
(page 1 only)

VACCINE INFORMATION STATEMENT

What You Need to Know

Many Vaccine Information Statements are available in Spanish and other languages. See www.immunize.org/vis.

Hojas de Información Sobre Vacunas están disponibles en Español y en muchos otros idiomas. Visite http://www.immunize.org/vis

1 What is HPV?

Genital **human papillomavirus (HPV)** is the most common sexually transmitted virus in the United States. More than half of sexually active men and women are infected with HPV at some time in their lives.

About 20 million Americans are currently infected, and about 6 million more get infected each year. HPV is usually spread through sexual contact.

Most HPV infections don't cause any symptoms, and go away on their own. But HPV can cause **cervical cancer** in women. Cervical cancer is the 2nd leading cause of cancer deaths among women around the world. In the United States, about 12,000 women get cervical cancer every year and about 4,000 are expected to die from it.

HPV is also associated with several less common cancers, such as vaginal and vulvar cancers in women, and anal and oropharyngeal (back of the throat, including base of tongue and tonsils) cancers in both men and women. HPV can also cause genital warts and warts in the throat.

There is no cure for HPV infection, but some of the problems it causes can be treated.

2 HPV vaccine: Why get vaccinated?

The HPV vaccine you are getting is one of two vaccines that can be given to prevent HPV. It may be given to both males and females.

This vaccine can prevent most cases of cervical cancer in females, if it is given before exposure to the virus. In addition, it can prevent vaginal and vulvar cancer in females, and genital warts and anal cancer in both males and females.

Protection from HPV vaccine is expected to be long-lasting. But vaccination is not a substitute for cervical cancer screening. Women should still get regular Pap tests.

3 Who should get this HPV vaccine and when?

HPV vaccine is given as a 3-dose series

1st Dose	Now
2nd Dose	1 to 2 months after Dose 1
3rd Dose	6 months after Dose 1

Additional (booster) doses are not recommended.

Routine Vaccination

- This HPV vaccine is recommended for girls and boys **11 or 12 years of age.** It *may* be given starting at age 9.

  Why is HPV vaccine recommended at 11 or 12 years of age?
  HPV infection is easily acquired, even with only one sex partner. That is why it is important to get HPV vaccine before any sexual contact takes place. Also, response to the vaccine is better at this age than at older ages.

Catch-Up Vaccination

This vaccine is recommended for the following people who have not completed the 3-dose series:
- Females 13 through 26 years of age.
- Males 13 through 21 years of age.

This vaccine *may* be given to men 22 through 26 years of age who have not completed the 3-dose series.

It is *recommended* for men through age 26 who have sex with men or whose immune system is weakened because of HIV infection, other illness, or medications.

HPV vaccine may be given at the same time as other vaccines.

Source: http://www.cdc.gov/vaccines/pubs/vis/downloads/vis-hpv-gardasil.pdf

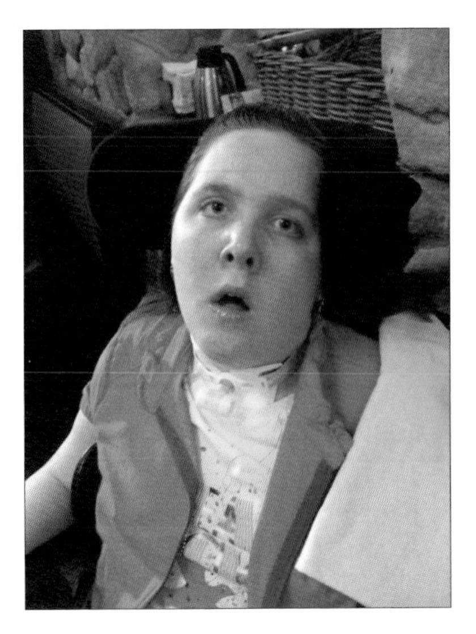

Zeda Pingel before and after HPV vaccination. She received the quadrivalent human papillomavirus vaccine, also known as Gardasil.

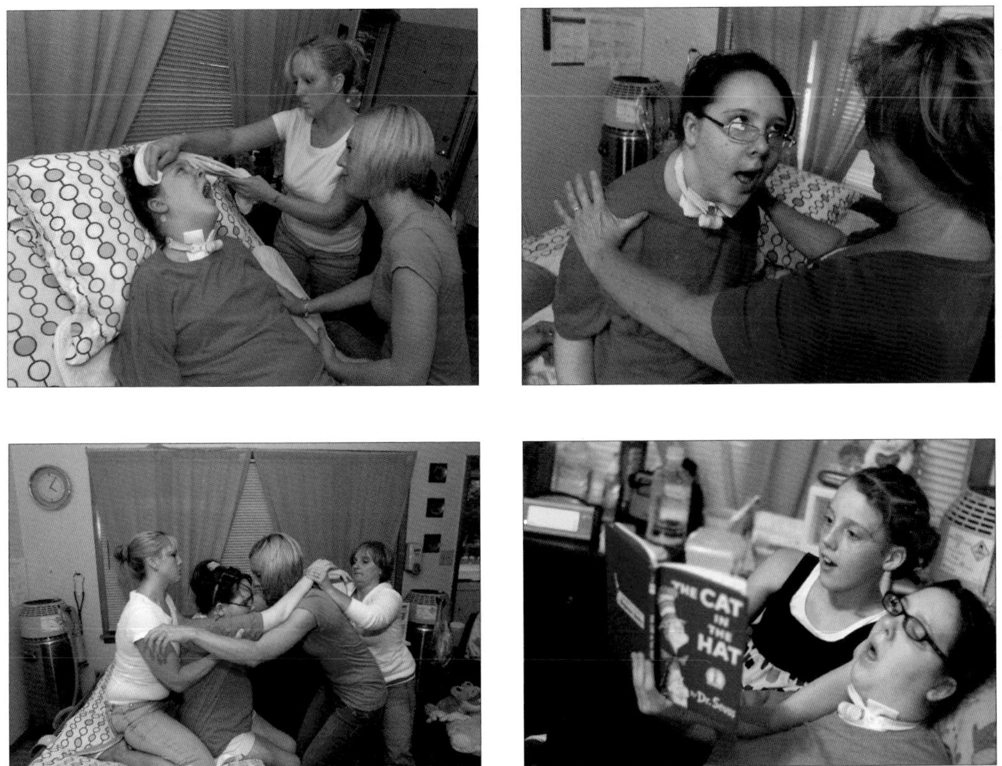

Zeda now lives in a hospital bed in her family's living room.

Photos courtesy of Jeffrey D. Nicholls/Post-Tribune of Northwest Indiana

ARE WE POISONING OUR KIDS

IN THE NAME OF PROTECTING THEIR HEALTH?

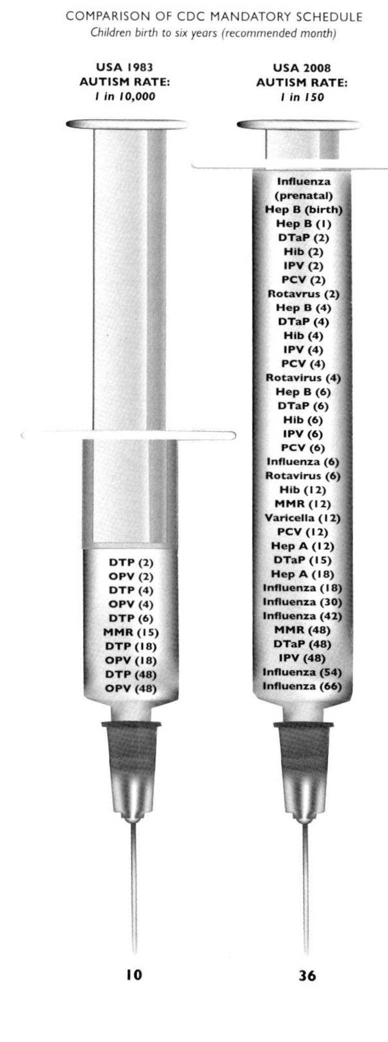

Green our vaccines. And administer them with greater care.

Mercury. Aluminum. Formaldehyde. Ether. Antifreeze. Not exactly what you'd expect—or want—to find in your child's vaccinations. Vaccines that are supposed to safeguard their health yet, according to our studies, can also do harm to some children.

The statistics speak for themselves. Since 1983, the number of vaccines the CDC recommends we give to our kids has gone from 10 to 36, a whopping increase of 260%. And, with it, the prevalence of neurological disorders like autism and ADHD has grown exponentially as well.

Just a coincidence? We don't think so. Thousands of parents believe their child's regression into autism was triggered, if not caused, by over-immunization with toxic ingredients and live viruses found in vaccines. The Centers for Disease Control and the American Academy of Pediatrics dispute this but independent research and the first-hand accounts of parents tell a different story.

Why are we giving our children so many more vaccines so early in life?
Why do we only test vaccines individually and never consider the combination risk of vaccines administered together? Given the dramatic rise of autism to epidemic levels, isn't it time for the scientific community to seriously consider the anecdotal evidence of so many parents? We urge the CDC and AAP to help us find the answers to these questions and learn why the increase in the number and composition of so many vaccinations has led to a surge in neurodevelopmental disorders. Our children deserve no less.

GENERATION RESCUE
www.generationrescue.org

We want to thank Jim Carrey and Jenny McCarthy for their generous support of Generation Rescue and their never-ending commitment to solving the growing challenges of autism.

Courtesy of Generation Rescue

Appeared in *USA Today*, on February 12, 2008

The Poling family when Hannah was 11 months old. Hannah's attention is appropriately focused on the same thing that is drawing both her parents' attention.

Hannah at 12 months old. She had the health, affect, and attention of a typical child.

Photos courtesy of the Poling family

Hannah at 19 months, shortly after receiving nine vaccines (in five injections) in one visit. Hannah developed hypotonia in her extremities. Immediately following her shots, she refused to walk and use the stairs.

Photos courtesy of the Cedillo family

A healthy and happy eight-month-old Michelle Cedillo, pictured with her father.

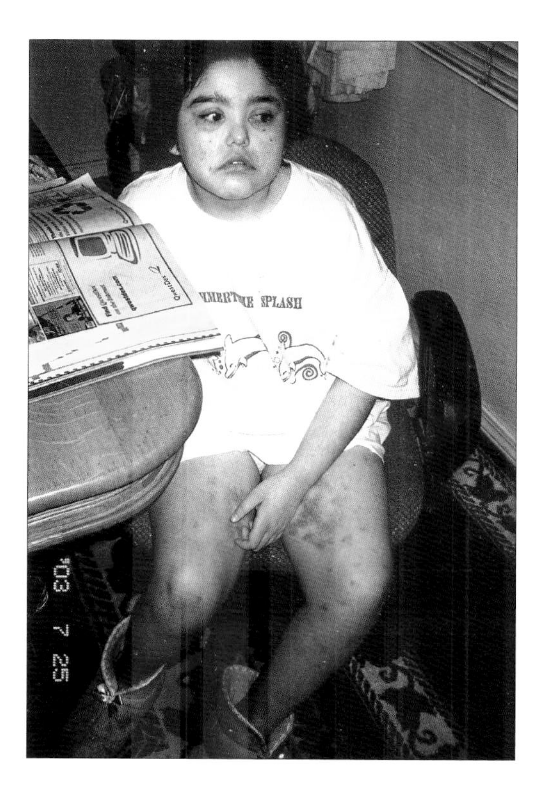

Michelle Cedillo was born on August 30, 1994. Like other children born in the United States at that time, she received twelve mandated vaccines containing ethylmercury.

At fifteen months, Michelle received the combination measles-mumps-rubella (MMR) vaccine. Seven days later, she developed a fever of over 105 degrees. Since then, she has suffered many severe medical problems—autism, mental retardation, life-threatening seizures, inflammatory bowel disease, uveitis, partial blindness, and arthritis. Michelle cannot talk, cannot walk unassisted, and cannot care for herself. She requires round-the-clock care.

"Green Our Vaccines" Rally—Washington, DC
June 4, 2008

Photo Courtesy of Manette Louden

"Freedom of Choice" Rally—Trenton, New Jersey
October 16, 2008
700+ protested four new mandates including a flu shot for daycare and preschool.

Rally for Vaccination Choice during New Jersey gubernatorial debate
William Paterson University, Wayne, New Jersey
October 16, 2009

Mary Holland, Russ Bruesewitz, Robie Bruesewitz, Louise Kuo Habakus

Press conference at Supreme Court to protest *Bruesewitz v. Wyeth* decision
Washington, DC
March 3, 2011

Photo Courtesy of Larry Kuo

Vaccine Epidemic book tour
Borders bookstore, Palo Alto, California
June 27, 2011

Front Lawn Rally for Senator Obama en route to DNC fundraiser
Habakus residence, Middletown, New Jersey
September 5, 2008

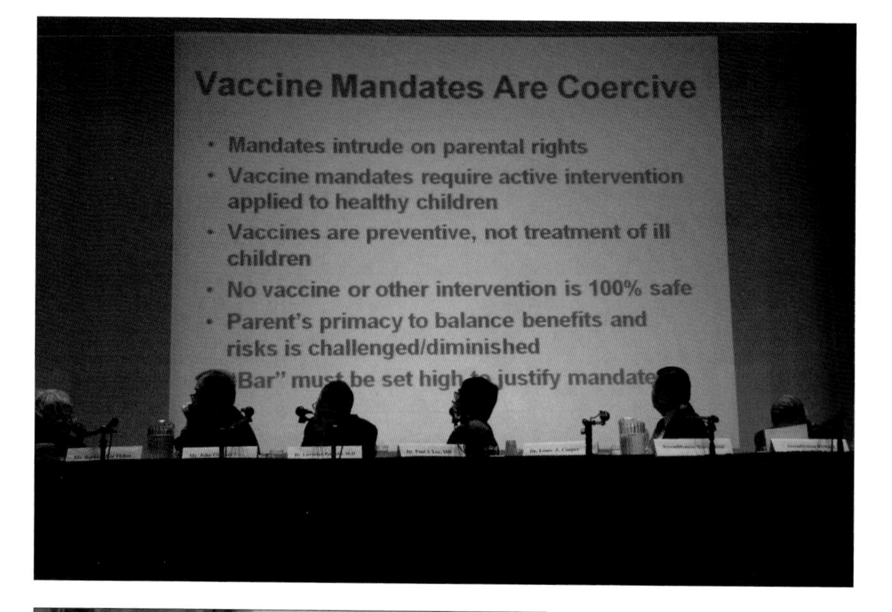

Photos courtesy of Christine Heeren

New York State Legislative Vaccine Roundtable—Stony Brook, New York
December 15, 2008

Speakers: Assemblyman Marc Alessi; Debra Blog, MD; Louis Cooper, MD; Barbara Loe Fisher; John Gilmore; Assemblyman Richard Gottfried; Louise Kuo Habakus, MA; Paul Lee, MD (AAP); Larry Palevsky, MD

Photos courtesy of Manette Loudon

Healthcare workers protest mandatory seasonal and H1N1 swine flu shots
Albany, New York
September 29, 2009

www.greatergoodmovie.org

Photo courtesy of Ronald Habakus

American Rally for Personal Rights
Grant Park, Chicago, Illinois
May 26, 2010

nursing homes, and hospices; it provided only a narrow exemption for medical reasons. The commissioner argued that voluntary immunization programs for HCWs had failed and that they were obliged to protect patients under their care.[61] New York State has been the only state to attempt to mandate flu vaccination for all HCWs.

The New York State regulation brought the state in direct conflict with individual HCWs, the New York Committee for Occupational Safety and Health, the New York State Public Employees Federation, several other unions, and the New York Civil Liberties Union.[62] HCWs and unions filed several lawsuits against Commissioner Daines for overstepping his administrative authority; the HCWs argued that such mandates required legislative authorization. HCWs held large protests in Albany, photographs of which are included in the photo insert of this book.

On October 13, 2009, a New York State trial court judge issued a temporary restraining order against the health commissioner and the state Department of Health requiring the state to stay the imposition of the mandate until there could be a full-scale trial on the merits, scheduled for October 30, 2009. On October 22, New York State rescinded the mandate.[63] Claiming that shortages of H1N1 vaccines would impede implementation of the mandates, Governor David Paterson announced that the state would not seek to impose the HCW vaccination requirement. It seems possible that the state's real reason for reversing policy was the very serious risk that it might have lost at trial. Since 2009, New York State authorities have not attempted to impose a flu vaccine mandate for HCWs. NVAC's recommendations may be an alternative solution for the failure of state HCW mandates.

Employer Vaccine Mandates for HCWs

Today, growing numbers of private hospitals and health care companies already impose annual flu vaccination mandates on HCWs as conditions of employment, ahead of NVAC's 2012 recommendations. Many major health care agencies and institutions have called for private mandates, including the Infectious Diseases Society of America, the American College of Physicians, the Association for Professionals in Infection Control and Epidemiology, and the National Foundation for Patient Safety.[64] Health care employers that have imposed mandates include the Hospital Corporation of America, The Johns Hopkins Health System, University of Iowa Hospitals, Hospital of the University of Pennsylvania, The Children's Hospital of Philadelphia, and the U.S. Department of Defense.[65]

Unlike The Cochrane Collaboration and other sources cited in this chapter, these institutions presumably assess the evidence of flu vaccine safety and efficacy in a positive light.

On the whole, courts have upheld private vaccination mandates for HCWs as a term of employment. While the NVAC recommendations note that there have been legal challenges for employer mandates and some setbacks, overall this incremental approach appears to be working to make vaccination mandates part of a new workplace environment for HCWs.[66] There is every reason to imagine that private HCW mandates will continue to proliferate. To the extent that such mandates become more common and accepted, it seems likely that other private institutions, including universities, community centers, and even apartment buildings, may follow suit. HCWs are undoubtedly on the front lines of the flu vaccine mandate battle, but it would be foolish to assume that they are the only ones likely to be affected. Some believe that flu vaccine mandates for teachers, police, firefighters, nursing home residents, and hospital patients are likely to be next.

CONCLUSION

Given serious safety concerns, the lack of independent research, undue pharmaceutical influence, and the reality of adverse reactions, it is essential that HCWs continue to have the right to decide for themselves about the flu vaccine. If the vaccine worked as well as the government says, HCWs would be first in line to get their shots. HCWs are among the most medically knowledgeable people in the country. The rest of us should be paying close attention to the governmental and private efforts to coerce them to receive vaccination or lose their jobs.

Chapter Twenty-Three

FORCED CHILD REMOVAL

Kim Mack Rosenberg, JD

You have a religious exemption from vaccination for your child at school. You get a call from Child Protective Services. The caseworker says she wants to meet with you to discuss whether your child is the subject of "medical neglect." Do you know your rights?

Few would dispute that Child Protective Services (CPS) serves a critical function in our society by protecting vulnerable children who have been harmed or are in harm's way in their homes. New York State, for example, created a network of child protective service agencies to investigate cases of suspected abuse and to provide swift and competent protection and rehabilitative services to the child and family.[1]

Despite good intentions, state agencies such as CPS, have long histories of abusing power[2] and failing to provide due process, to the detriment of children and their families.[3] In the absence of actual or imminent harm, CPS's role should not be to "protect" children from parents who make alternative healthcare choices.[4] In these circumstances, the state should not use CPS to usurp parental authority and to instead assert its view of the child's best interest.

What happens if CPS scrutinizes a parent's medical choice not to follow vaccination mandates? There are two basic questions:

1. Is the parent at risk for investigation of "medical neglect"?
2. Regardless of the reason for the child's removal, may the state vaccinate a child against the parent's wishes?

There are no simple answers to these questions. Child protection laws and their enforcement vary widely among jurisdictions. Parents should be aware, however, that the answer to each of these questions may be "yes." Through the misuse of CPS procedures, a state may be able to forcibly impose vaccinations.

This issue is of critical concern to parents who do not vaccinate according to government-recommended schedules. Recent surveys also have found that a majority of parents have serious concerns about vaccine safety. Parents are concerned that vaccinations may result in or contribute to autism, ADHD, sudden infant death syndrome, and other health problems. When families have a child with autism, for example, parents may provide their children with alternative medical and nutrition-based treatments. CPS has taken action against parents for medical neglect and has removed children because parents chose to make alternative healthcare choices for children on the autism spectrum. Many parents, including many parents of children with autism, make the decision not to vaccinate their children based on sincerely held religious beliefs. All but two states recognize religious exemptions from vaccination requirements but this may not, in every instance, protect a child from vaccination at the behest of CPS.

CPS and some courts have considered parents' choices for autism spectrum disorder treatments not "medically necessary." Many autism treatments—diets, auditory training, vitamin and mineral therapy, hyperbaric oxygen therapy, and chelation, to name a few—are hotly disputed.[5] Parents seeking to make decisions in the best interests of their children with autism are pawns in a broader dispute—involving parents, government, the pharmaceutical industry, and concerned scientists and medical professionals—about the safety of vaccines and the etiology of autism.

MEDICAL NEGLECT

CPS is required to promptly investigate allegations of medical neglect. But what is "medical neglect"? While laws vary, neglect is generally an absence of appropriate care. *The Merck Manual* defines neglect as

> the failure to provide for or meet a child's basic physical, emotional, educational, and medical needs. Neglect differs from abuse in that it usually occurs without

> intent to harm. . . . Medical neglect is failure to ensure that a child receives appropriate preventive care (e.g., vaccinations, routine dental examinations) or needed treatment for injuries or physical or mental disorders.[6]

Similarly, under New York law, for example, a neglected child is one

> whose physical, mental or emotional condition has been impaired or is in imminent danger of becoming impaired as a result of the failure of his parent . . . to exercise a minimum degree of care . . . in supplying the child with . . . medical, dental, optometrical or surgical care, though financially able to do so or offered financial or other reasonable means to do so.[7]

The above passages demonstrate that the definition of medical neglect is vague. If a parent does not vaccinate a child in accordance with state mandates, is the parent neglectful? There is not a clear-cut answer, in part because many sources fail to differentiate between parents with legitimate exemptions from vaccination requirements and parents who may have no objection but simply fail to vaccinate their children. Various documents from New York State, New York City, and other sources interpret the statutory definition of neglect to include the failure to vaccinate. For example, the New York City Children's Services "Parent's Guide to New York State Child Abuse and Neglect Laws" states:

> A parent or caregiver is required to supply adequate medical, dental, optical and surgical care for a child under 18 years old. This medical care includes seeking adequate treatment for conditions that impair, or threaten to impair, the child's mental, emotional or physical condition. You should follow the treatment prescribed for medical, psychiatric, and psychological care. *You should also obtain preventive care such as well-baby care checkups, and immunizations for polio, mumps, and measles.*[8] (emphasis added)

Similarly, New York University Child Study Center defined medical neglect as "a failure to provide or comply with prescribed medical treatment for a child (e.g., such as immunization, surgery, medications)."[9] A Pennsylvania presentation entitled "Keeping Your Children & Getting Them Back," designed to provide parents with information concerning child abuse and neglect laws in that state, gave "not taking a child for medical checkups or shots" as a straightforward example of medical neglect.[10]

While anyone may report suspected abuse or neglect to CPS, doctors, as a general rule, must report cases of medical neglect to CPS for investigation.[11] However, as noted above, what constitutes medical neglect is subject to interpretation. Doctors are the frontline in encouraging compliance with mandated vaccination schedules and, therefore, may view with suspicion parents who do not vaccinate in compliance with such schedules. For this reason, and because they are among the few people who have access to children's medical records, doctors may be the reporters to CPS in most vaccination-related medical neglect cases.

Simply put, a child's physician who believes that the parent's decision not to vaccinate is neglectful may report the parent to CPS. The American Academy of Pediatrics (AAP) has spoken to the issue of vaccinations and medical neglect and its opinions—while perhaps not appearing unreasonable on their face—are worrisome. The AAP in 2005 stated that:

> [c]ontinued refusal [to vaccinate] after adequate discussion should be respected unless the child is put at significant risk of serious harm (as, for example, might be the case during an epidemic). Only then should state agencies be involved to override parental discretion on the basis of medical neglect. Physician concerns about liability should be addressed by good documentation of the discussion of the benefits of immunization and the risks associated with remaining unimmunized. Physicians also may wish to consider having the parents sign a refusal waiver.[12]

Several years earlier, in 1997, the AAP's Committee on Bioethics published an article concerning religious exemptions to medical care. In that paper, the AAP called for the repeal of all religious exemption laws, proposing instead that parents should be better educated about their children's medical needs.[13] While the statement concerned exemptions from all forms of medical care, the committee took particular note of vaccines, stating that "some parents, acting in accord with state laws, refuse to have their children immunized because of religious beliefs. The AAP does not support the stringent application of medical neglect laws when children do not receive recommended immunizations."[14] Recognizing that "the risk to unimmunized individuals is relatively low" and encouraging "universal immunization," the AAP stated that it "supports the use of appropriate public health measures, such as mandatory mass vaccinations in epidemic situations, when necessary to protect communities and their unimmunized members."[15] In the 1997 statement, the AAP cautioned physicians that "threatening or seeking state intervention should be the last resort, undertaken only when treatment is likely to prevent substantial harm or death."[16]

These AAP statements raise several concerns. First, the undercurrent of both statements is paternalistic. Underpinning the AAP's position is the suggestion that not vaccinating a child is an illegitimate, scientifically or medically unsound choice,[17] despite laws that grant that right. The 2005 statement demonstrates this when it refers to failure to vaccinate as a "problem."[18] Further, the waiver that the AAP suggests doctors present to a non-vaccinating parent for signature contains language that arguably could later be used against the parent as a potential admission of neglect.[19] First, it indicates that the parent has been advised of the risks and benefits of vaccination and is written in such a way as to imply that the parent agrees with what he or she has been told.[20] The form then states, "I know that failure to follow the recommendations about vaccination may endanger the health or life of my child and others with which my child might come into contact."[21] While suggesting that "doctor knows best," the AAP's statements offer scant guidance about what constitutes an emergency that doctors should report. How does a doctor decide which emergency, which risk, which harm? With insufficient AAP guidance, there inevitably will be a lack of uniformity in application of this policy. Even the 2005 example of the epidemic is not clear-cut. One needs only to reflect on the 2009 H1N1 influenza outbreak to learn an object lesson on the subjectivity of identifying an epidemic. Finally, anecdotal evidence suggests that not all doctors are following the AAP's policy; some are reporting families who choose not to vaccinate to CPS in the absence of an emergency or threat to the child's health, such as a disease outbreak.

It is likely that only a small percentage of CPS referrals for medical neglect are based solely on noncompliance with vaccination mandates.[22] Nonetheless, for any family who has to face investigation by CPS, the social, emotional, and financial toll is incalculable, even if ultimately vindicated. In one case, the court found that a parent who "impressed the court as a diligent, loving and concerned parent" was neglectful for failing to immunize his daughter against measles during a measles outbreak.[23] It found that his objections to the measles vaccine, while sincere, were based on medical and scientific information, not religion. Because there was no other evidence of any "neglect," the court did not force vaccination of the child, at least in part, because the child was not yet attending school. The court might have decided differently if the child had been school-age and had determined that the parent's objections were not religious. This case also shows how time-consuming the judicial process can be—it took the court almost two years to decide the case after the initial petition was filed.[24]

As the number of parents exercising their rights to make vaccination exemptions grows, so does the opposition to that exercise. This is a contentious area and one that the Center for Personal Rights is closely monitoring.

FORCED VACCINATION

A child in foster care may be subject to vaccination unless his or her parent has made formal objections to vaccination, and even then, the parent's exercise of his or her rights is not guaranteed. Parents cannot rely on CPS to affirmatively enquire as to parental beliefs with respect to what are widely regarded as "routine" medical care practices and must take care to not waive their rights if they have valid exemptions against vaccination.

Vaccination according to federal recommendations and state mandates is a cornerstone of foster care, under both state and federal laws. Most children in foster care are eligible for federally-funded Medicaid.[25] In New York, for example, the Medicaid Early Periodic Screening Diagnosis and Treatment (EPSDT) requirements include "immunizations in accordance with the most current New York State or New York City Recommended Immunization Schedule, as appropriate."[26] Thus when a child enters the New York State foster care system, CPS attempts to obtain current immunization history.

The initial EPSDT medical assessment includes "preventive services, such as immunizations . . . appropriate for the child's age."[27] Following the initial assessment, pediatric visits include "immunizations consistent with current NYS/NYC DOH recommendations for age."[28] In fact, the EPSDT provider manual unequivocally states that "persons under 19 years of age must be immunized in accordance with the most current" state vaccination schedule. The manual also recommends "simultaneous administration of all routinely recommended vaccines appropriate to the age and previous vaccination status."[29] Of gravest concern to parents who elect not to follow vaccination mandates is the manual's encouragement to "catch up" on missed vaccinations. The manual states that "most of the widely used vaccines can generally be safely and effectively administered simultaneously. This is particularly important in scheduling children with missed immunizations."[30]

Given the emphasis on vaccination in foster care, parents should document any objections to vaccination. They should consider providing CPS with a copy of any vaccination exemption they previously obtained and making sure that any medical authorization[31] that they execute does not include consent for vaccination. If the child previously received vaccines, parents should consider providing CPS with a copy of the child's immunization history to avoid unnecessary revaccination. Depending on individual circumstances, a parent may also be able to participate in doctor visits while a child is in foster care.

Challenges to a parent's right to object to a child's vaccination on religious grounds when a child is in foster care vary by state. In New York, for example, courts have upheld exemptions for parents whose children are in foster care, but only if the parents meet all the usual requirements for vaccination exemptions. In New York, this means demonstrating a sincere and genuine religious objection to the court's satisfaction.[32] In many states, religious exemptions are not as automatic as they may seem and, if CPS seeks court intervention to vaccinate a child, a parent may even be required to "prove up" his or her exemption, even if previously accepted.[33]

In a decision upholding a parent's exemption, a mother with two children in foster care successfully defeated a CPS petition to compel vaccination.[34] Several years before, the mother had obtained a religious exemption for the elder child for day care, and neither child was vaccinated.[35] The agency's foster care paperwork noted the mother's opposition to vaccination on religious grounds.[36] The court held a hearing to assess whether the mother's religious beliefs were genuine and sincere and found that they were. The agency's application to vaccinate was denied.[37]

While the mother prevailed in this case, the facts of the case exposed a glaring hole in the system. The foster parent, acting on her own authority, had one of the children vaccinated "to protect her own family."[38] CPS then petitioned for the right to vaccinate the other child after the parent threatened to sue.[39] Even though the agency was aware of the mother's objections, a rogue action by the foster parent resulted in a child receiving a vaccination against the mother's religious beliefs. The foster parent's action violated the role of foster parents in New York. Foster parents "do not have legal authority to request or authorize immunizations."[40]

In another New York case, the court decided against a parent exercising a religious exemption. A court recently determined that a mother whose child was found to be neglected based on, among other things, failure to maintain a sanitary home and failure to provide medical care, did not support her religious objection to vaccinations by a preponderance of the evidence.[41] As a result, the court granted the Administration for Children's Services's motion to allow the child to be immunized in accordance with New York's vaccination mandates.[42]

In a recent Arizona decision, a court found that a mother whose child had been placed in foster care still was entitled to exercise her religious objection to vaccination.[43] The court was careful to note that such cases needed to be determined on their individual facts and stated "we would not hesitate to find a compelling state interest had the Department shown that [the child] was especially vulnerable to diseases prevented by immunization, due perhaps to malnutrition or some other medical condition."[44] Finding that the child's risks were no greater than the risk to other children, the court upheld the mother's right to continue to exercise her religious objection while her child was out of her custody.[45] This case demonstrates that parents must be prepared not only to support their religious objections to vaccination but also perhaps to address medical or scientific concerns the court may have.

In other cases, courts have held that when parents lose custody of their children, either temporarily or permanently, they lose the right for their religious beliefs to control medical decision making. In a 2002 North Carolina case, parents who were adjudged neglectful and whose children were placed in foster care were denied the right to exercise their religious objection to vaccination.[46] The court did not examine the sincerity or bona fide nature of the parents' beliefs, instead finding that they had lost their rights to exercise a religious exemption—though their parental rights were not terminated.[47] The court found that only the department of social services could make immunization and other healthcare decisions for the children.[48] In 2002, a Georgia court decided to permanently remove a child because it found the child to be "deprived." As part of that decision, the mother lost the right to make medical decisions.[49] The court granted custody, which included the right to make medical decisions for the child, to the child's grandmother.[50]

THE POTENTIALLY LIFE-ALTERING RESULTS OF FORCED VACCINATION

While this chapter has focused on parents' rights, children's rights are inextricably linked to their parents' rights. Forced vaccination of children, in violation of parents' rights, risks real physical harm to children. The results of forced vaccination while in foster care may be devastating.

Dan Olmsted, editor, *Age of Autism*, an online newspaper, wrote a story in 2008 about a young girl vaccinated while in foster care against her parents' religious convictions.[51] CPS removed an Amish child from her family when she was one year old because CPS thought the family displayed "medical neglect" and failed to treat her chronic ear infection properly. In foster care, she was vaccinated, apparently with multiple vaccines in two sessions, presumably to "catch her up" to the mandated schedule. The child now has autism and is in her parents' care. In the Old Order Amish community, parents do not vaccinate their children and autism is almost unknown.

CPS may protect children from child abuse and neglect, but it also poses a potential risk to families when parents choose not to follow vaccination mandates. Even in those jurisdictions where, by law, CPS and foster parents must respect parent choices about vaccination, they do not always do so. When parents are faced with the reality of child-removal proceedings, they must carefully document and communicate vaccination exemptions to CPS and know their rights.

The author wishes to acknowledge Malika Felix, graduating law student and CPR intern, for her research assistance. *This chapter is an introductory overview and is not legal advice. State law, especially on religious treatment exemptions, varies widely. Parents should consult a lawyer to discuss individual circumstances; the cases used here are examples only.*

Chapter Twenty-Four

THE GREATEST THREAT TO OUR COUNTRY

Julian Whitaker, MD

The forced childhood vaccination program endured by American families is, in my opinion, the single greatest threat this country has ever faced. I do not make this statement lightly. As a practicing physician, founder of the largest integrative medical clinic in the country, author of fourteen books, and editor of *Health & Healing,* a popular monthly newsletter, I am an outspoken critic of a number of widely accepted practices in medicine. Yet no one—not the government, the media, physicians, or even patients who are most at risk—seems to be alarmed by or even recognize the very real and dangerous blinds spots of conventional medicine.

For instance, at my clinic, prevention means helping patients adopt a good diet and exercise routine, optimize their nutritional status with targeted supplements, and boost their overall health. In the irrational world of conventional medicine, however, prevention often means the search for disease. Beginning in infancy, people are encouraged to see their physicians for regular exams and screenings whether they're sick or not. Scans, blood tests, and the latest screening technologies give doctors a plethora of information—the significance of which is far from definitive and open to interpretation. But rather than take a conservative course, most doctors equate the presence of "risk factors" with disease, immediately slap on diagnoses, initiate "treatments," and turn millions of essentially healthy people into anxiety-ridden patients.[1] Overdiagnosis inevitably leads to unnecessary treatment. Although some surgeries and invasive procedures are lifesaving, far too many are unwarranted. To give a few examples, studies suggest that up to 70 percent of all hysterectomies are inappropriate.[2] Many invasive procedures for prostate cancer and benign prostatic enlargement have complications and are questionable in terms of necessity and long-term success.[3] And more than a million patients per year are frightened into coronary artery bypass surgery or angioplasty—the bread and butter of invasive cardiology—although the scientific literature clearly shows that, in most cases, they should not even be considered.[4]

The greatest threat to our health and well-being, however, is the pharmaceutical industry, which has usurped the doctor-patient relationship, has driven up health care costs, and harms millions of people every year. Big Pharma has co-opted medicine. It begins in medical school, where budding physicians are taught how to prescribe by instructors who are compensated by drug companies. Students who do well in this environment become authority figures and opinion leaders, and they're paid handsomely to extol the benefits of a never-ending stream of pharmaceuticals. This unholy alliance has even tainted our most prestigious medical journals, which are dependent on pharmaceutical advertising for revenue and rife with conflicts of interest.

Drug companies spend nearly twice as much on marketing and promotion as on research and development, and this includes almost $5 billion a year in direct-to-consumer advertising.[5] The United States and New Zealand are the only countries in the world that allow pharmaceutical companies to advertise directly to consumers, and in the U.S. direct-to-consumer advertising has infiltrated our culture to such a degree that we don't even notice it. When is the last time you questioned the appropriateness of advertisements suggesting that you "ask your doctor if this drug is right for you"?

It has also permeated our government. Big Pharma spends nearly $250 million a year on lobbying expenses, is a top campaign contributor, and has notoriously close and compromising financial ties to the Food and Drug Administration, the agency tasked with monitoring drug approvals and safety.[6] This industry is also in cahoots with government-funded agencies that promote public health campaigns that aggressively "educate" us on the importance of screenings and medication compliance—efforts that have created millions of new customers for their products. For example, several years ago policymakers adjusted the "normal" blood pressure range downward; overnight, 13.5 million additional Americans became hypertensive and in need of a drug. Updated cholesterol goals made 40 million more people eligible for cholesterol-lowering drugs. New bone density thresholds increased the number of people with osteoporosis by an astounding 85 percent—yet another windfall for the pharmaceutical companies.[7]

Today, physicians whip out the prescription pad even before giving safer, less expensive alternatives a try. Between 1999—about the time direct-to-consumer advertising really took off—and 2009, the number of prescriptions dispensed in the United States went from 2.8 billion to 3.9 billion, for an average of 12.1 retail prescriptions for every man, woman, and child in America. That's a 39 percent increase during a decade in which our population grew by only 9 percent.[8]

If these drugs had a clean record of safety, I wouldn't be so concerned. But they don't. A landmark study published in *JAMA, The Journal of the American Medical Association*, estimated that 106,000 hospitalized patients die every year due to adverse reactions to prescription drugs.[9] That's 290 deaths from prescription drugs every day, 365 days a year—the equivalent of a 747 airliner going down every other day! These patients don't lose their lives as a result of an overdose or because a pharmacist dispenses the wrong drug. They die because they have an adverse reaction to a properly prescribed and administered drug.

The damage doesn't end there. Approximately 4.5 million visits to physicians' offices and emergency rooms every year are directly linked to adverse effects of prescription meds.[10] And a 2011 report by the U.S. Agency for Healthcare Research and Quality revealed that in 2008, 1.9 million patients in American hospitals (nearly 5 percent of all hospitalizations) had "medication-related adverse outcomes," "side effects of prescribed drugs that were taken as directed, unintentional overdosing by the patient, and medication errors such as incorrect prescribing and dosing."[11] To put this into perspective, *every day* nearly 13,000 people have adverse drug reactions severe enough to merit a trip to their doctor or to an emergency room, and 5,400 hospitalized patients are injured by medications that are supposed to help them.

I'm particularly concerned about the drugs marketed to children. A quarter of American children take at least one drug for a "chronic condition." One in ten boys has been prescribed Ritalin or similar addictive, amphetamine-like drugs. Pediatric use of antipsychotics, powerful, dangerous medications initially used to treat schizophrenia but now prescribed for depression and anxiety, doubled in a recent eight-year period. Antidepressants, which increase risk of suicide and uncharacteristically violent behavior, are now taken by millions of children.[12]

Nothing, however, compares to one class of drugs that is forced upon every child in the United States: vaccines. Of all the follies perpetuated in medicine, the government-mandated vaccination program is the most heinous because it is literally destroying our most valuable assets, our children. It is time for strong words. We are facing nothing short of a public health catastrophe. That this catastrophe is avoidable makes it doubly tragic.

In this chapter and elsewhere in this book, you will read about the damage vaccines have inflicted and continue to inflict upon our children. You'll learn of the toxicity of vaccines and the lack of valid scientific data supporting the vaccine program. However, this nightmare really has nothing to do with medical science, good or bad. It is the result of a breakdown in the checks and balances of our government and a suspension of what we Americans value most, our individual freedoms. There is a solution—but first, the problem.

THE VACCINE-AUTISM LINK

If a child is reported missing, everyone immediately joins in the search. Men and women alike have powerful intrinsic urges to protect and guard our young. But what about the millions of children who have "disappeared" into the abyss of autism? Where is the hue and cry?

Fifty years ago, autism was rare, identified in only 1 in 10,000 children. Rates began to creep up in the early 1980s, with the disorder appearing in about 1 in 5,000. By 2007, however, autism affected 1 in every 150 children. And today, 1 in 88 American children—and 1 in 54 boys—have an autism spectrum disorder. That's a stunning 78 percent increase in just five years![13]

These statistics, which were released in 2012 by the Centers for Disease Control and Prevention, are terrifying. If current growth curves hold steady, autistic children will soon outnumber healthy children. Unless something is done to reverse this insidious trend, virtually all male children will be diagnosed with autism spectrum disorder by 2032, followed by all girls in 2041. This frightening projection may seem implausible, but it is based on current government statistics and simple math.[14]

The prevalence of other neurological disorders affecting our children has soared as well. One in six American children is now identified as having a learning disability, and the numbers of students enrolled in special education increased by 30 percent in the past decade alone. What's going on? In my opinion, the major culprit—and the only obvious and absolute constant for these skyrocketing numbers of autistic and learning-disabled children—is the dramatic increase in federally recommended and state-mandated vaccinations.

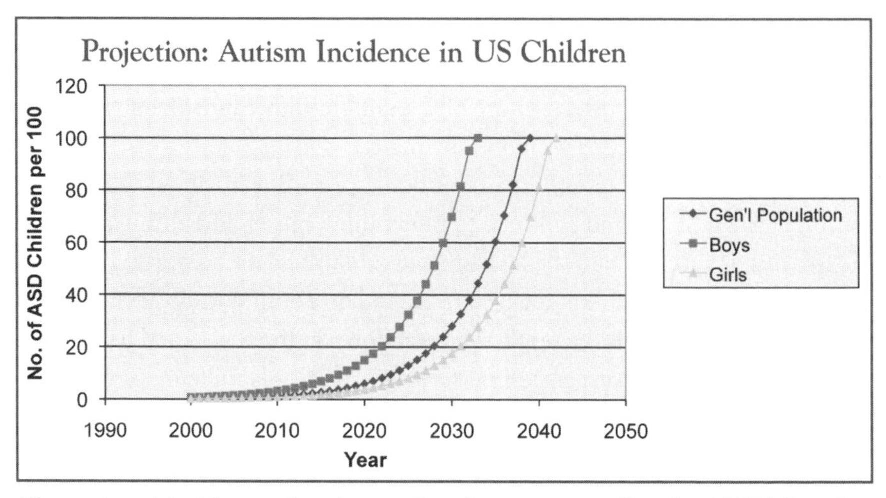

The projected incidence of autism and autism spectrum disorder (ASD) based on current statistics. Graph courtesy of Shawn Siegel.

In the 1940s and 1950s, our government recommended that American children receive the DPT (diphtheria, pertussis, and tetanus), polio, and smallpox vaccines; in 1971, measles, mumps, and rubella vaccines were added to that list. Then the "vaccine epidemic" of the 1990s and 2000s arrived, and the numbers of doses of vaccines forced upon our children exploded.[15] Today, twenty-five doses of nine different vaccines are given to our children by the time they are six months old, one administered hours after birth, and thirty-eight doses of fifteen vaccines are mandated by age two. (Compare this to Japan, which has a voluntary vaccination policy. The earliest vaccinations there are given at three months of age, and just ten doses of four vaccines are recommended by age six months with a total of seven vaccines in seventeen doses by age two.)[16]

It's no coincidence that autism spectrum disorder rates in the U.S. exploded at the same time. Autism is not a disease and it is misleading to call it a developmental disorder. Autism is diffuse, severe brain damage caused, in my opinion, almost exclusively by conglomerates of toxic substances put into vaccines and injected into pregnant women, newborns, infants, and children. Thirty years ago, most cases of autism were present at birth. Today, the majority of affected children develop normally until sometime between their first and second birthdays. These "late-onset" autistics hit all the usual milestones before they begin to regress—in many cases abruptly after their fifteen- to eighteen-month booster shots.

According to Mary Ann Block, DO, a Hurst, Texas, physician who has been treating autistic children for more than fifteen years, "Many parents have videos of their children taken before and after the vaccines, and a very real difference is clear.

They go from being happy, connected children to slipping away into their own worlds."[17] U.S. Representative Dan Burton, speaking at a 2002 Congressional hearing on vaccines, spoke of his "talkative, playful, outgoing, healthy" grandson Christian. "My only grandson became autistic right before my eyes—shortly after receiving his federally recommended and state-mandated vaccines. . . . Within a few days he was showing signs of autism."[18] These are not isolated incidents, yet vaccination proponents are loath to even consider that vaccines may be the cause.

TOO MANY UNNECESSARY, UNSAFE VACCINATIONS

A single shot can contain a brew of adulterated bacteria, viruses, aluminum, mercury, formaldehyde, hydrochloric acid, and/or numerous other multisyllabic chemical additives.[19] (For more information on vaccine ingredients and substances used in the manufacturing process, see page 3 of the photo insert.) To assume that repeated exposures to such a wide range of toxins have no cumulative adverse effects on a child's developing nervous and immune systems is irrational, naïve, or worse. You may find this hard to believe, but there is not a shred of evidence to support the safety of the increasing vaccine load foisted upon our children, and any doctor who assures parents that vaccines are completely safe is either misinformed or flat-out dishonest.

To ascertain vaccination safety and efficacy, large comparative studies of the health outcomes of statistically significant numbers of vaccinated and unvaccinated children would have to be conducted, and they would have to include variables such as administering as many as six or eight different vaccines at a time. Such studies have never even been attempted. Yet one vaccine after another has been added to the already overloaded list of required vaccinations.

Furthermore, many of the required vaccines are absolutely unnecessary. Hepatitis B, which is given hours after birth, is transmitted primarily by sexual contact or dirty drug needles. Why in the world would newborns of healthy moms need such a vaccine?[20] The same goes for diphtheria and tetanus, which are first administered at two months of age. Diphtheria may be a problem in developing countries, but its incidence in the United States since 1980 is one case in every 100 million people. And tetanus (lockjaw) isn't even contagious—it's usually contracted through deep wounds with contaminated objects. Granted, neonatal tetanus isn't uncommon in countries with unsanitary childbirth conditions, but it is virtually nonexistent here in America. Polio is another vaccine that is completely unwarranted. Although you may associate polio with the paralysis, iron lungs, and deaths of the mid-1900s epidemics, this disease was eradicated decades ago. There hasn't been a case of naturally occurring paralytic polio in the United States in more than thirty years. Today, polio is much more likely to be caused by the live virus vaccine than by "wild" poliovirus.[21]

The truth is that severe illness and death rates from virtually every infectious disease we vaccinate against were dropping rapidly and were, in fact, almost eradicated before vaccines even came onboard.[22] The real conquerors of the infectious diseases that killed millions in the past were not vaccines at all but improved standards of living, better sanitation, safer food, and elimination of waterborne pathogens. However, you can't charge billions of dollars every year for improved hygiene and living conditions. (See U.S. Disease Mortality Rates chart on page 5 of photo insert.)

Despite these well-documented facts, drug companies, government officials, medical policymakers, schools, mainstream media, and physicians constantly bombard parents and the public with fear tactics. They spout nonsense such as, "Without forced vaccination programs, there will be a return of major epidemics, and bodies will stack up in the streets," and "Polio might come back, so we need to vaccinate six million children every year." These statements are just plain absurd. Many Amish children, many homeschooled children, and millions of children who otherwise receive legal exemptions don't get vaccinated—and they don't drop dead or get polio either. And don't forget the adults. Most aren't vaccinated against all these diseases, and even if they were, vaccine immunity wanes over time. Where are the hundreds of millions of adults succumbing to deadly epidemics?

In an attempt to frighten us into submission, pro-vaccine propagandists make other unfounded presumptions with no hard evidence to back them up. A common argument is that unvaccinated children put others at risk of infectious diseases. My response is that if vaccines work, why do parents worry about children who are not vaccinated? They also offer up platitudes like, "Preventing one death makes it well worth vaccinating all children." How about the children who die or are injured by vaccines? Aren't their injuries or deaths worth preventing? If vaccinations prevent that one death but do irreparable harm to thousands or more, is that okay? Of course not. It's insane!

HPV VACCINE: AN UNMITIGATED FRAUD

Shifting gears, let's look at another fraudulent, wasteful, and dangerous vaccination program, one that targets preteen and adolescent girls, young women, and now boys and young men: the human papillomavirus (HPV) vaccine. The purpose of Gardasil (Merck's HPV vaccine) and Cervarix (GlaxoSmithKline's version) is, in theory, to prevent cervical cancer.[23] I predict, however, that the HPV vaccine program will not only be a colossal failure in terms of health benefits but will leave an expanding wake of young people suffering with vaccine-induced seizures, strokes, chronic headaches, and worse. (One of the "Gardasil girls" severely damaged by the vaccine is Zeda Pingel, whose story is chronicled by her mother Amy Pingel in Chapter 14.)[24]

Here's why: Regardless of what the drug companies, the CDC, or physicians say, there is no firm data to support the widespread contention that the HPV vaccine prevents cervical cancer. It has been shown to prevent *precancerous* changes of the cervix,[25] but to assume that these changes will progress to cancer is dishonest and manipulative.

The National Cancer Institute (NCI) reports that 90 percent of HPV infections clear up spontaneously within two years.[26] Recognizing this, in 2011 the U.S. Preventive Services Task Force issued new guidelines recommending that women be tested for cervical cancer with Pap smears (not HPV tests) every three years rather than annually because more frequent testing leads to overtreatment of low-grade changes that would in all likelihood not turn out to be cancerous. The Task Force further bolstered the irrationality of vaccinating young girls when they changed the screening guidelines to include only women ages twenty-one through sixty-five, noting that cervical cancer is exceptionally rare in women under age twenty-one.[27] So, why does the federal government recommend the HPV vaccine to eleven- and twelve-year-olds?

In any case, cervical cancer is very slow growing, so we won't know for twenty or thirty years whether mass vaccination reduces death rates. It's all presumption, a huge lottery in which there may well be no winners—other than the drug companies, aided and abetted by their cronies in government. (See Mark Blaxill's and Dan Olmsted's chapter 19, "A License to Kill," for more information on this public-private collaboration.)

SKEWED STATISTICS, WINDFALL PROFITS

The pro-vaccine camp underscores the need for mass HPV vaccination by trotting out government statistics showing that more than a quarter of American females ages fourteen to fifty-nine and nearly 45 percent of those ages twenty to twenty-four have been infected with HPV. But they fail to mention that there are forty sexually-transmitted HPV strains, and those targeted by Gardasil (types 6, 11, 16, and 18) and Cervarix (types 16 and 18) are rare. HPV types 6 and 11, which can cause genital warts, were detected in 1.3 and 0.1 percent of women, respectively, and types 16 and 18, which are linked with some cases of cervical cancer, were present in only 1.5 and 0.8 percent, respectively.[28] Furthermore, scientists acknowledge the real risk of serotype replacement, whereby new, untargeted cancer-causing strains increase in prevalence as those targeted by a vaccine decrease.

HPV FACTS

4,000: Cervical cancer deaths per year in the U.S., or 2.4 per 100,000 women. (Women are 75 times more likely to die of heart disease than of cervical cancer.)[30]

2.3 percent: Women ages fourteen to fifty-nine infected with the two types of HPV associated with cervical cancer. (Thirty percent of cervical cancer cases are not associated with HPV.)[31]

90 percent: HPV infections that clear up on their own.[32]

$15 billion: Cost of vaccinating all thirty million females in target age range.[33]

Thousands NNT: Number needed to treat to save one life (theoretical best-case scenario).[34]

The bottom line: very, very few women who have HPV are infected with high-risk strains, and far fewer get cervical cancer. Every year in the United States, twelve thousand women are diagnosed with this cancer, and four thousand die of it. Of course, any premature death is a tragedy, but we cannot lose sight of the fact that, according to the latest statistics from the NCI, only 0.68 percent of women will ever be diagnosed with, let alone die of cervical cancer. Why are we recommending this vaccine for all children? It's not as if HPV can be "caught" by sitting next to someone in the classroom. Nevertheless, this vaccine is mandated for school admission in Washington, DC, and the state of Virginia, and there is pending legislation to require it in more than twenty other states.[29]

To further underscore the absurdity of universal HPV vaccination, let's look at the concept of "number needed to treat," or NNT, an extremely useful statistic for evaluating any medical treatment. Simply stated, NNT tells us how many people need to be treated with a given therapy to get the desired benefit in one patient. The lower the NNT, the more effective the treatment. For example, peptic ulcers are primarily caused by *Helicobacter pylori* bacteria, and antibiotics that eradicate it are an extremely successful therapy. For every eleven patients with *H. pylori* who are treated with antibiotics, ten are cured of their peptic ulcers. Therefore, the NNT is 1.1 (10 divided by 11).[35]

So what is the NNT of the HPV vaccine in terms of preventing cervical cancer deaths? Even if the vaccine completely wiped out cervical cancer—which no one expects it to do—thousands upon thousands would have to be vaccinated in order to prevent one death. The others would obtain no benefits, yet would be needlessly exposed to the inherent risks of this vaccine. Most statisticians agree that an NNT over forty is no more than a crapshoot. An NNT in the thousands is an unmitigated fraud, especially when there's no evidence that the vaccine will save even one life.

We already have a system in place for preventing cervical cancer that works very well: regular Pap tests. Even the most vocal vaccine proponents admit the vaccine doesn't eliminate the need for Pap testing—or that most cervical cancer deaths occur in women who hadn't been screened in the previous five years. This system has reduced the incidence of cervical cancer from nearly 15 in 100,000 women in 1975 to 6.8 per 100,000 in 2009.[36] Why fix something that isn't broken?

The answer is obvious: Follow the money. In the United States, there are roughly 30 million females between the ages of nine and twenty-six who are "eligible" for HPV vaccination, which requires three doses spread out over six months at a retail price of about $130 each ($390 total). That's nearly $12 billion right into the pockets of Big Pharma. Now, let's add in physicians' fees and average the cost for the three-dose course at $500: $500 x 30 million = $15 billion. Imagine spending $15 billion on a vaccination program with no hard evidence that any lives will be saved.[37]

Let's take it a step further and *assume* this lavish blanket of presumed protection actually works and cervical cancer is eliminated. (Never mind that a miniscule percentage of HPV-infected women ever develop cervical cancer, that 30 percent of women with cervical cancer have *not* been infected with HPV, and that it will take decades before we know if the darned thing really does what it's predicted to do.) Guess how much it would cost per life saved in this best-case scenario? $7.5 million. If we took that $15 billion and put it towards improving nutrition, reducing obesity, and implementing other proven health interventions, we could save tens of millions of lives. But, incredibly, the powers that be prefer to waste it on a fraudulent vaccination program.

As if thirty million girls and young women weren't enough, Merck tried to get Gardasil approved for women up to age forty-five, but even the FDA recognized this absurdity for what it was. However, Big Pharma now has males in their sights. In February 2012, the Advisory Committee on Immunization Practices formally added routine HPV vaccination for eleven- to twelve-year-old boys, with "catch-up" shots for those aged thirteen to twenty-one and previously unvaccinated "immunocompromised" young men through age twenty-six.[38] The argument goes that they, too, must be vaccinated in order to prevent the spread of HPV to their sexual partners.

If vaccinating females makes no sense, going after males is completely illogical. HPV is almost always an inconsequential infection in males, so males themselves get very little to no benefit at all. Unfortunately, they are not immune to the adverse effects of the vaccine. In fact, they are likely at greater risk of damage—at least that's what we've learned from the standard childhood vaccinations, which negatively affect two to three times more boys than girls. As for the costs, don't get me started. We would spend another $15 billion vaccinating all males in the target age range—with zero benefits to them and absolutely no assurance or evidence that this "experiment" will ultimately reduce the rate of cervical cancer in women. There's only one given: Any amount of money spent on such an ill-advised campaign would be a 100 percent waste to the public, yet hugely profitable for vaccine manufacturers.

MANUFACTURERS HAVE NO LIABILITY

These pharmaceutical companies, which make billions of dollars on vaccines, have hoodwinked the government into requiring all these vaccinations *and* taking on the liability for vaccine-induced damages. These guys have nothing to lose and everything to gain.

This unprecedented turn of events occurred in 1986, when a reluctant President Ronald Reagan signed into law the National Childhood Vaccine Injury Act. This law decrees, "No vaccine manufacturer shall be liable in a civil action for damages arising from a vaccine-related injury or death."[39] By seeking this protection, the drug companies could be said to have acknowledged that the vaccines they created for children are toxic, that they are designed to make children sick in order to stimulate an immune response, and that they contain highly toxic ingredients that will cause serious side effects, including death. Therefore, they in effect told lawmakers that if the pharmaceutical industry's liability was not eliminated, they would never be able to implement a plan for mass vaccination of American children because they could never withstand the judicial scrutiny of vaccine damage in courts of law.

In short, liability for injuries and deaths caused by most vaccines was removed from the pharmaceutical companies because the vaccines are so dangerous. Just imagine 100,000 families suing Merck because their children were severely damaged after receiving MMR vaccinations. Their defense? "Well, your honor, they are probably protected from measles."

Isn't that why we need judicial scrutiny, to protect us from exactly what is now happening with these horrifically dangerous vaccines? In place of judicial scrutiny, this law created a special administrative program with an "Office of Special Masters" composed of lawyers with no scientific background who determine if harm has been done and if compensation is warranted. The decisions of these "Special Masters" carry the force of law, and they have passed out more than $2 billion to thousands of families of damaged children.[40] Guess who paid these billions? Not the drug companies, but vaccine consumers. This is on top of the tax dollars spent administering this program.

If this law had not eliminated all of Big Pharma's liability for vaccine-related injuries, there would be no forced vaccination program—and no autism epidemic.

CHILDREN ARE FORCED TO BE VACCINATED

The second reason for this horrific nightmare is the suspension of our civil liberties. Big Pharma has convinced the government to force parents to submit their children to whatever vaccines they produce, no matter how unnecessary or toxic. If you as a parent decided you did not want your child to be vaccinated, you couldn't just say no. You would have to go through a lengthy and time-consuming process of getting an exemption.

Such processes vary from state to state and in some states are unbelievably harsh. In 2007 in Prince George's County, Maryland, State Attorney Glenn F. Ivey and the chairman of the school board, Owen Johnson, Jr., implemented a vaccine roundup. They issued summonses to parents of more than 2,300 children who had not provided certificates of vaccination. Parents were told to appear in court on a Saturday to subject their children to on-the-spot, state-mandated vaccinations—up to 17 doses—or face imprisonment.[41]

According to Kathryn Serkes of the Association of American Physicians and Surgeons, "This campaign of intimidation to brutally enforce blanket vaccine mandates by the government agencies and the school district gives no consideration for the rights of the parents or the individual medical condition of the child." So why did it happen? Money, plain and simple. The school district stood to lose a substantial amount of funding if students didn't comply with the state's vaccine mandates. "Apparently the district wants that money," said Serkes, "even if it gets it off the backs of children."[42] What is the difference between this activity in the state of Maryland and tyranny elsewhere in the world? There is no difference.

The pharmaceutical industry also is going to equally great lengths in an attempt to force the HPV vaccination program down our throats. Several state governments have enacted laws that require schools to hand out information on the vaccine to sixth graders and their parents and insurance companies to provide reimbursement. For example, the state of New Jersey delivered HPV vaccine propaganda to parents via their children's backpacks. And in California, Governor Jerry Brown signed legislation that enables children as young as twelve to be vaccinated against HPV and hepatitis B *without a parent's knowledge or consent.*[43] Imagine, parents have to sign permission slips for their children to go on a school field trip, but twelve-year-olds, without parental knowledge or consent, can have toxic substances injected into their growing bodies.

WE MUST STOP THE MADNESS

There are only two things that can stop this madness. First, liability must be borne by the pharmaceutical companies. These companies make huge profits selling vaccines. Therefore, like other businesses in this country, they—not the American people—must be legally accountable for the safety of their products and financially responsible when injuries occur.

Second, parents must have the right to decide what is injected into their children. It is up to you to protect them, even if that means bucking the status quo and facing rabid opposition. The most important thing you as a parent can do is base your decisions regarding their well-being on accurate, unbiased information. In the current climate, that's easier said than done.

You can't count on your doctor for objective guidance. Virtually all pediatricians follow the dictates of the American Academy of Pediatrics, the CDC, and Big Pharma and vigorously support every vaccination program that comes down the pike. Their refusal to rock the boat by acknowledging the growing body of evidence about the dark side of vaccines is safe for the doctors, who also bear no liability, but terribly dangerous for your child. I encourage parents to spend the time to find a philosophically aligned physician. It's not easy, but they are out there. Because some of these doctors are afraid of being targeted and threatened by the state medical boards and departments of health, many won't advertise the fact that they accept parents who selectively vaccinate or refuse all vaccines. However, if you network with concerned parents individually or through support groups, you will find them. You deserve to have a relationship with a doctor who does not bully you or coerce you into doing something that violates your intuition or your beliefs.

Finally, take advantage of the excellent resources available in this book, which include scientific research, legal information, and guidelines for exemptions and alternative vaccine schedules. Do your homework and hold your ground. Because older children will likely be influenced by peers and others, keep the lines of communication open and make sure they understand the extreme risks and lack of proven benefits of the HPV vaccine. Freedom of informed choice, tarnished though it may be, is your right as an American and your responsibility as a parent.

It may take many years and, unfortunately, countless injuries and untold numbers of deaths, but we're not as naïve or stupid as Big Pharma takes us to be. Mark my words. The public will eventually wake up and see forced vaccination for what it is—a dangerous, money-grabbing, scientifically invalid fraud.

Chapter Twenty-Five

A DOCTOR'S VIEW OF VACCINES AND THE PUBLIC HEALTH

Sherri Tenpenny, DO, AOBNMM

According to the *Encyclopedia Britannica*, public health is "the science and art of preventing disease, prolonging life, and promoting health through organized community efforts." Its mission, as led in the United States by the Centers for Disease Control and Prevention (CDC) includes developing sound public health policy for the population at large. There can be little argument that public health has played a significant and valuable role in sanitation, injury prevention, environmental toxicology, food safety, and clean water. However, in recent years, one of public health's primary tools for disease prevention—vaccination—has come under attack. Its importance has been seriously brought into question.

Said to be the cornerstone of disease prevention, vaccination has become one of the most expansive areas of public health. All levels of government—local, state, national, and global—spend billions of dollars each year on the manufacture, procurement, distribution, and administration of vaccines. In 1983, the vaccination schedule from birth to eighteen years of age included twenty-four doses of seven different vaccines: polio, MMR (measles, mumps, and rubella), and DTP (diphtheria, tetanus, and pertussis). The federal government now recommends that children receive seventy doses of sixteen different vaccines by the time they graduate from high school. Parents have begun to question the necessity of so many vaccines. An ethical tug-of-war has arisen between the people's right to choose what is injected into their bodies (and the bodies of their children) and the presumed benefits of vaccines to society as promoted by health officials.

Discussions about the right to refuse vaccination can become contentious, even hostile. Those who firmly believe that vaccines are safe, protective, and mostly harmless stand a chasm apart from those who have had a different experience with vaccination, involving serious injury, chronic illness, and even death. The debate has intensified, and so have the stakes. As more parents investigate and weigh the risks and benefits of vaccines, many have chosen to opt out. These parents have decided to take personal responsibility for the health of their children, refusing to accept that health must come through a needle. Every vaccination has potential side effects and many parents feel that the risk of the vaccine is unacceptably larger than the risk of the illness.

Moreover, scientists cannot prove the efficacy of a particular vaccine in a particular individual because it is impossible to predict the susceptibility of an individual to a specific pathogen. Parents know of children who have remained well without being vaccinated and of persons who have contracted an illness even though they received their respective shots. Public health officials and parents who advocate vaccination downplay reports of outbreaks among the vaccinated, fearing the value of the vaccination paradigm would be called into question.

With more individuals refusing vaccines, alarmed health officials are concerned that the unvaccinated will cause a loss of "herd immunity"—a vaccination level assumed to provide community-wide protection. They point to outbreaks of chickenpox, mumps, and pertussis over the last few years as proof of waning herd immunity. Officials have lodged blame for these occurrences squarely on unvaccinated children, even though most of the children who contracted the illnesses were *fully vaccinated.*

Vaccine makers and medical professionals claim vaccines were responsible for eradicating a large number of diseases, but official statistics clearly document vaccine-preventable illnesses were on their way out or, in many cases, nearly eliminated prior to the widespread use of the vaccines designed to prevent them. However, scientific facts rarely seem to change the perspectives of either side. Reasons for vaccination choice often reside in personal convictions. Many individuals insist that the separation between their person and the state begins at their skin. They have strong convictions that mandatory injections are an infringement of personal rights. The question, then, is whether society has a right—or perhaps a duty—to override personal choice for the good of the whole country. This argument has sparked debates for more than a century.

A fact that cannot be debated, however, is that we have a national epidemic of sick children. Today, parents schedule occupational therapy, physical therapy, and speech therapy into their weekly routines, as though these activities were as ordinary as soccer practice and piano lessons. Pediatricians have started to say, "two years of age is when kids get asthma," or "children get autism before age three," as though these were typical growth milestones.

Is there a common factor contributing to the epidemic of illness among children today? It's not genetics: Different families have different genes. It's not exercise: Some kids are athletes, others are couch potatoes. It's not environmental exposures: Some sick kids live in unhygienic public housing projects, some live in pristine gated communities. It's not food: Some eat only organic, some eat mostly fast food. What touches almost all children today and likely contributes to the plethora of chronic illnesses are seventy doses of sixteen vaccines—which include measurable amounts of chemicals, excipients, detergents, adjuvants, and heavy metals—by eighteen years of age. Science has not proven these ingredients to be harmless as administered. Why do we accept that vaccination cannot be a contributor to illness and poor health?

We have come to accept asthma, allergies, eczema, chronic ear infections, and learning difficulties as "normal" childhood conditions. In our zeal to eradicate a few infectious diseases, we have traded typically mild illnesses of childhood, such as chickenpox, mumps, and measles, for a lifetime of poor health and drugs. We have exchanged temporary illnesses for pervasive, lifelong diseases, disorders, dysfunctions, and disabilities. A chronic illness held in check by medication is not health. Society must gauge the wellbeing of children by more than high vaccination rates and low infection rates caused by a specific short list of pathogens.

ECONOMICS OF PUBLIC HEALTH: A SNAPSHOT

Since the 1970s, health economists have justified vaccination through cost-benefit analysis, a technique for quantifying and measuring the value of an intervention by weighing the likely costs of an illness against the positive benefit of the procedure. According to public health officials, vaccination has one of the best cost-benefit profiles for reducing medical care expenditures.[1] However, these calculations are based at the population level and do not include the potential costs to the individual. Policy makers admit that side effects occur, but consistently use large epidemiological studies to minimize the number of reactions and their catastrophic results. The story below, unfortunately, well illustrates this point.

> Several years ago, I had a young patient in my practice who was a virtuoso violinist. She had won an impressive list of awards and competitions and was about to enter a prominent school of music. She had grown up healthy and unvaccinated in a state where her mother had exercised a philosophical right to refuse vaccines throughout her twelve years of primary education. The university where she intended to matriculate was unbending in their vaccination requirements, particularly insisting that she receive an MMR (measles, mumps, rubella) vaccine. While this was disconcerting to her parents, she was otherwise healthy and they agreed to the demands of the college. Within a few days of the inoculation, her health began to deteriorate and within months, she was unable to lift her arms due to excruciating pain in her joints and extreme weakness in her muscles. Despite the help of a long list of conventional and integrative medical practitioners, this aspiring musician was unable ever again to play her instrument. Today, she is on disability due to progressive neurological deterioration and chronic pain.[2]

Physicians and officials begrudgingly admit that vaccination side effects can occur. The CDC has written a fact sheet on each vaccine called a Vaccine Information Statement (VIS). The VIS provides basic information about the risks and presumed benefits of the vaccine. By federal law, all vaccination providers must give patients, or their parents or legal representatives, the appropriate VIS prior to administering a vaccine.[3]

Each VIS is written in clear and simple language so it is easily understood by parents. A VIS contains 1) a description of the illness the vaccine is designed to prevent; 2) a basic description of the vaccine; 3) a definition of who should receive the vaccine and when it should be administered; 4) the risks of the vaccine; and 5) a list of possible mild, moderate, and severe reactions to the vaccine.

A VIS is given to reassure parents they are making the correct decision to vaccinate. Depictions of the diseases are strong, and include incidence rates for hospitalization and death. This makes every infection seem serious and potentially deadly—a frightening perspective for a nonmedical person possessing no experience with which to place the statistics in context. On the other hand, a VIS tends to minimize the potential for serious vaccine side effects. Nearly all VIS forms state, in one way or another, that severe reactions are "exceedingly rare, less than one out of a million doses. These are so rare it is hard to tell if they are caused by the vaccine."[4]

However, severe reactions are not nearly as rare as many doctors and public health officials would have us believe. In 1986, Congress passed the National Childhood Vaccine Injury Act, which included a national system to collect and evaluate the reports of possible adverse events. The Vaccine Adverse Event Reporting System (VAERS) is a passive surveillance system that relies on physicians, healthcare providers, parents, and vaccine manufacturers to submit reports of adverse reactions, typically within thirty days following vaccination. Since 1999, VAERS has received 11,000 to 12,000 individual reports per year. Approximately 15 percent of filed reports are classified as "an event that is fatal, life-threatening, requires or prolongs hospitalization, results in permanent disability, or in the judgment of the physician could lead to such an outcome in the absence of medical intervention."[5] In 1993, Dr. David Kessler, former head of the U.S. Food and Drug Administration, stated that only about one percent of serious adverse drug events (reactions) are reported. This could mean that there are potentially 1,200,000 adverse events from vaccines occurring *each year.*

In addition to filing a report, severely injured parties can file a claim for compensation through the Vaccine Injury Compensation Program (VICP). As of November 3, 2010, 13,613 claims have been filed and the VICP has committed to, or has paid, more than $2 billion in awards to 2,541 persons, with more than 5,800 remaining to be adjudicated. The federal government grants awards for extreme reactions to vaccination, such as chronic arthritis, anaphylaxis, encephalopathy, and death. Mild to moderate disabilities and side effects are not eligible for damage awards.

The medical literature is rife with reports of autoimmune illness, hearing and vision loss, skin disorders, blood disorders, and a long list of neurological disorders occurring postvaccination. The positive cost-benefit analysis lauded by public health bureaucrats does not account for economic and personal losses sustained by the many thousands of affected persons who have been subjected to millions of dollars of tests searching for the cause of an illness that was sourced from a vaccine. If a true accounting of the vaccine program included these expenses—and the cost of drugs used to treat illnesses arising from vaccine side effects and reactions—the program's economic cost-benefit ratio would seriously be called into question.

PUBLIC HEALTH: PULL OR PUSH?

Public health officials strike a delicate balance between cooperation, cajoling patients into vaccination compliance, and coercion, using subtle and not-so-subtle threats of serious consequences when people refuse. It is the classic carrot-and-stick approach applied to healthcare. A patient has the right to refuse medications and elective surgeries because an adverse outcome will affect only that individual's health. However, health officials fiercely challenge patients and parents when they refuse vaccination because officials fear that outbreaks of contagious diseases can affect the welfare of a community.

The negative attitude toward adults who choose not to vaccinate their children began almost a century ago. During the 1920s, the tenor of health officials' rhetoric transformed from gently urging the public to cooperate with diphtheria vaccination programs into a tone of stern authority. Strong language was used in public service announcements, pamphlets, and medical brochures. Officials insinuated that parents had culpability if their unvaccinated children became ill. In 1926, the American Child Health Association charged that "the time will come when every case of diphtheria will be an indictment against the intelligence of the parents and it will not be many years before every death from diphtheria will be referred to a coroner's jury for investigation to fix criminal responsibility."[6]

While no parents have yet been charged with criminal liability because their unvaccinated child contracted or died from a so-called vaccine-preventable disease, there have been coercive attempts by public health officials to forcibly vaccinate children against the wishes of their parents. For example, in 2007, when more than 2,300 children did not comply with the newly mandated chickenpox and hepatitis B vaccinations, the state of Maryland threatened parents with legal action as a last resort. Children were absent from the Prince George's county school district for nearly two months because of failure to meet vaccine requirements, creating what officials deemed a "truancy issue." Circuit Judge C. Philip Nichols, Jr., who handles juvenile matters, and Maryland Attorney General Glenn F. Ivey sent letters to the families who were not compliant, ordering them to appear for a hearing. The letter warned that their children's unexcused absences may subject them to criminal charges, which included a $50-per-day penalty and up to ten days in jail. The state ordered parents to arrive at the courthouse with children in tow. If parents did not provide proof that their children had received all of the state-mandated vaccinations or hand over a valid vaccination-exemption letter, the children would be vaccinated on site.[7] Many parents scrambled to update their children's records privately through their family physicians' offices and sent the records directly to the school system. However, nearly 1,700 children appeared with their parents at the courthouse on Saturday morning, November 11, 2007. The judge verbally reprimanded the parents, many children were immunized at the courthouse, and the students were told to return to school on Monday morning.

Another more serious example occurred in New York in 2008 when three children were removed from their mother's custody by the Department of Social Services while the court decided, among other issues, if the mother's opposition to immunizing her children as a school requirement qualified as a "genuine and sincere religious belief." In New York, parents have the burden to prove by a "preponderance of the evidence" that their opposition to immunization is "personal and sincere." In this instance, the court found that the mother had met her burden. The state granted her an exemption. The experience was no doubt terrifying to this family and sent a strong message to others that the state of New York enforces vaccination requirements with a heavy hand.[8]

Compulsory vaccination laws have generated disagreements between officials and parents since the widespread use of the first vaccine for smallpox in the 1800s. As more vaccines have been developed, the question is, increasingly, whether all licensed vaccines are truly necessary to protect the public's health. For example, many states are debating school requirements for human papillomavirus (HPV) vaccination, Gardasil and Cervarix, which have been licensed and recommended for girls and boys as young as nine. HPV is not an airborne pathogen; it is spread by sexual intercourse and sexual contact. Is it reasonable to make this vaccine mandatory for sixth graders in exchange for a public education?

VACCINATION EXEMPTIONS: THE LEGAL RIGHT TO REFUSE

By 1968, nearly 50 percent of U.S. states had laws requiring one or more vaccinations for public school admission. By 1974, the number of states with vaccination laws had increased to forty.[9] Today, all U.S. states and territories have vaccination requirements for entering public schools. Notably, when state governments first passed vaccination regulations, significantly fewer vaccines were in general use. Since 1985, when the only vaccines required for school were polio, MMR, and DTP, most state governments have added multiple doses of nine additional vaccines and more boosters to school requirements.[10]

Almost all state laws allow parents to refuse vaccinations for their children. However, the ease of obtaining exemptions varies significantly by state. School and public health officials are seldom forthcoming about the availability of vaccination exemptions. Each fall, parents receive a letter from their local school districts reminding them to have their children's vaccination records up-to-date before the first day of school. An exemption form is seldom, if ever, offered unless requested. Many parents are unaware that their state laws provide a mechanism to exempt their children from vaccination requirements. Currently, twenty states allow parents to opt out based on philosophical disagreement with the principles of vaccination, and forty-eight states, all but West Virginia and Mississippi, allow parents to refuse for religious reasons.[11]

All fifty states allow medical exemptions from vaccination, but in practice, they are difficult to obtain. As a general rule, only medical exemptions signed by a medical doctor (MD) or doctor of osteopathic medicine (DO) are accepted. The letter must state that, in the opinion of the physician, the administration of one or more vaccines would be detrimental to the health of the child. States primarily grant medical exemptions on a vaccine-by-vaccine basis rather than as a blanket exemption to all vaccines. In some state health departments, medical directors have the authority to override the opinion of the patient's personal physician if they do not think the exemption is justified. Even when physicians agree that a child should be medically exempted, they are often hesitant to put their opinions in writing, fearing repercussions to their private practice. State health departments occasionally rebuke physicians who write too many medical-exemption letters.

The final type of exemption, rarely discussed, is the "proof of immunity," or serological, exemption. Establishing proof of immunity requires a blood test called a titer level. If the results demonstrate the presence of an acceptable level of protective antibody, this serves as proof that the person is protected, and no further vaccines for that disease are required. Acceptance of this type of exemption varies by state. It is often honored by military commanders for members who wish to avoid overvaccination.

VACCINE INGREDIENTS

There are more than sixty chemicals contained in measurable amounts in various vaccines. With just one dose of each vaccine, an individual receives a number of potentially toxic and carcinogenic substances. Skeptics claim that the concentration of chemicals is negligible, and people routinely consume many, such as gelatin. Scientists understand there is a significant difference between *ingestion* of a substance through the gastrointestinal tract and *injection* of a substance into the bloodstream, which can have a significantly different outcome.

For example, vaccine makers use gelatin in twelve different vaccines; it functions as a heat stabilizer.[12] For most people, consuming gelatin is safe. An odorless, tasteless, and colorless protein, it forms a jelly when heated and then cooled. Manufacturers produce gelatin by boiling the connective tissues, bones, and skins of animals, usually cows and pigs.[13] Researchers document that gelatin in vaccines may cause systemic allergic reactions, including anaphylaxis, a massive and potentially deadly response.[14] Additionally, a further cross-reaction between gelatin in vaccines and gelatin in food may occur, making once-safe foods potentially dangerous.[15]

The medical literature reports serious reactions to gelatin found in the MMR and chickenpox vaccines.[16] The reason for these reactions may be due to the large amount contained in each shot. The chickenpox vaccine contains 12,500 micrograms (12.5 milligrams) of gelatin per dose. The MMR vaccine contains 14,500 micrograms (14.5 milligrams) of gelatin per dose.[17] To put that into perspective, the volume of venom delivered by a wasp sting ranges from 2 to 15 micrograms. A bee sting, generally estimated to contain 50 micrograms of venom,[18] can cause anaphylaxis and death. With thousands of micrograms of gelatin present in a single injection, there can be little surprise that some people have violent, even deadly, reactions to these vaccines.

While gelatin can cause severe reactions in some people, it is not the only ingredient of concern. Many have heard about thimerosal, a preservative containing mercury. Parents are told that childhood vaccines now are "mercury free." This is misleading. While most childhood vaccines no longer contain thimerosal as a preservative, thimerosal is still used in the manufacturing of vaccines and is still found in many vaccines in trace amounts. Mercury is a potent neurotoxin and any amount, even a trace amount, is worrisome,[19] especially when injected.

Most flu vaccines, including those encouraged for pregnant women and very young children, still contain thimerosal. A multidose flu shot contains 25 micrograms of mercury per dose. By comparison, the EPA announced in March 2004 that nearly all seafood contains traces of mercury[20] and the FDA warns that a serving of fish is unsafe if it contains more than one microgram of mercury per gram of fish. A normal serving of fish is about three ounces (18 grams). Both the FDA and EPA advise women who may become pregnant, pregnant women, nursing mothers, and young children to avoid shellfish and some types of fish due to elevated mercury content. It appears government officials are more concerned about ingesting 18 micrograms of mercury in a serving of fish than they are about injecting the 25 micrograms of mercury in a vaccine.

Many vaccines contain the adjuvant aluminum, in the form of aluminum phosphate, aluminum sulfate, or aluminum hydroxide. Aluminum is a potent neurotoxin used to strengthen the effect the antigen (the viral or the bacterial fragment) has on the immune response. An adjuvant stimulates the immune system in a nonspecific manner, leading to the development of high antibody levels. At the time of this writing, aluminum-based salts are the only adjuvants approved for commercial use in the United States. Although vaccine makers have used aluminum in vaccines for more than fifty years, the mechanism of action of the aluminum adjuvant is complex and not completely understood.

Aluminum is primarily eliminated from the body through the kidneys. Infant kidney function, called the glomerular filtration rate, is low at birth and does not reach full capacity until at least one year of age. Infants, therefore, may not be able to excrete aluminum effectively, and the cumulative amount of aluminum from repeated vaccination may contribute to significant, unrecognized neurotoxicity.[21] Notably, the American Academy of Pediatrics (AAP) admits that "aluminum has been implicated as interfering with metabolic processes in the nervous system and in other tissues."[22]

Aluminum is harmful enough on its own, but it is known to work synergistically with mercury. Where both are present, their effects are exponential, as Dr. Boyd Haley explains in chapter 19. Children's exposure to aluminum has increased as the number of vaccine doses has increased. Examples of vaccines with aluminum are DTaP, pneumococcal conjugate (Prevnar), hepatitis A, some forms of Hib, and HPV (Gardasil). For example, the prescribing information for Gardasil[23] states,

> Each 0.5-mL dose of the vaccine contains approximately 225 micrograms (0.225 milligrams) of aluminum (as amorphous aluminum hydroxy-phosphate sulfate adjuvant), 9.56 mg of sodium chloride, 0.78 mg of L-histidine, 50 mcg of polysorbate 80, 35 mcg of sodium borate, less than 7 mcg yeast protein/dose, and water for injection.

The Prevnar vaccine package insert[24] states,

> Prevnar is manufactured as a liquid preparation. Each 0.5 mL dose is formulated to contain: 2 µg of each saccharide for serotypes 4, 9V, 14, 18C, 19F, and 23F, and 4 µg of serotype 6B per dose (16 µg total bacterial wall saccharide); approximately 20 µg of CRM197 carrier protein (diphtheria); and 0.125 milligrams of aluminum per 0.5 mL dose as aluminum phosphate adjuvant.

The FDA states that the amount of aluminum in an individual dose of a biological product shall not exceed:

> (1) 0.85 milligrams if determined by assay;
> (2) 1.14 milligrams if determined by calculation on the basis of the amount of aluminum compound added; or
> (3) 1.25 milligrams determined by assay provided that data demonstrating the amount of aluminum used is safe and necessary to produce the intended effect.[25]

Multiple vaccines are often given together. If DTaP (Infanrix—0.625 micrograms), hepatitis A/B (Twinrix—0.850 micrograms), and pneumococcal conjugate (Prevnar—0.125 micrograms) vaccines are given at the same time, the infant will receive 1.6 milligrams of aluminum, thereby exceeding the FDA-recommended allowable dose of aluminum on a given day. This increases the risk of both short- and long-term complications, which is especially worrisome because the effects of cumulative doses of aluminum injected into babies have not been studied.

DOCTORS AND VACCINATION

It is a rare doctor who supports the rights of parents and patients to refuse vaccination. Doctors who challenge the medical dogma and the value of vaccination are professionally castigated by their peers, personally embarrassed within their communities, and can be legally silenced by their government and state medical boards. On the other hand, physicians who promote vaccination have begun to pressure patients into vaccinating by the use of coercion.

Physician coercion can take many forms. One coercive technique is threatening to report parents of unvaccinated children to state child protective services for medical neglect. A more common method of coercion is to post an office policy that includes discharging entire families from a medical practice for non-compliance with the AAP's vaccination recommendations.[26]

The one-size-fits-all, inflexible stance on vaccination increasingly assumed by physicians seems to be having an unintended effect: It is strengthening the resolve of parents to refuse.

Parents simply don't want to be told what to do. Many parents today want to be actively involved in healthcare decisions for their children, especially decisions about vaccination. Parents who inquire about vaccine safety are often more familiar with the ingredients and the potential side effects than their doctors. Some doctors perceive this involvement as opposition, and many do not like having their authority questioned. Unfortunately, most physicians know little about vaccines beyond what they are told by the pharmaceutical sales representatives who sell them. Doctors accept that vaccines are safe and provide protection with few side effects—this is an assumption made with little or no personal investigation beyond the statements made by the CDC. They read medical literature about their specialty but do not scrutinize research about the vaccines they administer. Few, if any, medical school courses discuss vaccination in detail. Even pediatric residencies focus on only the perceived benefits of vaccines, how to administer them, and how to catch children up if doses are missing.

In general, doctors wholly accept that vaccines single-handedly eradicated polio and smallpox. They believe that vaccines are harmless and five or more shots can be safely given at the same time. Medical histories are infrequently assessed for contraindications to vaccination, and detailed information on vaccine package inserts is ignored. In the rush to vaccinate and marshal patients through time-limited office visits, doctors do not discuss vaccination options. In fact, when parents want to discuss non-standard pediatric vaccine options, pediatricians are encouraged to charge patients an extra fee to compensate for their time.[27]

When an adverse event occurs, doctors undertake an extensive and expensive array of medical tests to find the cause of a new problem that seems to have randomly materialized out of nowhere. The rare physician will admit that the symptom or illness—that began shortly after an injection—could be a side effect of a vaccine. More disturbing, physicians will seldom know enough about vaccine side effects to make the connection.

There is a double standard when it comes to vaccine reactions. Doctors readily acknowledge a serious reaction to a shot of penicillin. They will place prominent drug alert labels on the patient's chart and advise the patient to wear a medical alert tag. This is not the case with vaccination. Doctors consider a vaccine reaction to be an anomaly. If a patient is allergic to a substance in a vaccine, the medical literature advises that it is safe to administer a vaccine containing that substance, as long as it is "under supervision of a physician,"[28]—as though the mere presence of a doctor in the room will eliminate a serious, possibly deadly, side effect. It is disquieting that physicians would risk a reaction just to be sure a patient was vaccinated.

Doctors must adhere to their promise to first, do no harm. If medical doctors will not stand behind a patient's right to refuse vaccination, then patients must stand up for themselves. Vaccination choice must become a legally guaranteed personal right. Infectious diseases come and go. With an immune system supported by clean water, exercise, adequate sleep, decent living conditions, reasonably good food, and a 25-hydroxy vitamin D level of greater than 60 milligrams per milliliter,[29] few will be at risk of dying from an infection. In fact, most won't get sick at all. Health, after all, does not come through a needle.

Chapter Twenty-Six

A HOLISTIC HEALTH PERSPECTIVE

Annemarie Colbin, PhD

"I regard it as fundamental that any doctor applying a remedy to a patient should be conversant, so far as possible, with its ill effects as well as with its good effects."
—Sir Graham S. Wilson, twentieth-century British microbiologist

"Science . . . will snatch the babies from the water sprinkling and inoculate them with disease to save them from catching it accidentally."
—George Bernard Shaw, *Man and Superman*

"None [are] so blind as those that will not see."
—Matthew Henry, eighteenth-century Presbyterian minister

OVERVIEW

Vaccination is generally accepted as a given: "Of course you vaccinate your children—that's how you protect them from disease." In some communities, however, vaccination is controversial. Many consider it a dangerous practice that threatens the integrity of the immune system. While the media, medical professionals, lawmakers, and the general public largely accept vaccine theory, many researchers and the holistic health community (including chiropractors, homeopaths, and anthroposophic physicians) increasingly believe that the risks outweigh the benefits.

Contemporary societal mores dictate that children receive all of their shots according to the recommended schedule. Proponents of the practice state that vaccines are a safe and effective way to avoid childhood diseases and thereby improve public health. They consider adverse effects rare and believe the benefits of vaccination far outweigh the risks. In order to attend public school, children must provide evidence of vaccination. The practical effect of these requirements is a government mandate. Although neither the state nor the federal government directly presses the needle through the skin without consent, they promote vaccines, pressure families to vaccinate, impose mandates as a condition of day care and school admission, and do not readily and fully disclose potential adverse effects or the availability of exemptions. Therefore, it could be said that the government practices medicine without a license.

In this chapter, I examine vaccination theory, safety and efficacy, politics and public pressure, and offer suggestions on what parents can do, including their options and the possible consequences.

THEORY OF VACCINES

In the Middle Ages, the method of treating an illness by the use of its own products, such as pus or other exudates, was known as "isopathy." Evidence suggests that this practice dates back to the druids in Great Britain. It was based on observation that those exposed to a certain illness either never got the illness or, after having it, never got it again.

Edward Jenner, a barber and podiatrist, is credited with establishing the smallpox vaccination program in 1796. He heard from a farmer that milkmaids suffering from cowpox did not get smallpox. He assumed it might be true for horsepox as well, and started to inoculate people directly from diseased horses. He claimed that once successfully vaccinated, a person would not contract smallpox ever again. Vaccination against smallpox became a popular and common practice during the nineteenth century in Europe; yet smallpox became an epidemic disease in spite of, or perhaps because of, the vaccination programs.[1] Today, smallpox is no longer an active disease—a fact that modern medicine credits to twentieth-century vaccination practices. On the other hand, we do not have the black plague any more either, and there was never a vaccine for that.

Louis Pasteur, a nineteenth-century French chemist, clarified the role of microorganisms in the fermentation of beer, wine, vinegar, and milk products and eventually extended this research to human and animal diseases. Focusing on the role of microorganisms, he concluded that they were the exclusive cause of infectious diseases (this is known as the germ theory). In his view, the condition of the host played no part. From there, it was a simple step to decide that the best way to eliminate diseases was to eliminate the microorganisms. Conversely, his contemporary, the biologist Antoine Béchamp, maintained that the microorganisms were not external but arose from the disease process itself, and that diseases had more to do with the host's condition (this is known as the pleomorphic theory). Therefore, he believed that the way to eliminate disease was to strengthen the body's immune system. Once that was accomplished, the bacteria would simply disappear.

Modern vaccination theory is, at its core, a hybrid of Jenner's and Pasteur's models. It claims that microorganisms cause disease, and the way to avoid a particular disease is to introduce the disease artificially so that the body can set up defenses in the event that the "real" disease shows up later. A healthy immune system creates antibodies in the blood to specific diseases, and these antibodies depend on what we have been exposed to in the environment since birth. Vaccines introduce the microorganisms, either live or inactivated, to provoke the body's production of antibodies to that disease. Vaccination theory posits that a vaccine offers "protection against an infectious disease by inducing a mild attack of the disease beforehand."[2] The theory also assumes that the body will react to the introduction of the disease by setting up a defense mechanism, or immunity, to that same disease *only*; the theory ignores or minimizes other reactions.

Proponents of the vaccination schedule also argue that aggressive vaccination is necessary to maintain "herd immunity." This theory holds that if a high percentage—between eighty-five percent and ninety-six percent, according to the Centers for Disease Control and Prevention (CDC) and the World Health Organization (WHO)—of the population is immunized, they will protect the unimmunized.[3]

This theory ignores many crucial facts: In many outbreaks, a significant portion of those contracting the disease have been vaccinated; for certain vaccines, recipients may shed the virus following vaccination, risking the spread of disease; and vaccination does not prevent individuals from being carriers of a disease although they may remain asymptomatic.

VACCINE INGREDIENTS

We always should read labels and know what is in whatever we eat, drink, or take by mouth or injection. It may give us pause, push us to do more research, or cause us to make a different choice. Yet, most people fail to ask, "What's in the shot?" as the pediatrician injects the child.[4]

A review of the vaccine "excipients" lists, available on the CDC's website,[5] is eye-opening. You can review the information by vaccine or by ingredient. Vaccine ingredients include several types of aluminum, antibiotics, benzethonium chloride, dye, DNA, albumin, bovine calf serum, formaldehyde, gelatin, monosodium glutamate (MSG), polysorbate (20 and 80) and other surfactants, thimerosal (mercury), yeast protein, and urea.

The two with which most people are probably familiar are aluminum, used as an adjuvant to provoke a stronger immune response, and thimerosal, a mercury salt used as part of the production process or added as a preservative to some vaccines, such as flu vaccines. Aluminum and mercury are both potent neurotoxins. It may take a lifetime for the body to deal with these heavy metals. Mercury is a toxic heavy metal, and there is great concern about mercury in dental fillings. In addition, fears of mercury contamination in seafood have been widespread since the 1970s.

While mercury and aluminum may be the best-known ingredients of concern in vaccines, there are many unanswered questions surrounding other ingredients, as well. The potential long-term impact on babies and children of these vaccine ingredients is unknown. The truth is that we do not have any firm longitudinal data about their safety from proper randomized, controlled clinical trials or official epidemiological tracking.

We also must ask questions about the safety of vaccine "active ingredients." Where do scientists find the virus or bacteria necessary to create a vaccine? The primary place that these organisms are found is in the victims of the diseases. Recently, some vaccines have been produced through genetic engineering, a new and unproven modality whose safety will be tested ultimately by time.

The hepatitis B vaccine is produced by genetic engineering. Its pre-licensure testing was short and based on two questionable assumptions: that the older, serum-based hepatitis B vaccine had been used successfully for years and that the genetic-engineering process could not possibly be a risk factor. However, the antigen for the hepatitis B vaccine is produced by recombinant technology in yeast (*saccharomyces cerevisiae*), and hypersensitivity to yeast is therefore a contraindication to vaccination.[6]

Many parents believe that the government tests vaccines rigorously before licensure and marketing. This is not the case. For many vaccines, pre-licensure testing is brief and undertaken by the manufacturer primarily to look at efficacy and at only acute reactions. Testing protocols sometimes follow vaccine recipients for only days or weeks. The manufacturer tests safety, but with an obvious stake in a successful outcome. Vaccines come to market with scant evidence of their safety or efficacy. Instead, only after vaccines are on the market and thousands begin receiving them in routine practice are their effects monitored in what is known as "post-marketing surveillance." We need long-term, controlled studies to ascertain the effects of injecting genetically-engineered products into infants and toddlers. In the interim, children are essentially used as experimental human subjects in this practice.

ARE VACCINES EFFECTIVE? ARE THEY SAFE?

The prevailing wisdom is that vaccines have kept children and others from getting certain diseases. However, some researchers disagree. Thomas McKeown states in *The Modern Rise of Population* that vaccinations were introduced after infectious diseases, such as whooping cough and measles, had declined by about ninety percent.[7] The physician and noted researcher on human longevity Leonard Sagan points out in *The Health of Nations* that the level of generalized resistance to disease may be more important in controlling disease mortality than the eradication of any infectious agent.[8]

EFFECTIVENESS

Vaccination may not be truly effective in keeping illness at bay. In some instances, the opposite may be true. Researchers implicate vaccines as a cause in some epidemics. For example, in the 1870 to 1871 smallpox epidemic in Germany, as many as ninety-six percent of those who had the disease had been vaccinated at least once before they became ill. The same happened in an 1881 smallpox epidemic in Great Britain. Diphtheria increased by thirty to fifty percent in Europe after the introduction of mass, compulsory vaccinations.[9] Measles also went up. For example, before a vaccination program was established, there were 4,056 cases of measles recorded in the Los Angeles County Health Index; after a vaccination program was established, there were 13,912 cases.[10] There were at least eighteen reports of measles outbreaks during the 1980s, all among school populations where 71 percent to 99.8 percent of the children had been fully immunized.[11] In 1994, there was a measles outbreak in Cincinnati, even though eighty percent of the children had had at least three doses of the vaccine.[12] In 2009, the department of health confirmed over two dozen cases of pertussis in Hunterdon County, New Jersey; all the infected children had been vaccinated prior to contracting the disease.[13] In 2010, there was a mumps outbreak in New York City, primarily among Orthodox Jews. According to the CDC, most of those who contracted mumps had been vaccinated against the disease. The CDC conceded that the "mumps portion of the vaccine is less effective than the other parts" of the MMR.[14]

Do vaccines really prevent epidemics? European authorities at the end of the nineteenth century decided they did not, so they began to rely more on isolation and improved hygiene to combat disease. These methods met with surprising success and led to the decline in the incidence of smallpox. Smallpox incidence declined everywhere, among vaccinated and unvaccinated populations.[15] After the WHO's final smallpox vaccination effort in 1978, mandates ended in the 1980s. People credit vaccines with the eradication of this once-feared disease. On the other hand, smallpox became epidemic after, not before, vaccination began. Did the illness lead to the vaccine, which, in turn, increased the illness that increased vaccination? Did the illness decrease because vaccination decreased? The smallpox virus supposedly does not exist anymore; however, there are new viruses associated with pox diseases, such as monkeypox and white pox, which are indistinguishable from the variola (smallpox) virus.[16]

Dr. Robert Mendelsohn, the late pediatrician and author, was adamantly opposed to mass vaccination. He noted that while we attribute the end of polio epidemics in the 1940s and 1950s to vaccination programs, polio ended simultaneously in Europe where people did not use the vaccine extensively.[17] Diseases are known to be cyclical, and some, such as bubonic plague, and scarlet fever, have disappeared without any vaccinations whatsoever. This fact weakens the claims about vaccines' effectiveness in preventing epidemics. When epidemics do not appear, the preventive factor could have been the vaccine, or increased hygiene, or better food supplies, or simply just the times.

We cannot prove why something does not happen. If a child gets a polio shot and does not get polio, there is no way to prove that he or she would have or would not have gotten polio in the absence of the vaccine. We do not know whether it was the child's immune strength, herd immunity from vaccination, or just good luck. Believing that one child did not get polio because of a vaccine is simply an assumption, as there are other children who did not get the vaccine who did not get polio either.

Vaccines do not necessarily lower overall mortality rates. Dr. Sagan notes that, even though smallpox deaths in the early 1800s fell after the introduction of vaccination, "the overall death rate was relatively unaffected as deaths from gastrointestinal and other diseases subsequently rose. One cause of death was replaced by another."[18]

Dr. McKeown flatly states that the contribution of medical intervention to the prevention of sickness and premature death was minimal. Of much higher impact was the increase in the available food supply leading to improved nutrition, as well as "the purification of water, efficient sewage disposal, and improved food hygiene, particularly the pasteurization of milk, the item in the diet most likely to spread disease." He notes that, according to the WHO, the best inoculation against common infectious diseases is an adequate diet.[19] Dr. Ignaz Semmelweis,[20] best known for the eponymous Semmelweis effect,[21] would surely add that the simple act of thorough and consistent hand washing can drastically cut the spread of disease.

What are the long-term effects of vaccination on the possibility of epidemics? It may turn out that in the long run, diseases become more severe and widespread. For example, Dr. David L. Levy created a computer model to forecast the effects of a measles elimination program on the number of people susceptible to the disease. He found that in the pre-vaccine era, approximately ten percent of the population was susceptible to measles, most of whom were children under ten. According to the computer model, with the institution of the immunization program, the proportion of susceptible people fell at first, to 3.1 percent in 1981, but then began to rise about 0.1 percent per year until it reached 10.9 percent in 2050. Levy predicts that, at that time, susceptibility will range across the entire population. He concludes that long-range projections indicate that the proportion of people susceptible to measles may eventually be greater than in the pre-vaccine era.[22] We may already have an example of this effect. The WHO instituted a polio eradication program in Oman, achieving a ninety-eight percent vaccination rate. Six months later, the community experienced a massive polio epidemic.[23]

Biologist Robert May postulated that, according to chaos theory, the cycles of epidemics will change with the introduction of a vaccination program. Chaos theory, also known as "nonlinear dynamics," looks at humans and other biological systems as dynamic, complex systems that move continuously over time. Therefore, linear predictions of behavior rarely bear out. In the public health field, once vaccine programs are implemented, instead of changing smoothly downward, as is the expectation of vaccination theory, May predicted that the downward trend would be dotted with large peaks of the same illness.

He mentioned that data from campaigns such as the one to eliminate rubella in Great Britain show precisely such peaks.[24] There is also strong evidence that the incidence of shingles, a painful disease generally associated with the elderly and caused by a reactivation of the same virus that causes chickenpox, is increasing, perhaps due to people's inability, as a result of the chickenpox vaccination, to develop antibodies to the virus because of exposure to a different but related strain of wild chickenpox in the air. Moreover, shingles is now even seen in children who have been vaccinated against chickenpox. In other words, vaccine programs often do not eliminate a disease; they simply rearrange its occurrence, causing it to mutate, or rise in related areas. Rather than recognizing the frightening implications of the mutating disease strains, vaccine developers have taken this "opportunity" to develop new vaccines to address new strains (e.g., new pediatric pneumococcal vaccines for thirteen strains instead of seven) or to vaccinate against diseases previously not vaccinated against (e.g., shingles). However, the vaccine developers will never have the upper hand on diseases; illnesses always will be one step ahead. New vaccine strains are just an invitation for viruses to mutate. In short, manufacturers are creating their own market for new vaccines.

POST-VACCINATION ENCEPHALITIS

According to medical historian Harris Coulter, PhD, the occurrence of encephalitis, or brain inflammation, after vaccination is much more common than we think. Different vaccines (i.e., measles, polio, and rabies) have been shown upon autopsy to produce both myelin destruction and encephalitis.[25] Myelin is the sheath that covers and protects the nerves; it is inexplicably destroyed in diseases such as multiple sclerosis and adrenoleuko-dystrophy[26] (ADL). Autopsies show that as a result of post-vaccination encephalitis, there are "many small yellowish-red lesions in the white matter of the cerebrum, cerebellum, brain stem, and spinal cord," indicating destruction of the myelin sheaths.[27]

Encephalitis, among other causes, may come from an allergic reaction to foreign protein. Vaccines may provide some of these damaging proteins, because they contain traces of monkey kidney cells and chick embryo cell cultures, as well as casein, or milk protein. The presence of these foreign proteins injected in the blood may sensitize the child to develop allergic reactions that may later manifest as allergies.

How would the foreign proteins get into the brain? The body protects the brain very carefully from unregulated fluctuations of elements in the blood (such as hormones, amino acids, mineral ions, and others) with the blood-brain barrier. Endothelial cells inside the capillaries that bring blood to the brain establish this barrier. The cells are so tightly packed that only certain necessary nutrients, such as glucose, are able to pass through the capillary walls into the brain. Experiments have shown that a strong solution of sugar in the blood loosens the tightly packed endothelial cells and thereby increases the permeability of the blood-brain barrier. Interestingly, once the sugar concentration disappears, the barrier returns to normal.[28]

High-pitched crying, inconsolable screaming, head banging, rocking, seizures, convulsions, and staring episodes (which may be absence seizures) are all indications of brain inflammation. The aftermath of brain inflammation may include speech, hearing, and vision difficulties; hyperactivity; retardation; learning disabilities; dyslexia; antisocial behavior; personality disorders; and "flat affect," or the inability to respond to emotion. Disturbingly, there has been an increased incidence of children having encephalitis after having a childhood disease and after receiving a vaccination against that childhood disease, for many decades.[29] Anna-Lisa Annell pointed out in 1953 that common childhood diseases showed an increased tendency to attack the nervous system since the 1920s, when only rare individual cases affected the nervous system.[30] Coulter blamed the increase on "the emergence of an allergic dimension that had not been there before." That is, there are more neurological complications with both diseases and vaccinations because a new element has been introduced that has rendered children more susceptible to allergic reactions. This new element, Coulter postulated, is the increased number of vaccinations children receive.

We really do not have enough data on what the subtle long-term effects of vaccinations are. We cannot predict which children will respond with serious reactions and which will not. As the title of the 1982 documentary so aptly identified, "vaccine roulette"[31] may be a dangerous game.

GENERATIONAL EFFECT

Can vaccines damage DNA? Is vaccine damage cumulative? Can it be inherited? Dr. Robert Mendelsohn called immunization "a medical time bomb" and urged parents to reject all vaccinations for their children. He stated that there is no convincing scientific evidence that mass inoculation eliminated any disease. He further stated that there are significant risks and numerous contraindications that make giving vaccines dangerous. He believed that the mass vaccination program might be at the root of the increase in autoimmune diseases in this century.[32] Harold E. Buttram, MD, predicted in 1982 that "there may be large scale immune malfunction following current childhood vaccination programs . . . a weakening which may be cumulative from one generation to the next."[33]

Live or attenuated viral vaccines may be particularly dangerous, according to Professor Richard DeLong, a retired microbiologist.[34] He has been warning the medical community since 1960 of the dangers of live or attenuated viral vaccines. These vaccines can cause mutations, chromosomal aberrations, birth defects, cancer, and new diseases; they may revert to virulence; and they may be contaminated with other viruses and other microbes.[35] Professor DeLong emphasizes that live viral vaccines may originate new diseases in four different ways:

1. Gene recombination occurring among two or more different viruses infecting the same cell;
2. Adventitious viruses in the live vaccine that come from the cells that were used to produce the vaccine;
3. The vaccine viruses having gene changes occurring during reproduction in the *in vitro* phase of vaccine production; and
4. The vaccine viruses having gene changes during *in vivo* reproduction in the vaccine recipient.

Mass vaccination increases the probability of all these things. He considers attenuated and live vaccine viruses nothing less than uncontrolled, genetic manipulation, which may be transmitted from one generation to another.[36]

Professor DeLong's theory is receiving further support from information that came to light in the 1990s concerning disturbing aspects of the polio vaccination program. We have known since the 1950s that the polio vaccine, grown on kidney cells of the African green monkey, was contaminated with a monkey virus named simian virus 40 (SV40) that had cancer-causing effects in hamsters. Microbiologist Howard Urnovitz has suggested that the human immunodeficiency virus (HIV) could have originated from the contaminated polio vaccines through recombination with human genes.[37]

In short, instead of eradicating childhood illnesses with vaccines, we may have simply shifted their timing and virulence. We used to have a population that generally contracted mild versions of childhood diseases when young and then became naturally immune for life. Now we have a growing worldwide population that is artificially immune for a short time, and is becoming susceptible to diseases that were mild in children but that pose dangers to adults. The population may also be at risk for a host of new viral and immune system diseases that did not exist before the vaccination era. We cannot assume that these events are unrelated.

WHO IS AT RISK FOR ADVERSE EFFECTS?

Some children react violently to vaccines and others do not. Could there be a connection for some children between nutritional status and the likelihood of an adverse reaction? Certainly, the status of maternal nutrition and the type of infant feeding (breast or bottle, home-cooked food or commercial and fast food) play a major role in the basic immunocompetence of human beings. For example, in Australia, it was found that malnourished aboriginal children who received routine immunizations had a death rate as high as 500 per 1,000. Once their nutritional status was supplemented with vitamin C, their reactions to the vaccines diminished and the death rate went down.[38] We should consider carefully what this hypothesis means for our population.

What are the variables that would allow vaccines to cause neurological damage and encephalitis? According to the natural healing model that I follow, the following list names the worst risk factors because they weaken the entire system:

- Parents' malnutrition and use of drugs and alcohol
- Mothers' weak nutritional status
- Bottle feeding
- High intake of milk products
- High reliance on commercially processed foods
- High intake of sugared snacks and juices
- A lack of fresh vegetables and whole grains
- Insufficient good-quality protein foods

The more of these factors that are present in a child's life, the higher the risk would be of susceptibility to adverse reactions. Additional risk factors include a personal or family history of allergies, convulsions, seizures, food or mold sensitivities, or any condition that has compromised the immune system.

To prevent adverse reactions, it is important to be aware of contraindications. Vaccine manufacturers, the American Academy of Pediatrics (AAP), and the CDC's Advisory Committee on Immunization Practices, and individual pediatricians establish contraindications. I suggest paying special attention to the manufacturers' warnings and contraindications listed on the package inserts for vaccines, as they will be the most detailed. It is reasonable and certainly within a parent's legal right to firmly refuse vaccines on medical grounds when manufacturers' contraindications are present.

Many vaccinated children do not show obvious adverse effects. Even if no overt reactions appear, I must assume that vaccines seriously stress the immune system. I believe that there have to be consequences when, over the first two years of life, a child is injected with many viruses and viral particles, plus their respective preservatives and adjuvants. By definition, vaccines make a healthy child sick.

We do not know if or how the vaccines influence illnesses later in life. According to the chaos theory, human beings are complex and dynamic systems that show a "sensitive dependence on initial conditions."[39] Therefore, vaccines administered early in life present a small variation, which could spread out into a considerable change later in life. As nature makes no value judgments, this change could be for good or ill in human terms. We do not know if vaccinated people would have contracted the diseases in the absence of vaccination. One cannot prove why something does not happen. The only way we can attempt to demonstrate the efficacy of these practices is through rigorous, unbiased scientific research.

Parents who decide to immunize should make sure their children are well nourished and should review all manufacturer contraindications for each specific vaccine to minimize the risk of injuries. Careful pediatricians would follow this same protocol.

PUBLIC PRESSURE AND VACCINATION CHOICE

Vaccination choice is a thorny issue. Vaccination is demanded by schools, day care centers, doctors, and even other parents, including friends and neighbors. Regardless of their position on vaccines, all parents ultimately have a common goal: healthy and happy children. Because they have been told so by the public health system, many parents believe that vaccinations help achieve this goal. They follow the systems set up by their culture and society. Often these systems work; sometimes they do not. They surely are not working when well-meaning parents follow our prescribed healthcare model scrupulously and then find themselves with a neurologically impaired child who requires lifelong care. This is a tragedy.

The marketing technique used most extensively to promote vaccination is fear. Both the media and the health establishment whip up irrational fears of disease in the population, so that people buy the products recommended without hesitation. One excellent book on this subject, debunking the fear of viruses as well as the practice of vaccination, is *Fear of the Invisible*, by investigative reporter Janine Roberts (Impact Investigative Media Productions, Bristol, U.K., 2008). For example, she uncovers that evidence had been presented to the U.S. Congress that linked summer polio epidemics to the common use of heavy metal or organophosphate pesticides in the summer. If that is true—and it possibly is, as these pesticides work by irreversibly inactivating enzymes essential to nerve function in insects as well as in humans—this dreaded illness may have had more to do with pesticides than the polio virus.

RISKS OF THE DISEASE VERSUS RISKS OF THE VACCINE

Is the risk posed by the disease higher or lower than the risk posed by the vaccine? This is a difficult question to answer, because there are risks on both sides. Vaccine promoters downplay the risks of adverse events—both their frequency and their severity. The risks of vaccination, however, have been shown to be real, and should not be ignored.

For many years, it has been known that the statistics concerning exposure to "childhood" disease risks are questionable. One symposium addressed the issue of "Immunization, Benefit Versus Risk Factors." W. Schumacher, one of the presenters, stated the following:

> Inoculations are applied to healthy persons in order to protect them from a disease to whose risk of infection the inoculated subject will perhaps never be exposed. The risk is a mere statistical value. As no medical treatment and no inoculation are free from the risk of undesired side effects, one must judge in every case whether the risk of catching the disease is considerably higher than the risk of vaccination side effects . . . The subject affected by an inoculation has, without doubt, made a special sacrifice in the interest of the general public.[40]

According to a leading medical ethicist,[41] if we are to consider the "best interest of the child," healthy children should not have to be vaccinated to protect others. "We don't require adults to serve as kidney donors or even blood donors," so it is discriminatory to require children to be "splendid Samaritans . . ." for the health of others.[42]

It is important to understand the facts about the childhood illnesses against which we vaccinate. Measles, mumps, chickenpox, and whooping cough were in decline well before vaccines became available.[43] These decreases were largely attributable to improvements in nutrition, hygiene, sanitation, the advent of running potable water, the refrigeration of food, and the invention and increased popularity of indoor flushing toilets. In addition, we have many more treatment options today than we had at the beginning of the century when these diseases were common, including countless drugs and antibiotics, as well as "alternative" medicine such as homeopathy, herbs, dietary management, and acupuncture. In countries with good public hygiene and access to healthcare, these diseases are rarely fatal. In fact, most childhood diseases do not have as many adverse effects as the vaccines for those diseases. While there is the risk of severe injury or death for any disease, millions worldwide have survived measles and mumps without any lasting damage and at the same time have acquired lifelong immunity.

There are several diseases that are more serious, such as diphtheria, polio, tetanus, and hepatitis B. However, good hygiene, good nutrition (especially a diet low in sugar and milk products, high in vegetables and whole grains, and providing sufficient protein), healthy vitamin D levels, and clean living conditions can go far in preventing those illnesses.

Since at least 1979, polio has been largely an iatrogenic illness: most of the cases since then have been vaccine-induced, a condition called vaccine-associated paralytic polio or VAPP.[44] Tetanus and hepatitis B are not contagious childhood illnesses. Tetanus is very rare in the United States and affects more adults than young children.[45] It may be a danger if one steps barefoot on a rusty nail in a horse stable and has a weak resistance to disease. This fails to justify the insistence of most emergency medical departments to administer tetanus shots to everyone without any wound at all. The best prevention for this disease is proper wound hygiene, including allowing the wound to bleed (for cleansing) and to be in contact with the air (the tetanus microorganism is anaerobic; that is, it develops in the absence of air). Tetanus is treatable with antibiotics.[46]

Since the early 1990s, most newborns in the United States receive their first dose of hepatitis B vaccine before they leave the hospital. The reason for this, which is the same reason for vaccinating at well-child visits, is simply because the babies are there, within reach. The vaccine is useless yet poses real risks for the vast majority of infants (an exception might be those born to hepatitis B surface-antigen positive mothers). Hepatitis B is a disease communicable only by the exchange of bodily fluids; it is rarely contracted before the age of sexual maturity. Moreover, vaccinating newborns against hepatitis B does not necessarily protect them later. Because immunity from the vaccine fades, children receiving the vaccine at birth may need boosters before they become teens and are actually at risk for the disease. Outside the United States, most countries with a low prevalence of chronic hepatitis B virus infection do not administer a dose of the hepatitis B vaccine at birth.[47]

THE REMEDY MAY BE WORSE THAN THE DISEASE

Those who promote vaccinations make the major assumption that infectious diseases are a bad thing and should be avoided at all costs. The system we have in place for avoiding a disease actually provokes the disease in mild form; it does not avoid it. In other words, a person who receives a vaccine against measles gets the measles, but usually without the visible symptoms (i.e., the rash). The medical assumption is that this is preferable to getting the real thing, and that the antibodies created by the vaccine exposure to the disease will help protect the vaccinated against the disease in the event of later exposure.

Very few studies have compared "real" disease with vaccine-induced disease. With respect to the measles virus, however, a Danish study showed a marked relationship between the presence of measles antibodies, such as those elicited by the measles vaccine, the absence of a history of an actual measles rash, and certain serious illnesses in adulthood.[48] These diseases, manifested in adulthood, included immune diseases, skin diseases, bone and cartilage disorders, arthritis, lupus, connective tissue disease, Crohn's disease, lung disease, nerve paralysis, and various tumors. The author states, "It is assumed that individuals who are at risk of developing non-measles associated disease are those who have been infected with measles virus, but who never manifested a rash." These diseases were not considered adverse effects of a vaccine, but they were more frequent in the subjects that had not had the measles rash.

Is getting measles naturally, with the rash, so bad? Naturopathic doctors, especially those who follow the anthroposophic philosophy of Rudolf Steiner,[49] believe it is not. Some believe that, through these diseases, the body cleanses itself of unnecessary matter.[50] In fact, when children get minor diseases early in life, they are considered a form of natural vaccination, because they prevent the same disease from reappearing later.[51] Doctors have long known that, when a childhood illness is acquired naturally, its immunity lasts for life. This is particularly important because childhood diseases such as chickenpox, measles, and mumps are much more severe in adulthood. When allowed to vanquish the disease on its own, a child's immune system will have a sort of "workout." Just as for all other body systems, the immune system is strengthened by use and repeated exposure. It will get stronger after each bout of a minor illness. Each illness a child overcomes naturally provides training for the immune system and increases its effectiveness later in life.

The natural healing model posits that the proper way to deal with childhood diseases is to allow a disease to complete its course. Fasting, rest, and natural remedies help this process. In some instances, using drug treatment to "stop" the symptoms of a disease interferes with the immune system and causes the illness to get worse, or leads to other illnesses.

I believe strongly that vaccines are worse than natural disease. The possibility of catching a natural disease is contingent upon the body's resistance and strength of its immune system; we can be exposed to an illness and not catch it because of a strong immune system that gives us resistance. With a vaccine, a pathogen is injected directly into the bloodstream and, because we cannot predict what our reactions may be, there is no foolproof way to protect ourselves or our children against any possible adverse effects. These risks may or may not be statistically small, but they are risks nevertheless, and some of them are serious, especially those for potential neurological damage. Frankly, we don't really know exactly how often adverse events occur, because there is no active tracking system for adverse effects. If properly informed, I doubt there are many parents who would willingly subject their children to the risks of neurological damage, no matter how small they are. I certainly would prefer that my children get measles or whooping cough naturally and thereby probably develop lifelong immunity, rather than be exposed to the threat of brain damage, convulsions, Guillain-Barré syndrome, juvenile diabetes, uncontrollable seizures, and other serious adverse reactions of a vaccine.

UNVACCINATED CHILDREN

How is the health of unvaccinated children? Do they die of childhood diseases in droves? The answer is no, they don't. There is a general notion that unvaccinated children get sick more often. Indeed, it may be the opposite.

When I interviewed anthroposophic physician Dr. Philip Incao in the 1990s, he treated about 500 children in his Harlemville, New York practice,[52] and a large number of them were not immunized. I inquired about the health of these unimmunized children and asked if he saw more or less incidence of asthma, allergies, hyperactivity, learning disabilities, and seizures among them. He said he saw much less—almost none—and if he had the money to do a study of the patient populations of the several different anthroposophical doctors in the country, he felt sure there would be far fewer of those problems among them as well.[53]

Serious epidemiological study of unvaccinated people would be highly beneficial. There are pockets of them here and there—among the anthroposophists, the Christian Scientists, and similar groups. A most interesting field study was done by a journalist, Dan Olmsted, who went looking for autism cases among the Amish, many of whom do not vaccinate their children. Statistically, he should have found at least 50+ in the population size he was studying; instead, he found three cases—an adopted girl from China, and two children who were local and had been vaccinated.[54]

Among those who favor vaccinations, the thought of unvaccinated children in their midst stirs up fears usually reserved for contagious epidemics. One colleague told me that when she tried to put her daughter in a playgroup and the other parents discovered that her child was unvaccinated, they asked her to remove her child. "I felt for the first time what it is like to be discriminated against," she said. The thought that unvaccinated children pose a danger to the vaccinated population is totally irrational if the vaccines are really effective. Allowing an unvaccinated population to exist next to a vaccinated one is not dangerous to the latter; it may be dangerous to the former, as in the cases where adults contracted paralytic polio from recently vaccinated infants.[55] When there is a disease "outbreak" it is always assumed that it was started by an unvaccinated person. But unless "patient zero" is positively identified as such, it is always possible that a vaccinated person started the outbreak.

If unvaccinated people get a disease naturally, they can self-quarantine. Another colleague mentioned to me that his five unvaccinated children were at one time exposed to measles. The measles vaccine offers no protection after one has been exposed, so when the school authorities gave them the choice of vaccines or quarantine, he chose the latter. All of his children had a mild case of measles and overcame it easily. "They were sick and unhappy," he says now, "but in the long run I believe their immune systems got stronger."[56] Moreover, they are immune to measles for life.

If vaccination programs were halted, the major concern of those who promote vaccines is that we would face epidemics of childhood diseases. However, such a scenario is unlikely. Current data show that these diseases appear among the vaccinated populations anyway.[57] In some parts of the country, a high proportion of children may be unimmunized or partially immunized. Most adults are not routinely vaccinated and their childhood vaccines have likely worn off. Where, then, are these epidemics? Herd immunity cannot explain these inconsistencies.

POLITICS AND PUBLIC PRESSURE

Public health authorities promote the idea that vaccinating is what *good* parents and *good* physicians do; therefore, if you do not vaccinate, you are probably a *bad* or *neglectful* parent. In my experience, however, parents who do not vaccinate tend to be much more health-conscious and mindful than average parents. The irony is that the social healthcare system is stacked against them. There have been cases in which a parent's conscious refusal to vaccinate, for health reasons, has been used against him or her in custody disputes. There are numerous cases in which the parents have refused to vaccinate, for any reason, and then health authorities have removed the children from the legal custody of their parents and, under custody of the courts, vaccinated them, alleging child abuse or medical neglect. Then the children are returned to the parents, who are left to deal with adverse effects, if there are any. (See chapter 20 for more information on vaccination and forced child removal.)

OFFICIAL INFORMATION

In April 1992, the U.S. Department of Health and Human Services (HHS) began distributing a series of informative booklets concerning vaccines to pediatricians and public health clinics. While the booklets insisted that the benefits of vaccines outweighed the risks, they also clearly identified the potential adverse effects and encouraged parents to communicate any adverse effects to their doctors. Parents were required to read the booklets carefully and sign a form before submitting their children to vaccination.[58] This procedure implied that parents could choose not to vaccinate their children after receiving information about the potential adverse effects. Doctors complained that it took too long for patients to read the booklets and that parents were asking "too many" questions.[59] Doctors also thought that excessive emphasis on the adverse effects would discourage parents from complying with the vaccination schedule. As a result, by 1995 the CDC replaced the booklets with a single sheet for each vaccine, offering a more upbeat tone about the benefits of vaccination, and including cute drawings of children. These sheets that are used today, usually entitled, "What you need to know," provide incomplete risk information, minimize adverse effects, and require no signature. Parents simply do not receive enough information to give truly informed consent for the vaccination of their children.

PEDIATRICIANS AND HEALTHCARE WORKERS

Imposing healthcare upon people against their will is dictatorial. It is reasonable to expect that our democratic political system would allow individuals to make healthcare choices for themselves and their children based on free and informed consent. Even if vaccinations are supposed to protect the public from epidemics and disease, the acknowledgement that this protection entails the sacrifice of some members of the public should give us pause.

If parents do not wish to vaccinate, and they voice their concerns to a pediatrician, one of the following scenarios will happen:

1. The doctor, who may agree or not, will believe the parents have a right to choose what type of medical care they want for their child and will encourage the parents to make choices in accordance with their conscience.
2. The doctor, who believes immunization is not a choice, will protest strongly and may even refuse the child as a patient. In extreme cases, the doctor also may alert authorities and potentially raise concerns about the parents' failure to vaccinate as evidence of child neglect.

Parents are entitled to use legal exemptions to opt out of vaccination mandates. Depending on the state, parents may be able to obtain religious, medical, or philosophical exemptions. Parents should also inquire about the possibility of "documented immunity," whereby a lab test can demonstrate proof of immunity (in the form of antibody titers) showing the vaccination is therefore not required. The difficulty with which parents can obtain exemptions varies from state to state.

Obtaining natural immunity through disease must be an allowable choice. Society has developed the strange notion that getting sick with a disease for which there is a vaccine should be punishable by law, fine, or child removal for parental neglect. This idea is dangerous. Physicians of earlier times noted that after a bout of illness, children entered growth spurts and became healthier. In any case, there are available treatments for most diseases against which we vaccinate. For example, the administration of high doses of vitamin A, which will alter the course and severity of measles, is used for the treatment of measles in developing countries. Deciding not to vaccinate or to vaccinate selectively must be legitimate choices in a society that respects human rights.

WHAT TO DO? THE PARENTS' CHOICE

Good health is not always due to good luck: It is very often the result of our choices, even if we are not aware of which choices caused what. There are many educated, well-informed parents who believe that the best and safest way to keep their children healthy is to breastfeed them, to give them natural, organic food with plenty of vegetables, and to avoid drugs and vaccines.

Parents who elect not to vaccinate or to vaccinate selectively need to make their own choices about how much they will discuss the issue with others, in light of the possible social repercussions. It should be noted that parents who decide not to vaccinate their children generally do so as a conscious choice, after much research and soul-searching. These parents often give one or more of the following reasons for refusing vaccines.

1. **Religious arguments.** They belong to a religious or spiritual group that is opposed to the introduction of foreign substances such as drugs into the human body (e.g., Christian Scientists), as the body is considered holy and the temple of the soul. Many parents also have a strong, valid, personal spiritual belief system, which may or may not be officially sanctioned by an official religious group.

2. **Health-related arguments.** One or more of their children suffered from the adverse effects of a vaccine, and they do not wish to risk further damage to the health of this child, or to risk harming their other children. They may also have seen firsthand incidents of vaccine-induced injury in other children.

3. **Philosophical and health theory-based arguments.** While Western medicine promotes vaccination, there are other healing systems, such as the naturopathic or anthroposophic, which believe it is better for the immune system to fight off the real illness naturally, rather than to deal with vaccines' inactivated or live viruses and toxoids. These health systems assert that vaccination causes more damage than it prevents.

4. **Research-based arguments.** After researching and studying popular and scientific literature, especially the many papers on adverse effects, a number of parents simply conclude that childhood diseases are rare, the related vaccines carry too many serious risks, and the theory behind their widespread use does not hold up. They decide to avoid the risk of lifelong seizures and brain damage, and accept the risks of childhood diseases, which, when treated with natural remedies, fasting, and rest, usually run their course.

FACING THE ISSUES

As the number of parents voicing concerns about vaccine safety and policy increases, and as more parents refuse to vaccinate their children in accordance with federal recommendations and state mandates, the rhetoric of vaccine promoters escalates as well. I recommend the following course of action when facing these issues.

1. **Become informed.** Read the scientific literature and books questioning vaccines, talk to parents, and ask your healthcare practitioners.
2. **Make a clear choice.** You can change your mind later on if new information comes your way. The choices are:

    - Vaccinate
    - Vaccinate partially or later than recommended
    - Do not vaccinate

3. **Learn the law.** Know the operative vaccination or immunization laws in your state. Study them carefully.
4. **Act according to your principles, legally.** Find the mode of action that allows you to follow your choice within the law. If you are just trying to take care of your family in the best way you know how and if your community follows the medical system and you don't, the prudent approach is to not argue about it—just do what you want to do. If you choose to become an activist, then it is worth the effort to speak up and to be willing to accept the consequences.
5. **Demand your right to informed consent.** Under the doctrine of informed consent, adults are allowed to refuse vaccination. On the other hand, parents' right to make the same choice for their own children is severely constrained by state laws and school regulations. The doctrine of informed consent legally requires physicians to explain the following points, among other things.

    - The benefits of the recommended treatment or procedure;
    - The risks associated with the recommended treatment or procedure; and
    - Alternatives to the treatment or procedure.

Health consumers have a legal right to make choices regarding their health and medical care needs. Informed consent to what type of healthcare to accept, including vaccination, is a human right.

A FINAL VIEW

Regardless of what we do, children do get sick from time to time. Even those who promote vaccination will admit that it does not protect against all sickness. The choice we have is between the risk of man-made, iatrogenic sickness, and the risk of natural disease. Let us be very clear that this is our own, personal choice—and not the choice of the government or of healthcare practitioners.

Chapter Twenty-Seven

WHAT SHOULD PARENTS DO?

Louise Kuo Habakus, MA

"Who shall decide when doctors disagree?"
—Alexander Pope, eighteenth century English poet

WHERE CAN PARENTS TURN FOR MEDICAL ADVICE REGARDING VACCINATION?

Many parents are deeply concerned about vaccine safety. The Center for Personal Rights (CPR) upholds the rights of individuals to make vaccination decisions with their healthcare practitioners.

The following list illustrates the broad spectrum of medical perspectives on vaccination. It starts, at one end, with a recommendation to give all scheduled vaccinations, and concludes, at the other end, with a world-famous pediatrician's argument against vaccination. Many inquiring parents find it helpful to review a wide range of views. We have gathered and summarized the views of nine board-certified pediatricians and other licensed professionals. These views concur with and differ from the government's recommendations. Many of these professionals support vaccination choice and raise important questions and concerns about Centers for Disease Control and Prevention (CDC)-recommended vaccination protocols. A medical association or medical doctor authorized to give medical advice has published each of the following perspectives.

1. The CDC and AAP Childhood Vaccination Schedule

Most doctors follow the CDC's recommended vaccination schedule for children. The American Academy of Pediatrics (AAP) urges its 60,000 member pediatricians and pediatric specialists to endorse and adhere to this schedule:

In 2008, the AAP published a letter in its official monthly newsmagazine, *AAP News*, from All Star Pediatrics, a member pediatric practice in Exton, Pennsylvania. Excerpts from this letter echo the medical advice the CDC and AAP offer. The following are excerpts from its press release dated October 2010:

> We firmly believe that all children and young adults should receive all of the recommended vaccines according to the schedule published by the Centers for Disease Control and the American Academy of Pediatrics. . . .
>
> Please be advised . . . that delaying or "breaking up the vaccines" to give one or two at a time over two or more visits goes against expert recommendations, and can put your child at risk for serious illness (or even death). . . .
>
> Finally, if you should absolutely refuse to vaccinate your child despite all our efforts, we will ask you to find another healthcare provider who shares your views. We do not keep a list of such providers, nor would we recommend any such physician. . . . [B]y not vaccinating you are putting your child at unnecessary risk for life-threatening illness and disability, and even death.[2]

Immunization providers and departments of health receive significant financial support from CDC-funded programs at the state and local level. Through its Section 317 Immunization Program, the CDC supports the "activities needed to assure vaccination of a population."[3] These include the "integration of new vaccines into routine medical care" and "strategies to increase vaccination coverage rates."[4] Section 317 funding has exceeded $500 million annually over the past five years. The American Recovery and Reinvestment Act of 2009 provided an *additional* one-time appropriation of $300 million that "provides a historic opportunity to make the benefits of vaccination available to more Americans by providing a 40 percent increase in funding over the two-year funding period."[5]

The AAP is similarly concerned with maintaining high vaccination rates. The following are excerpts from their press release dated October 2010:

> Deadly diseases are making a comeback. . . .
>
> At the heart of all these outbreaks: Parents who have chosen not to vaccinate their children. . . .

> "We are at a crossroads," said Dr. [Paul] Offit. "If population immunity continues to break down, the door will open wider to diseases that we thought had been vanquished by vaccines."[6]

In 1982, the AAP rejected a resolution of its members to give comprehensive information to parents about vaccine risks and benefits.[7] In 1993, the AAP led the successful effort to substantially reduce the amount of information that doctors were required to give patients under the 1986 National Childhood Vaccine Injury Act.[8] Public health and mainstream medicine prioritize vaccine compliance above the individual's right to free and informed consent.

2. Vaccines: What You Should Know

Paul A. Offit, MD, FAAP and Louis M. Bell, MD, FAAP (John Wiley & Sons, Inc., 2003)

Drs. Offit and Bell are pro-vaccine medical doctors who believe that individuals should follow the recommendations of the experts, namely, the CDC and AAP. Offit and Bell are board-certified pediatricians. This revised edition, which updates a 1999 book by the authors, covers vaccines for children and adults. In simple language, it addresses the need for vaccines, how they work, how they are made, safety, and who recommends them. It also covers special situations, including vaccines for travelers, bioterrorist agents, rabies, Lyme disease, and AIDS. There is a chapter for each vaccine, with a discussion of the disease and its incidence, the vaccine and its side effects, and the AAP's recommendation for number of doses and the ages at which they should be administered. In a chapter on "Practical Tips," they offer brief observations regarding the fear of shots, who shouldn't get shots, vaccinating children who are ill or who have allergies, premature babies, giving vaccines simultaneously, missed vaccines, and vaccinating children who are adopted from other countries.

Offit and Bell state no concerns about death or severe injuries from vaccines given to children and nearly all vaccines given to adults. They make bold statements about the safety and absence of adverse events of many vaccines.[9] The authors explain that vaccines are not "completely free of any possible negative effects." However, they list many other "gambles" individuals unconsciously make every day, including bathing, eating, being struck by lightning, and even contamination from money which may harbor bacteria and viruses.[10] When asked if vaccines are safe, they believe the more relevant question is whether the benefits outweigh the risks. Their answer is a resounding yes.[11]

The authors support the use of multiple vaccines in one visit: "All routinely recommended vaccines can be given simultaneously. There is no evidence that giving one vaccine significantly interferes with the immunity caused by another. Nor is there any evidence that any of the vaccines increases the rate of side effects of another."[12] They welcome the arrival of new combination vaccines that make it easier for children to get vaccines without getting so many shots: "The good news is that help is on the way. Several companies are now working together to combine vaccines."[13] (See the appendix to this chapter for a list of combination vaccines available today, plus a discussion of issues related to their safety and efficacy.)

> The authors also maintain that "it is safe to give all of the recommended vaccines to children with "minor illnesses." Minor illnesses include low-grade fever, ear infections, cough, runny nose, diarrhea, or vomiting. Several studies "found [these children] were not at greater risk of side effects and had similar immune response after immunization as children without minor illnesses."[14]

In the "Common Concerns about Vaccines" chapter, the authors give the background for Dr. Offit's public statement that a child could safely receive 100,000 vaccines at once. They offer the theory of two immunologists from the Salk Institute in San Diego:[15]

> The number of microorganisms to which a body can respond depends on the number of cells in blood that can make antibodies sufficient to recognize all the relevant parts of the microorganism. . . . the number of microorganisms to which one responds depends on size. [They] estimated that elephants can produce immunity to about a hundred times more microorganisms than humans... The scientists estimated that even young infants could respond to about 100,000 different organisms at one time. Therefore, the eleven vaccines required for all children will use up only about 0.01 percent of the immunity that is available.

Offit and Bell acknowledge that Americans are entitled to refuse vaccines by law but question the wisdom of exemptions: "One could argue that an individual's rights should not include the right to catch and spread contagious and potentially fatal diseases."[16] They offer two examples of outbreaks that occurred within communities that refused vaccinations on religious grounds.

3. *The Vaccine Book: Making the Right Decision for Your Child*

Robert W. Sears, MD, FAAP (Little, Brown and Company, 2007)

Dr. Sears is a pro-vaccine medical doctor who endorses vaccination choice. He is board-certified in pediatrics. This book is for parents. In his words, "vaccination isn't an all-or-nothing decision" and he urges doctors to give medical care to children whose parents choose all, some, or no vaccines. He does not believe that all vaccines are equally important, and explains that some diseases are more common and are more deadly. In his book, Sears reviews each childhood disease and vaccine, and offers information on combination vaccines, vaccine safety research, side effects, ingredients, and exemptions. He is a leading proponent of alternate vaccination schedules and offers his own schedule, which is guided by the following principles:

- No more than two vaccines at any one time to limit potential side effects
- No more than one live virus vaccine at a time
- No more than one aluminum-containing vaccine at a time (aluminum-free if possible)
- Avoid combination vaccines where possible
- Avoid mercury-containing vaccines where possible
- Start at two months of age instead of at birth

Sears tells parents exactly what he thinks in the "The Way I See It" section for each vaccine. He lists what he considers the most important vaccines (DTaP, rotavirus, Hib, pneumococcal), recommends delayed vaccination for diseases that are usually mild for infants (hepatitis A, rubella, chickenpox), and supports parents who choose to skip vaccines for diseases that children are unlikely to get (hepatitis B, polio, and flu). He advocated giving infants the individual measles, mumps, and rubella vaccines until sole manufacturer Merck decided in 2008 to stop making them. More information is available at www.thevaccinebook.com.

4. *What Your Doctor May Not Tell You About Children's Vaccinations (revised)*

Stephanie Cave, MD, FAAFP (Wellness Central, 2010)

Dr. Cave is a medical doctor who supports the use of some vaccines and endorses vaccination choice. She is a board-certified family physician. This book is for parents. She is concerned about the rise in chronic childhood medical conditions that mirror the rise in recommended and mandated vaccination. "There's much we don't know, but this much I am certain of: parents who blindly listen to their doctors can no longer afford to do so. It's time for parents to ask questions and learn all they can about vaccines before a shot is scheduled."[17] Her book discusses vaccine safety, autoimmunity, autism, and mercury in vaccines. She reviews each childhood disease and its corresponding vaccine. Cave lists questions for parents to ask the doctor and questions parents should ask themselves before vaccinating. She offers tips to reduce vaccine reactions including:

- Ensure the child is in good health and, in particular, does not have a fever or recent bacterial or viral infection.
- Give the doctor your child's complete family and personal medical history.
- Supplement with vitamins A and C to protect against adverse reactions.
- Request single-dose vials and thimerosal-free vaccines.

Cave is also specific in her recommendations. Her top-priority vaccines are the ones she believes are "safer for the general population": Hib, polio, and DTaP.18 The hepatitis B vaccine is only recommended for an infant if the mother has tested positive for the disease. The combination MMR vaccine is not recommended. She asks parents to contact Merck and ask them to resume manufacture of the single measles, mumps, and rubella vaccines. Cave does not recommend the remaining childhood vaccines— varicella, influenza, hepatitis A, rotavirus, meningococcal, and human papillomavirus19—and her counsel is instead designed to help parents meet state requirements for day care and school. She advocates postponing the mandated vaccinations as long as possible to reduce the possibility of adverse reactions while still staying within state laws. Cave urges parents to get informed and not be satisfied with one-sided answers.

5. *Vaccinations: A Thoughtful Parent's Guide*

Aviva Romm, MD (Healing Arts Press, 2001)

Dr. Romm supports vaccine use and she also supports parents' informed and educated choice. This book is for parents. Romm is a medical doctor and a recognized expert in midwifery and pediatric herbal medicine. Her vaccination book reflects these multiple perspectives. She believes no single approach is appropriate for every child and, therefore, does not make specific vaccine recommendations in her book. She addresses each disease at length and includes data on incidence, symptoms, risks and complications, and conventional treatment. She also provides a section called "food for thought" that provides related information to further support a parent's decision. For example, she describes pertussis as "a generally self-limiting disease that causes discomfort and inconvenience but little damage in a well-nourished, healthy population. It also confers permanent immunity."20 In the section on vaccine risks, Romm observes:

> [Vaccines] can have immediate short-term consequences that are both mild and serious. However, we also saw that vaccines are implicated in a number of chronic health problems, such as allergies, asthma, diabetes, and rheumatoid arthritis. When we consider the consequences of such insidious and pernicious problems as mercury toxicity and learning disabilities, and the rise in chronic autoimmune conditions, we must consider whether avoiding certain diseases is a trade-off courting others. Our children deserve long and healthy lives. Vaccines offer benefits, but they are also a bit of a gamble. Perhaps vaccinations are only one small part of disease prevention and more emphasis needs to be placed on enhancing immunity through other (and safer) means.[21]

In addition to reviewing each childhood vaccine, she covers vaccine history, research on safety and efficacy, and the risks of adverse events from vaccines. She outlines personal choices, public policies, and natural approaches to health and immunity. Her book is evenhanded and offers herbal and homeopathic options for parents who do not vaccinate. She respects many vaccination choices and encourages parents to do what is appropriate. This includes an identification of the circumstances in which they might be able to be delayed or avoided. More information is available at http://avivaromm.com/books.

6. *The Vaccine Guide: Risks and Benefits for Children and Adults (revised)*
Randall Neustaedter, OMD, LAc (North Atlantic Books, 2002)

Dr. Neustaedter is a medical doctor who opposes vaccination and endorses vaccination choice. He holds a doctorate in oriental medicine and has practiced holistic medicine for over thirty years, specializing in children's health. This book is for anyone who is considering vaccination. He begins his book with four facts:[22]

1. Compulsory vaccine laws exist in every state.
2. All vaccines carry risks of adverse reactions.
3. Every parent has the legal right to exempt children from vaccinations.
4. The vaccination campaign has traded diseases of childhood for chronic immune and autoimmune diseases that afflict both children and adults.

Neustaedter's book is divided into two parts: Choices and Vaccines. Part I covers adverse vaccine reactions, contaminated vaccines, chemicals in vaccines, conventional studies, the role of alternative medicine in achieving health, building a strong immune system, and the legal requirements of vaccination. Part II addresses the diseases and their vaccines. He presents the details needed to make an informed choice and acknowledges that people "must decide what is best based on their philosophy, comfort level, and understanding of the real issues surrounding vaccinations. Only then will your choice be based upon intelligent reasoning and not blind faith."[23] In his preface, he mentions the political goal to garner popular support for vaccinations and notes the actions of four presidents: Ford's ill-fated 1976 swine flu program; Carter's campaign that increased federal vaccination funding nearly sevenfold; Reagan's passage of the 1986 Act to protect vaccine manufacturers from lawsuits; and Clinton's passage of the Comprehensive Child Immunization Act of 1993.[24] Neustaedter addresses the role of natural medicine, the treatment of vaccine reactions, and legal requirements, including compulsory vaccination, exemptions, immigration, and military service. In addition to the childhood vaccines, he covers anthrax, smallpox, and Lyme disease vaccines. More information is available at www.cure-guide.com.

*7. **Make an Informed Vaccine Decision: A Parent's Guide to Childhood Shots***
Mayer Eisenstein, MD, JD, MPH with Neil Miller (New Atlantean Press, 2010)

Dr. Eisenstein is a medical doctor who opposes vaccination and endorses vaccination choice. He is board-certified in preventive medicine and holds graduate degrees in law and public health. This book is for parents. Since 1973, his medical practice has offered medical care to 35,000 children who received few or no vaccines; this unvaccinated population has no autism and virtually no asthma, allergies, respiratory illnesses, or diabetes. When vaccinated children contract so-called vaccine-preventable diseases, Eisenstein says, people should direct their outrage towards the drug company that manufactured and falsely promoted an ineffective product, not towards unvaccinated children.[25] He says alternative schedules are "probably more sensible than taking all of the shots at once. . . . However, you're merely mitigating the potential damage. I don't believe you're going to eliminate it altogether."[26] In response to claims that vaccines reduce the incidence of disease, he retorts: "Several diseases—tuberculosis, scarlet fever, plague—infected thousands of people every year but virtually disappeared without any vaccines. How do we explain this?"[27]

With Neil Miller, a medical research journalist, Eisenstein reviews each disease and its corresponding vaccine. They offer detailed information on the incidence and complications of the diseases, and on issues of manufacturing, safety, efficacy, and adverse events associated with each vaccine. They discuss the role of vitamin supplementation, such as vitamin D3 for influenza; vitamin A for measles; and vitamin C for cancer. There is a chapter devoted to autism, a neurodevelopmental disorder that was "extremely rare prior to 1943" and now affects one in ninety-one children and one in fifty-eight boys.[28] He quotes Congressman Chip Pickering: "More children will be diagnosed with autism this year [2006] than AIDS, diabetes, and cancer combined."[29] The book closes with a discussion of multiple vaccines given simultaneously, the use of aluminum and other vaccine ingredients, and questions about social obligation and herd immunity. Additional books and resources about vaccination are available at www.thinktwice.com.

8. *Saying No To Vaccines: A Resource Guide for All Ages*

Sherri Tenpenny, DO, AOBNMM (NMA Media Press, 2008)

Dr. Tenpenny is a doctor of osteopathic medicine who opposes vaccination and endorses vaccination choice. She is board-certified in emergency medicine (through 2006) and in neuromusculoskeletal medicine. This book discusses vaccines for children and adults. Tenpenny compiles medical, governmental, and scientific sources that demonstrate that vaccines are less safe and less protective than advertised. In addition to a discussion of childhood vaccines, Tenpenny addresses the vaccines recommended for adolescents, adults, healthcare workers, soldiers, and international travel. The book refutes twenty-five common arguments used to promote vaccination. The table of contents allows readers to easily find the answers to their questions. Topics include:

- Safety and efficacy studies for approved vaccines;
- The smallpox vaccine and myths regarding smallpox's "eradication;"
- A detailed discussion of pertussis and its vaccine;
- Antibody production as an inaccurate measure of protection against infectious diseases; and
- The impact of stray viruses, heavy metals, chemicals, and other vaccine components.

Tenpenny argues for the importance of fevers and the need to allow a disease to run its course. Unwarranted or excessive treatment hinders recovery and makes the original disease appear more severe. She offers practical advice on how to manage a fever at home and when a doctor's care is necessary. Tenpenny explains influenza viruses, the politics of mass vaccination policies, and the relationship between war and malnutrition in global flu outbreaks. The book contains more than 300 references to medical journals. The final 100 pages contain cross-referenced tables, detailed charts, and other resources, such as a list of ingredients in the most commonly used vaccines, examples of refusal forms for newborns, school exemption letters, and suggested websites. More vaccination information is available at www.tenpennyvaccineinfo.com.

9. *How to Raise a Healthy Child in Spite of Your Doctor*

Robert Mendelsohn, MD (Ballantine Books, 1984)

The late Dr. Mendelsohn, a professor and renowned pediatrician in practice for thirty years, opposed vaccination. President Lyndon Johnson appointed him national medical director of Project Head Start, a program of the federal government that provides comprehensive health education and programming to low-income families. Mendelsohn was also chairman of the Medical Licensure Committee for the state of Illinois, and associate professor of Preventive Medicine and Community Health in the School of Medicine of the University of Illinois. He believed "every vaccine causes neurological damage."[30] He vaccinated patients early in his career, until he realized they were causing more harm than good.[31] His book stops short of advising parents to reject vaccination (in 1984) because "parents in about half the states have lost the right to make that choice."[32] Mendelsohn argued:

- There is no convincing scientific evidence that mass vaccination eliminated any childhood disease. If vaccines eradicated disease in the U.S., why did the diseases disappear simultaneously in Europe, where mass immunization did not take place?
- There are significant known risks and numerous contraindications associated with every vaccine, yet strong reluctance to abandon vaccination persists within the medical community. For three decades before ending its use, the smallpox vaccine remained the only source of smallpox-related deaths. "Think of it! For 30 years, kids died from smallpox vaccinations even though no longer threatened by the disease."
- No one knows the long-term consequences of injecting foreign proteins into children. There are growing suspicions that vaccination may be responsible for the dramatic increase in autoimmune diseases since mass inoculations were introduced.[33]

His book covers mumps, measles, rubella, pertussis, diphtheria, chickenpox, scarlet fever, meningitis, tuberculosis, sudden infant death syndrome (SIDS), poliomyelitis, and infectious mononucleosis. In his own words:

> Scarlet fever is another example of a once feared disease that has virtually disappeared. If a vaccine had ever been developed for it, doctors would undoubtedly credit that with the elimination of the disease. Since there is no vaccine, they give the credit to penicillin, despite the fact that the disease was already disappearing before the first antibiotics appeared. In all probability, as with other diseases, the true reason for its waning incidence is improved living conditions and better nutrition.[34]
>
> In view of the rarity of [diphtheria], the effective antibiotic treatment now available, the questionable effectiveness of the vaccine, the multimillion-dollar annual cost of administering it, and the ever-present potential for harmful, long-term effects from this or any other vaccine, I consider continued mass immunization against diphtheria indefensible. I grant that no significant harmful effects from the vaccine have been identified, but that doesn't mean they aren't there. In the half-century that the vaccine has been used no research has ever been undertaken to determine what the long-term effects of the vaccine may be![35]

Mendelsohn's 1984 book, which remains in print, highlights the views of a prominent pediatrician just prior to passage of the 1986 Act. He described the "bitter controversy over immunization" that raged among the medical professionals, the media, and parents:

- Commenting on the AAP's rejection of a resolution to make available clear and concise information that a reasonable parent would want to know about the risks and benefits of vaccination, Mendelsohn wrote: "Apparently the doctors assembled did not believe that 'reasonable parents' were entitled to this kind of information."[36]
- As an increasing number of parents were rejecting vaccination, parents of vaccine-injured children were filing malpractice suits against the manufacturers and doctors who administered the vaccines, and some vaccine manufacturers stopped making vaccines.[37] "Many doctors are becoming nervous about giving [the DPT vaccine], fearing malpractice suits. They should be

nervous because in a recent Chicago case a child damaged by a pertussis inoculation received a $5.5 million settlement award."[38]

- Doctors were attacking their own. In 1982, Mendelsohn stated in an NBC television documentary: "The danger [from the DPT vaccine] is far greater than any doctors here have ever been willing to admit." The *Journal of the American Medical Association* (*JAMA*) criticized the television program and questioned the credentials of the experts, including Mendelsohn's. The same *JAMA* issue stated, "[the DPT vaccine is] relatively crude and toxic, and the advent of a safer vaccine is eagerly awaited. Almost from the inception of widespread DPT immunization, severe reactions have been reported, beginning with Byers's and Moll's study of vaccine-associated encephalopathy in 1948."[39]

Mendelsohn fought against the passage of the 1986 Act that provided liability protection to vaccine makers. The 1986 Act paved the way for a three-fold increase in the number of federally recommended vaccines for American children during the past twenty-five years.

• • •

Most pediatricians and family doctors endorse and follow the CDC's recommended schedule. However, there are well-respected physicians who depart from this regimen and support vaccination choice. While they do not agree on every point, these physicians suggest that parents and individuals consider the following, to guide their vaccination decisions:

1. A family history of autoimmune, chronic, neurological, or developmental disorders, including adverse reactions to vaccines.
2. The child's medical history, including recent illness, and fevers, and use of prescription medications.
3. The number of vaccines received at a single visit, with special attention to live virus, aluminum- and mercury-containing, and combination vaccines.
4. Any indication of an adverse reaction or developmental delay is a reason to stop or slow down vaccination.
5. The use of vitamins, herbal medicine, and homeopathic remedies before, after, or instead of vaccination.

6. Case-by-case decision making, based on the individual, the vaccine, and timing. Delaying vaccines is a common strategy of many doctors who employ alternate vaccination schedules.
7. A review of the scientific and medical literature and manufacturers' package inserts regarding vaccine adverse events.

CPR offers one additional perspective. No country in the world vaccinates as early or as often as the United States. In the 1950s, we had the third best infant mortality in the world; more babies lived past twelve months of age in only two other countries. In the 1960s, the United States began requiring universal, mandatory vaccination for school admission and has gradually and significantly increased the number of required vaccines since then. The United States' infant mortality ranking has dropped precipitously over this same time frame; today, America ranks forty-ninth, behind every developed country but one (Poland), and many developing countries.[40] This trend holds true for other serious pediatric disorders, including asthma, autism, and ADHD. CPR does not assert that there is a definitive link because comprehensive research has not been done. But it is reasonable to ask and investigate what role, if any, vaccines may be playing.

In the absence of definitive science, CPR advocates the precautionary principle and urges doctors to support parents in their right to exercise vaccination choice. Medicine is for the best interest of the individual patient. Public health is for the best interest of the population. When the state demands that doctors prioritize public health requirements above the needs of an individual patient, and there is no free and informed consent, the consequences are severe: Doctors are no longer practicing medicine, and the sacred trust between physician and patient is broken.

LEGAL EXEMPTIONS TO MANDATORY VACCINATION

The CDC publishes a *recommended* schedule. Until your child enters publicly funded day care, preschool, or school, specific vaccination decisions—if, when, and which the child should receive—remain the privilege and responsibility of parents, guided by the healthcare practitioners of their choosing.

Each state has its own mandatory school vaccination schedule. The National Conference of State Legislatures offers a summary of state laws on its website, www.NCSL.org. Type "School Immunization" in the search box and click on the link "School Immunization Exemption State Laws."

The United States offers a "no-choice choice" in practice. State laws provide for vaccination choice, but they rarely disclose this information, and exemptions can be difficult to obtain. All fifty states offer medical exemptions, if the letter is signed by a medical doctor or doctor of osteopathy; however, the state departments of health can and do override medical exemptions at their discretion. Forty-eight states offer religious exemptions. Twenty states offer philosophical or personal belief exemptions.

CPR works to obtain full vaccination choice in every state. This means that vaccination choice is both available and honored. Please visit our website, www.centerforpersonalrights.com, for more information. We provide resources to help individuals and parents make the best choices for themselves and their families.

Chapter Twenty-Eight

WHO IS DR. ANDREW WAKEFIELD?

Mary Holland, JD

If you've heard Dr. Wakefield's name—and you probably have—you've heard two tales. You've heard that Dr. Wakefield is a charlatan, an unethical researcher, and a huckster who was "erased" from the British medical registry and whose 1998 article on autism and gastrointestinal disease was "retracted" by a leading medical journal. You've also heard a very different story, that Dr. Wakefield is a brilliant and courageous scientist, a compassionate physician beloved by his patients, and a champion for families with autism and vaccine injury. What's the truth?

BACKGROUND ON THE CONTROVERSY

Dr. Wakefield graduated from St. Mary's Hospital Medical School of the University of London in 1981; he was one in the fourth generation of his family to study medicine at that teaching hospital. He pursued a career in gastrointestinal surgery with a specialty in inflammatory bowel disease. He became a Fellow of the Royal College of Surgeons in 1985 and was accepted into the Royal College of Pathologists in 2001. He held academic positions at the Royal Free Hospital and has published over 140 original scientific articles, book chapters, and invited scientific commentaries.

TABLE 1. LANCET STUDY SIGNED CONSENT FORMS[A]

Patient No.	Colonoscopy[b]	Reference[c]
01	07-21 to 07-26-96	Day 1, p.10
02	09-01 to 09-09-96	Day 1, p.08
03	09-08 to 09-13-96	Day 1, p.11
04	09-29 to 10-04-96	Day 1, p.12
05	12-01 to 12-06-96	Day 1, p.17
06	~11-01-96[d]	Day 1, p.14
07	01-26 to 02-??-97[e]	Day 1, p.22
08	01-19 to 01-25-97	Day 1, p.21
09	11-17 to 11-22-96	Day 1, p.15
10	02-16 to 02-19-97	Day 1, p.23
11	NA[f]	NA[f]
12	01-06 to 01-10-97	Day 1, p.19

[a] *Parental consents were obtained by August 24, 1995 (TA Reed & Co., Note 23)*

[b] *Royal Free Hospital admission to discharge dates (TA Reed & Co., Note 8)*

[c] *Transcript of GMC hearings (TA Reed & Co., Note 8)*

[d] *"On or about" November 01, 1997 (TA Reed & Co., Note 8)*

[e] *Discharge day in February 1997 was unrecorded (TA Reed & Co., Note 8)*

[f] *Patient 11, a U.S. citizen, was not subject to the GMC investigations*

In the early 1990s, Dr. Wakefield began to study a possible link between the measles virus and bowel disease. He published a 1993 study, "Evidence of persistent measles virus infection in Crohn's disease"[1] and coauthored a 1995 article published in *The Lancet*, "Is measles vaccine a risk factor for inflammatory bowel disease?"[2] At roughly the same time, Dr. Wakefield wrote an unpublished 250-page manuscript reviewing available scientific literature on the safety of measles vaccines.[3] He was rapidly emerging as one of the world's experts on measles vaccination.

In 1996, an attorney, Solicitor Barr of the law firm Dawbarns, contacted Dr. Wakefield to ask if he would serve as an expert in a legal case on behalf of children injured by vaccines containing the measles virus. The lawyer was bringing the suit on behalf of parents who alleged that vaccines had caused their children's disabilities, including autism. Six months before this, and independent of the litigation effort, parents of children with autism and severe gastrointestinal symptoms began contacting Dr. Wakefield because of his publications on the measles vaccine, asking for help for their children's pain and suffering, which they believed was vaccine-induced. Dr. Wakefield made two major, but separate, decisions at about this time—to try to help the families dealing with autism and gastrointestinal problems, and to become an expert in the legal case regarding vaccines and autism.[4]

Barr asked Dr. Wakefield to study two questions: (1) whether measles could persist after measles infection or the receipt of the MMR vaccine; and (2) whether the measles virus could lead to complications, such as Crohn's disease or autism. Due to bureaucratic delays at his hospital, however, Dr. Wakefield did not begin this litigation-related study until October 1997.[5]

By July 1997, Dr. Wakefield and his colleague, Professor John Walker-Smith, had already examined the "Lancet 12"—twelve patients with autism and gastro-intestinal symptoms that were the basis for the case study in the 1998 article published in *The Lancet.* Dr. Wakefield and others had recommended the referral of these patients to Prof. Walker-Smith, an eminent physician described by his peers as one of the world's leading pediatric gastroenterologists.[6] Prof. Walker-Smith had recently moved to St. Mary's Hospital from a different institution and brought with him the same clinical privileges and ethical clearances that he enjoyed at his previous hospital. He, a colleague, Dr. Simon Murch, and a team of other physicians, did extensive clinical workups on these sick children that Prof. Walker-Smith deemed "clinically indicated," while Dr. Wakefield coordinated a detailed research review of the tissues obtained at biopsy. The clinical tests included colonoscopies, MRI scans, and lumbar punctures to assess mitochondrial disorders. "Clinically indicated studies" did not require permissions from The Royal Free Hospital ethics committee because the tests were required for the benefit of the individual patients.[7] Dr. Wakefield's research was covered by an appropriate ethical approval.

In 1998, to announce the publication of *The Lancet* article coauthored by Dr. Wakefield and twelve other scientists, the dean of St. Mary's Medical School called a press conference. While this was not standard practice, the dean presumably was seeking to enhance the school's visibility in cutting edge research. The article was labeled in the medical journal as an "early report," stating that it "did not prove an association between measles, mumps and rubella vaccine and the syndrome described. Virological studies are underway that may help to resolve this issue."[8]

At the press conference, Dr. Wakefield was asked about the safety of the MMR vaccine. In 1992, two different combination MMR vaccines had been withdrawn from the U.K. marketplace because they were unsafe, so MMR vaccination was already a hot topic before *The Lancet* article was published. Dr. Wakefield responded that, given the paucity of combination MMR vaccine safety research, and until further safety studies were done, the vaccines should be separated into their component parts. He had previously informed his colleagues that this was his view and that he would express it if asked.[9]

The 1998 press conference set off a media firestorm, with large numbers of parents raising uncomfortable questions about the safety of the "triple jab" and requesting single measles, mumps, and rubella vaccines. In the midst of the controversy, in August 1998, the British government took an extraordinary step. It made separate measles, mumps, and rubella vaccine components *unavailable*,[10] thereby forcing the hand of concerned parents. At that point, measles vaccination rates among children in the United Kingdom fell significantly. Measles disease outbreaks became more prevalent, and included a handful of cases of serious complications and deaths. Some sought to blame Dr. Wakefield for irresponsibly scaring parents and triggering a public health crisis.[11] The British government had a big problem on its hands—one that would soon make its way to the United States.

The controversy surrounding Dr. Wakefield simmered. In February 2004, it reached a boiling point when Dr. Richard Horton, editor of *The Lancet*, held a news conference to declare that the 1998 article was "fatally flawed" because Dr. Wakefield had failed to disclose financial conflicts of interest with the litigation-related study he conducted. British reporter Brian Deer published the story in the *Sunday Times*, detailing alleged undisclosed conflicts of interest. Immediately following publication, Mr. Deer sent a detailed letter to the British General Medical Council (GMC), which regulates the practice of medicine.[12] The GMC then initiated proceedings against Dr. Wakefield that culminated in Dr. Wakefield's delicensure[13] in May 2010 and the retraction of the 1998 article from *The Lancet*.[14]

THE ALLEGATIONS AGAINST DR. WAKEFIELD

The highly publicized, multi-year, multi-million dollar prosecution against Dr. Wakefield alleged that:

- Dr. Wakefield was paid 55,000 British pound sterling (about US $90,000) by litigators for the study published in *The Lancet*, and he failed to disclose this conflict of interest;

- He and his colleagues performed medically unnecessary tests on the children in the 1998 study and lacked appropriate ethical clearances;
- The children in the 1998 study were selected for litigation purposes (as described in the *Sunday Times* article) and not referred by local physicians; and
- He drew blood from children at his son's birthday party for control samples in the 1998 study with callous disregard for the distress that this might cause children.[15]

Based on its findings, the GMC concluded that Dr. Wakefield had engaged in "serious professional misconduct," and "dishonest," "misleading," and "irresponsible" behavior, warranting the sanction of his removal from the medical profession.[16]

Let's examine the GMC's charges and the evidence.

Failure to Disclose Payment from Litigators

Dr. Wakefield accepted 55,000 pounds to conduct a study for the class action suit regarding vaccines and autism. This was a research grant from which Dr. Wakefield personally received no compensation. Dr. Wakefield did not start this study until after the case series for the Lancet 12 had been submitted. Legal documents prove that Dr. Wakefield's hospital knew about this study and knew about the amount of money he received, most of which went to pay the salary of a designated laboratory technician. Documents further demonstrate that Dr. Wakefield disclosed in a national newspaper over one year before publication of the 1998 article that he was working with the litigators.[17] Dr. Horton, editor of *The Lancet*, had been informed and should have been well aware of Dr. Wakefield's role in the vaccine-related litigation before the publication of the 1998 article.[18]

"Medical Necessity" and Ethical Clearances

The Lancet 12 were sick. Each child was administered tests with the intent to aid that child. The hospital administration was fully aware of the tests being conducted and made no objections. Because all of the tests were "clinically indicated" and not for research purposes, no ethical clearance beyond what Prof. Walker-Smith already possessed was required. Notably, no patient, parent, or guardian has ever made accusations against Dr. Wakefield or testified against him for ethical violations or medically unnecessary procedures. Dr. Wakefield and his colleagues reject the GMC's ruling that the tests for the Lancet 12 were unnecessary.

The Lancet 12's Referrals

The GMC charged that the children were referred through the litigation effort and not through ordinary medical channels. This is incorrect. Parents started contacting Dr. Wakefield long before the litigation started, and independently of it. Since the litigation study was not yet started by the time *The Lancet* study was completed and submitted to the journal, this finding is false. Dr. Wakefield and his colleagues reject that claim; the families contacted them directly because of their medical expertise.[19]

Control Blood Samples from a Child's Birthday Party

Dr. Wakefield arranged for control blood samples from healthy, typically developing children to be taken at his son's birthday party. Most of the children's parents were medical colleagues and friends. He did this with the children's and parents' fully informed consent and gave the children 5 pounds each for their trouble. The procedure was undertaken by an appropriately qualified doctor using a standard technique. The children were happy to be helpful and went on to enjoy the birthday party. While this is admittedly an unconventional method of collecting control blood samples, it hardly amounts to "serious professional misconduct" or an ethical breach warranting delicensure. The GMC's description of this incident as an example of "callous disregard" for children's distress seems to be a gross exaggeration.[20] Indeed, the U.K. High Court of Justice exonerated Professor Walker-Smith in March 2012,[21] and the *Lancet* journal has suggested that it is considering reversing its retraction.[22]

The GMC failed to prove its case against Dr. Wakefield. Using Brian Deer's reporting as evidence, the GMC appears to have purposefully conflated the Lancet 12 study and the subsequent litigation study to create the appearance of a financial conflict of interest.[23] Similarly, the GMC appears to have wrongfully applied ethical research standards to tests that were "clinically indicated" for severely ill children.[24] Conflating treatment and research not only grievously harmed Dr. Wakefield and his colleagues, but set a threatening precedent for the practice of medicine. The government's medical regulators (of uncertain expertise) second-guessed Prof. Walker-Smith, the world's preeminent authority on pediatric gastroenterology, on his clinical judgment about what tests were necessary.[25] Which medical decisions will regulators second-guess next?

The press, and specifically reporter Brian Deer, tried Dr. Wakefield in the court of public opinion while the GMC was prosecuting him in its regulatory court. Deer alleged that Dr. Wakefield had a pending patent application for a separate measles vaccine and hoped to "cash in" by urging parents to forego the MMR for separate measles vaccines. The evidence proves that Dr. Wakefield was not a patent holder for a separate measles vaccine. St. Mary's Hospital held a patent for a *therapeutic* single measles vaccine using the beneficial immune properties of transfer factor, intended for people already infected with the measles virus. This measles vaccine was not a preventive product for people unexposed to the virus; in other words, there was no possible financial competition between the MMR vaccine and the single measles vaccine for which the hospital, and not Dr. Wakefield, held a patent.[26]

In 2009, Deer made additional allegations that Dr. Wakefield fabricated data.[27] The GMC never made this charge, but the media picked it up and, notably, the U.S. Department of Justice used it frequently in the Omnibus Autism Proceeding in the U.S. Court of Federal Claims. In those proceedings to determine whether families could receive compensation for MMR-induced autism, the US Department of Justice went out of its way to depict Dr. Wakefield as a scientific fraud, although he was not directly relevant to the proceedings.[28] In his 2010 book, *Callous Disregard*, Dr. Wakefield shows Deer's allegations of fraud to be fabrications.[29]

CPR finds no evidence of Dr. Wakefield's scientific fraud. On the contrary, many scientists and laboratories around the world have confirmed Dr. Wakefield's findings regarding severe gastrointestinal inflammation and symptoms in a high percentage of children with autism.[30]

In its February 2, 2010 retraction, *The Lancet* did not allege fraud.[31] Relying solely on the GMC proceeding, it retracted the article, asserting that the authors had not referred the patients as represented and the study team had not received the hospital's ethics committee's approval. The GMC's conclusions and *The Lancet*'s reliance on them appear unfounded.

THE MEANING OF THE WAKEFIELD PROSECUTION

What, then, was this high-profile prosecution really about? If there was no scientific fraud, no undisclosed financial conflicts of interest, no ethical breaches in performing tests on sick children, and no complaints from patients or their families, then what was the big deal? Did the international scandal and multi-million dollar prosecution proceed merely to chastise a doctor for drawing blood from children at a birthday party, with their consent and their parents' consent? Of course not.

Dr. Wakefield was, and remains, a dissident from medical orthodoxy. The medical establishment subjected him to a modern-day medical show trial for his dissent.[32] Dr. Wakefield's research raised fundamental doubts about the safety of vaccines and the etiology of autism. Dr. Wakefield was punished for his temerity to caution the public about vaccine risks and to urge them to use their own judgment. *Dr. Wakefield was punished for upholding vaccination choice.*

The purpose of the proceeding, as in any show trial, was to communicate to other doctors and scientists, and to the public, the error of the perpetrator's ways. A show trial offers a veneer of due process but, at its core, displays naked power. The apparent intent of the prosecution was to intimidate others from following Dr. Wakefield's footsteps and to teach the lesson that anyone in the medical or scientific community who dares to publicly question the safety and efficacy of vaccines will be punished with utmost severity. The GMC appears to have decided that if the price of such a lesson was scientific ignorance about vaccine-autism links and the suffering of severely ill children, then so be it. Dr. Wakefield was made an example.

The GMC destroyed Dr. Wakefield's professional reputation and livelihood, and *The Lancet* and other publications confiscated his professional accomplishment through retraction. The GMC colluded with *The Lancet*, the media, the British Department of Health, the pharmaceutical industry, and even with the U.S. Department of Health and Human Services and the U.S. Department of Justice, to discredit Dr. Wakefield.

The Center for Personal Rights is confident that the world will look back at the prosecution of Drs. Wakefield, Walker-Smith, and Murch with shame and remorse.

Dr. Wakefield has joined in a long, honorable tradition of dissidents in science and human rights. The world has benefitted profoundly from other courageous dissidents in science—Galileo, who argued that the sun is the center of the universe; Semmelweis, who reasoned that doctors must wash their hands to prevent transmission of infection; Needleman, who proved that lead exposure causes neurological damage in developing children; and McBride, who demonstrated that thalidomide caused birth defects.[33] As Thomas Kuhn explained, changing scientific paradigms is a revolutionary process, with the wrenching upheaval that revolution brings.[34]

In due course, the world has paid tribute to human rights dissidents, as well—Nelson Mandela moved from prison in South Africa under apartheid to become its most beloved President; Andrei Sakharov left Russia's internal exile to become its moral beacon; Vaclav Havel left a Czech prison to become its first post-communist President; and Liu Xiabo, a Chinese human rights advocate, received the 2010 Nobel Peace Prize in absentia because he remains incarcerated. In time, China will embrace Mr. Liu and look to him to help create a better future.

Before long, the world will likely recognize that it was Dr. Wakefield, not his detractors, who stood up for the practice of medicine and the pursuit of science. Dr. Wakefield remains an unbowed dissident in the face of a repressive medical and scientific establishment.

To learn more about his work and why it is controversial, see chapter 7, "An Urgent Call for More Research on Science," chapter 30, "The Suppression of Science," and chapter 29, "The Exoneration of Professor John Walker-Smith: A Great Wrong Partly Righted."

Chapter Twenty-Nine

THE EXONERATION OF PROFESSOR JOHN WALKER-SMITH

A Great Wrong Partly Righted

David L. Lewis, PhD

My responsibilities at the National Whistleblowers Center (NWC) in Washington, DC, include serving on its board of directors and managing the Research Misconduct Project (www.researchmisconduct.org). On a voluntary basis, I investigate cases of "institutional research misconduct" in which government, industry, and academic institutions use false allegations of research misconduct against scientists whose research threatens government policies and industry practices. In 2011, philanthropist Claire Dwoskin, who helped organize a vaccine safety conference in Jamaica, West Indies,[1] invited me to attend as an outside observer and comment on my experiences involving the suppression of scientific research (see appendix to this chapter). Dr. Andrew Wakefield was one of the speakers. He, with Professor John Walker-Smith and eleven other colleagues, wrote the controversial 1998 *Lancet* article in which parents linked autism in eight of twelve children with MMR (measles, mumps, rubella) vaccination.[2]

During the conference, news coverage of Brian Deer's articles in the *British Medical Journal* (*BMJ*) alleging research fraud against Dr. Wakefield broke in the international media.[3] I gave Dr. Wakefield my contact information, and he sent me a copy of his book, *Callous Disregard*. Afterward, he granted me full access to his personal files, which included many important documents that had never been made public. In November 2011, the *BMJ* published my "Rapid Response" about the documents and accompanied it with an editorial by its editor-in-chief, Dr. Fiona Godlee, a feature article by Deer, and two external commentaries.[4] After reviewing the documents, the author of one of the commentaries, Ingvar Bjarnason at King's College Hospital, told *Nature* that he did not believe Dr. Wakefield fabricated the diagnoses published in *The Lancet*. The documents "don't clearly support charges that Wakefield deliberately misinterpreted the records," he said.[5] Later, I filed a complaint with the UK Research Integrity Office (UKRIO) over Deer's and the *BMJ*'s handling of the evidence I uncovered.[6] This article is based on those experiences and my investigation of Dr. Wakefield's case.

BACKGROUND

Brian Deer's Allegations:

The controversy over the *Lancet* study began in 2004 when Brian Deer, a reporter working for Rupert Murdoch's *The Sunday Times*, alleged that the study lacked proper ethics approvals and imposed clinically unnecessary medical procedures on the *Lancet* children for research purposes.[7] He claimed that solicitor Richard Barr funded the study with a grant from the Legal Aid Board (LAB) to provide a basis for suing manufacturers of the MMR vaccine on behalf of parents claiming that the vaccine caused their children's autism. According to Deer, it was all part of a scheme in which the lead author, Dr. Andrew Wakefield, stood to profit from a patent on his own, safer measles vaccine.

Reluctant to publish the allegations, *The Sunday Times* required Deer to obtain on-the-record quotations supporting the allegations from *The Lancet*'s chief editor, Richard Horton. Deer met with Horton and other *Lancet* editors in February 2004. He was accompanied by Evan Harris, an outspoken supporter of the MMR vaccine in the British Parliament.[8] Horton, not surprisingly, refused to supply any quotations without first looking into the matter himself. After meeting with authors of the *Lancet* article and administrators at University College London (UCL) and the Royal Free Hospital, Horton and other editors published Deer's allegations with their response just three days after the meeting with Harris and Deer.[9]

The editors disagreed with Deer that medical procedures used in the *Lancet* study lacked proper ethics approvals and that the patients were not consecutively referred. They did agree, however, that Dr. Wakefield failed to inform *Lancet* editors about the LAB grant and his work with Richard Barr. Dr. Wakefield maintained that the LAB grant was for a separate study and that he had complied with *The Lancet*'s disclosure rules operating at that time. The *Lancet* editors decided that the alleged nondisclosures did not warrant retracting the paper. In short, the *Lancet* editors fully cleared Prof. John Walker-Smith and only chastised Dr. Wakefield over his alleged conflicts of interest. Walker-Smith, one of the thirteen coauthors of the 1998 Lancet article, is widely regarded as the "father of pediatric gastroenterology."[10] Unfortunately, before Dr. Wakefield could fully respond because he had moved to the United States, *The Lancet's* support for some of the allegations gave Deer instant credibility in the scientific community. This outcome created nearly insurmountable obstacles to Dr. Wakefield's ability to defend himself later. UCL, the only academic institution to evaluate Deer's allegations, disputed all of them.[11] Normally, all it takes to clear a scientist of fraud allegations is an investigation by the institution where the alleged acts occurred. But, because of the government's interest in the case, *The Lancet*'s findings triggered a debate in the House of Commons. Harris wanted the Crown Prosecution Service to hold Dr. Wakefield and his coauthors responsible for the deaths of unvaccinated children and consider charging them under criminal statutes.[12] In the end, the matter was turned over to the U.K. General Medical Council (GMC).

Deer's Nondisclosures:

Further crippling Dr. Wakefield's chances of a fair hearing, Deer either failed to obtain key evidence or chose not to disclose it to the GMC and others. For example, letters exchanged between Barr and Horton nine months before the *Lancet* study was published show that Horton, in fact, had been informed about Dr. Wakefield expert witness work with Barr.[13] Deer also quoted Barr saying (wrongly as it turned out) that he paid for the *Lancet* research.[14] Documents gathered after *The Lancet*'s investigation, however, show that A.J. Zuckerman, dean of the Royal Free Hospital Medical School, had sequestered all of the LAB funds until well after the *Lancet* study was submitted for publication. In May 1997, when it became apparent that the funds were in limbo, Dr. Wakefield requested that they be returned to the LAB.[15] Zuckerman later confirmed to the GMC that LAB funds were not used to support the *Lancet* study.[16]

Most damning of all, documents show that Deer obtained copies of the ethics committee approvals covering the research component of the *Lancet* study but apparently never disclosed them to the GMC. According to a Freedom of Information Act (FOI Act, a.k.a. FOIA) response from the London Strategic Health Authority of the National Health Service (NHS), Deer, in 2004, obtained copies of correspondence exchanged between Prof. Walker-Smith and the Royal Free Hospital's Ethical Practices Committee.[17] One memo from Prof. Walker-Smith to the committee on February 27, 1997, stated: "We currently have formal approval to take research biopsies during colonoscopy (Code 162-95) and I am writing to organize formal approval for research biopsies to be taken during upper biopsies."[18] (Approvals for upper biopsies were later designated as Code 70-97.) In another memo, a member of Prof. Walker-Smith's group transmitted a report to the ethics committee, which was titled "1999 Annual Report on ethical submissions 162-95 and 70-97." It states, "Samples, with fully informed parental consent (using the consent forms as detailed in the submissions), were obtained from upper and lower endoscopies. . . ."[19] In all cases, colonoscopies were performed on the *Lancet* children only after parental consent was granted on or before August 24, 1995 (Table 1). (The signed consent forms and other key documents, including the ethics approvals for 162-95, are available online at *www.VaccineEpidemic.com.)*

To substantiate his allegations that the *Lancet* study lacked any ethics approval for research, Deer provided the GMC with an Ethical Practices Committee (EPC) approval from the Royal Free Hospital that was coded EPC 172-96. This approval was for a study of 25 children with autism spectrum disorder and intestinal symptoms. It described clinical investigations involving ileocolonoscopy and upper gastrointestinal endoscopy with biopsies, barium (enemas), lumbar punctures, and various blood tests as part of "normal patient care."[20] This proposed study included research investigations intended to examine the possible causes of the children's disorders. It was approved on December 18, 1997, *after* seven of the *Lancet* children had already been investigated clinically for their intestinal and neurological symptoms.[21] Absent the EPC 162-95 approvals, the appearance was created that Prof. Walker-Smith collected research biopsies from seven of the *Lancet* children without any ethics approvals. It also made it appear that the parental consent forms should have been included with EPC 172-96 (instead of EPC 162-95) and that none of the research in the *Lancet* study was covered under any ethics approvals. Finally, it made it appear that Dr. Wakefield and his coauthors were dishonest when they stated in the

Lancet article, "Investigations were approved by the Ethical Practices Committee of the Royal Free Hospital NHS Trust, and parents gave informed consent."

The GMC Hearings:

The GMC, which regulates U.K. medical practice, held public hearings into allegations of ethical misconduct against the *Lancet* article's three senior authors, Dr. Wakefield, Dr. Walker-Smith, and Dr. Simon Murch. The hearings, which were based on allegations Deer first published in *The Sunday Times* in February 2004, began in July 2007 and lasted until May 2010. Prof. Walker-Smith's legal counsel, Stephen Miller, tried to correct Deer's apparent omission of the ethics approvals for the research component of the *Lancet* study (EPC 162-95) by introducing Exhibits 86(a), (b), and (c).[22] Dr. Wakefield and his coauthors described the research component as a "pilot study."[23] Miller quoted from Exhibit 86(a), a memo Prof. Walker-Smith sent to the chairman of the Royal Free Hospital dated August 24, 1995:

> For some years... we have had ethical permission to take two extra mucosal biopsies for research purposes. During colonoscopy children routinely have multiple biopsies taken for diagnostic purpose (4-6). The parents have signed a [consent] form as attached granting permission. These biopsies are used for a variety of 'research' investigations such as cytokine production where on occasion information of direct and immediate importance to the child's illness has been obtained as well as of research importance. I would be very grateful if you would grant permission for this to continue after our move to the Royal Free.[24]

Opposing counsel responded that this was the first time that the GMC's counsel had ever seen any of these documents concerning EPC 162-95. In a recent interview, Deer claimed that the correspondence he obtained under the U.K.'s FOIA concerning the research component of the *Lancet* study (EPC 162-95) was also entered into evidence by the GMC;[25] however, it was not. Deer did produce these documents in the defamation lawsuit Dr. Wakefield filed after Deer's *Sunday Times* articles of 2004,[26] which Dr. Wakefield later voluntarily withdrew. Their significance to the GMC proceedings only recently came to light after a comparison of Deer's disclosure to the GMC with a response from the NHS providing the documents Deer obtained under the FOIA in 2004.

Although the three documents introduced by Prof. Walker-Smith's attorney clearly demonstrated that the research component to the *Lancet* study had been preapproved by the Ethics Committee, they were not nearly as comprehensive as the EPC 162-95 approvals Deer had obtained from NHS. In other words, while Dr. Wakefield struggled to put all the facts together at the beginning of the hearings,[27] Deer apparently sat on the evidence Dr. Wakefield needed. In fact, there is no evidence that Deer *ever* disclosed any of the EPC 162-95 research approvals to the GMC, which simply disregarded the few meager exhibits introduced by Prof. Walker-Smith's attorney. In the end, the GMC sanctioned Dr. Wakefield and Prof. Walker-Smith for not complying with the "conditions for approval and the inclusion criteria" for Project 172-96 (i.e., investigating children before the approval date and failing to keep copies of the signed parental consent forms with EPC approvals for Project 172-96).[28] Dr. Wakefield and his senior coauthors, however, steadfastly maintained that EPC 172-96 had nothing to do with the *Lancet* study.[29]

Later on in the GMC hearings, Dr. Wakefield's attorney submitted correspondence exchanged between Richard Horton, editor in chief of *The Lancet*, and Richard Barr. Deer had failed to produce this critically important evidence, which proved that Horton was made aware of Dr. Wakefield's work with Barr nine months prior to publishing the *Lancet* article. But, instead of dropping the charge that Dr. Wakefield had not disclosed the work to Horton and apologizing for falsely accusing Dr. Wakefield, the GMC panel railed at Dr. Wakefield's attorney for introducing the evidence. Dr. Wakefield's attorney remarked to the panel, "Rather than [receiving] an apology, Dr. Wakefield has been the recipient of a vigorous attack by the Council for having had the temerity to bring these relevant documents to the Panel's attention."[30]

Ignoring key exculpatory evidence, the GMC found in 2010 that Prof. Walker-Smith and Dr. Wakefield had falsely claimed that the *Lancet* children were consecutively referred. They also found it false that certain medical procedures, including lumbar punctures and endoscopic biopsies, and collection of blood samples were clinically necessary and that these procedures had been approved by the hospital's ethics committee. But, the endoscopic biopsies were specifically approved (Table 1); and even British health authorities had employed lumbar punctures to collect spinal fluid in order to determine whether or not children were infected with the mumps virus from the MMR vaccine. They reported, "In Nottingham all children with febrile convulsions were lumbar punctured. . . ."[31] Moreover, Evan Harris acknowledged that British medical guidelines in effect at the time permitted the collection of blood samples for research purposes without ethical approval.[32] Separately, the GMC found that

Dr. Wakefield failed to disclose the LAB grant and other purported conflicts of interest to the *Lancet* editors and that he had exhibited "callous disregard" toward some of the *Lancet* children by collecting blood samples at a children's birthday party. Based on its findings, the GMC revoked Dr. Wakefield's and Prof. Walker-Smith's licenses to practice medicine.[33]

Targeting Dr. Wakefield in the U.S.

In the wake of the GMC's 2010 decisions, Deer and the *BMJ* began directing their attacks at Dr. Wakefield's safe haven in the United States, where they characterized him as a criminal attempting to flee justice.[34] In an interview with CNN, Deer called upon the U.S. Department of Homeland Security to "take a close look at Dr. Wakefield's visa application and how he got into the United States."[35] Fiona Godlee and others at the *BMJ* complained that Dr. Wakefield was allowed to continue his work in the United States and called on the news media, government agencies, and others to respond. "[H]e continues to push his views. . . . Meanwhile, the damage to public health continues, fueled by unbalanced media reporting and an ineffective response from government, researchers, journals, and the medical profession."[36]

Taking up the gauntlet, CNN's Anderson Cooper drew upon the *BMJ* articles to reject exculpatory evidence in Dr. Wakefield's book *Callous Disregard* and denounce him as a liar: "But, sir, if you're lying, then your book is also a lie. If your study is a lie, your book is a lie."[37] Cooper went on to say that a growing number of children have died from not being vaccinated, then asked, "Do you feel any sense of responsibility for that?"

Bill Gates spoke even more plainly with CNN's Dr. Sanjay Gupta. Alluding to the *BMJ* series, he said:

> Dr. Wakefield has been shown to have used absolutely fraudulent data. He had a financial interest in some lawsuits, he created a fake paper. . .it's an absolute lie that has killed thousands of kids. Because the mothers who heard that lie, many of them didn't have their kids take either pertussis or measles vaccine, and their children are dead today. And so the people who go and engage in those anti-vaccine efforts—you know, they, they kill children.[38]

BRIAN DEER AND THE *BMJ*

In January 2011, the *BMJ* published a series of articles several months after the GMC's decision to delicense Dr. Wakefield and Prof. Walker-Smith. It criticized Dr. Wakefield for his "involvement with a lawsuit against manufacturers of the MMR vaccine" and singled him out as a lone fraudster with a get-rich scheme.[39] In the now-discredited trial carried out by the GMC, one of the government's most damaging charges centered around Dr. Wakefield's "failure to disclose [a] patent" on a transfer factor used to *treat* measles infections.[40] These charges were based on Deer's allegations that Dr. Wakefield was attempting to undermine the MMR vaccine in order to market his own measles vaccine.[41] It is a common practice for universities to exploit the commercial potential for academic advances in science to allow the public to benefit.[42] Researchers are required to assign all patent rights to their universities who, in turn, file patent applications and create commercial development projects. This arrangement, however, makes researchers vulnerable to being seen as profiteers, even unethical, despite the fact that advances in science and medicine cannot be commercialized absent patent protection.

The measles transfer factor was first described in a provisional patent application filed by the Royal Free Hospital in 1997.[43] The application process was finally completed after the *Lancet* article was published.[44] Dr. Wakefield risked having the U.K. Patent Office void the application if he publicly described the transfer factor in writing before the application process was completed. In the end, it didn't matter anyway. The patent application was eventually abandoned,[45] and it became increasingly difficult to proceed to clinical trials as Deer and the GMC pursued the fraud allegations.

The patent application referred to two possible applications of Dr. Wakefield's work: a "measles *transfer factor*" and a "safer measles *vaccine*." Transfer factors confer cell-mediated immunity in immune-compromised patients.[46] They cannot be used to vaccinate the general population since they do not induce antibody immunity, which is considered a prerequisite for population-based vaccines. Hence, transfer factors cannot compete with the MMR vaccine manufactured by Merck and GlaxoSmithKline; and Dr. Wakefield, therefore, never stood to gain financially from the demise of the vaccine. Vaccines made from live viruses, such as the MMR vaccine, often cannot be safely administered to immune-compromised patients. In that sense, treatments derived from transfer factors for immune-compromised patients could be considered a "safer vaccine."

But, for Deer and the *BMJ* to claim that Dr. Wakefield created an MMR scare to sell his own patented measles vaccine in its place is simply false.

DR. WAKEFIELD'S FILES

In Deer's 2010 *BMJ* article titled "Wakefield's 'Autistic Enterocolitis' Under the Microscope," he accused Dr. Wakefield of exaggerating the information in pathologists' grading sheets in order to diagnose most of the *Lancet* children with colitis.[47] The pathologists, Professor Amar Dhillon and Dr. Andrew Anthony, examined the biopsy samples as part of a blinded independent analysis that included healthy (control) biopsy samples from another hospital. But grading sheets, Deer argued, "don't generate clinical diagnoses such as colitis." The "ultimate proof" of Dr. Wakefield's innocence or guilt, according to Deer,[48] is the biopsy slides described in the grading sheets, which turned up missing by the time the GMC began its investigation.

Dr. Wakefield's files contain the GMC's copies of many of the grading sheets.[49] They show that Dhillon included boxes to check for various diagnoses, such as Crohn's disease and ulcerative colitis. Consistent with the *Lancet* article, both pathologists found that only one child showed no evidence of inflammation. Unfortunately, the only grading sheets belonging to Anthony were completed shortly after the *Lancet* study was published as he continued to examine the biopsy slides. Still, they showed that he indicated the diagnosis of colitis in marginal notes. Dr. Wakefield's files also contained photomicrographs taken by Dhillon and Anthony for six of the children (Patients 2-6, and 9).[50] The various cellular structures described in Table 1 of the *Lancet* article are clearly visible on the photomicrographs.

Other important documents in Dr. Wakefield's files include Anthony's *PowerPoint* presentation in which he meticulously illustrated the approach he and Dhillon used to grade inflammation levels in the biopsy slides.[51] They also include copies of Dhillon's and Anthony's sworn statements to the GMC in which they described reviewing and approving Dr. Wakefield's summary of their grading sheets published in Table 1. In my opinion, these and other documents in Dr. Wakefield's files prove that he faithfully reproduced Dhillon's and Anthony's blinded analysis reported in Table 1 of the *Lancet* article.

Although the *BMJ* and Deer took pains to portray Deer as the origin of the analytical underpinnings of his *BMJ* article,[52] it never made sense that a freelance journalist with a lackluster career and no scientific or medical background would fully comprehend the medical records and histological data upon which the *Lancet* article was based. Thus, it was both shocking and revelatory to discover a confidential report comparing routine pathology reports from the Royal Free Hospital with the *Lancet* article in Dr. Wakefield's files.[53] The report was prepared for the GMC in 2006 by one of its experts, Professor Ian Booth, who was a pediatric gastroenterologist. It was the perfect intellectual blueprint for Deer's subsequent *BMJ* articles.

The discovery of Booth's report raises serious questions about whether or not Deer obtained the information from the GMC, directly or indirectly, and whether or not he used it as a template for his own articles in the *BMJ*. Professor Booth confirmed the report's authenticity to me in an email, stating:

> Yes, this is my document, although my understanding is that its contents remain confidential between myself and the GMC's solicitors to whom I submitted it. My analysis of the case records of the children presented in the Lancet publication was carried out specifically at the request of the GMC's solicitors and it formed part of the basis of the case brought against Wakefield et al by the legal team acting on behalf of the GMC.[54]

Shortly before the GMC sanctioned Dr. Wakefield and Prof. Walker-Smith in May 2010, the *BMJ* published Deer's autistic enterocolitis article suggesting Dr. Wakefield fabricated the diagnosis of nonspecific colitis from Dhillon's and Anthony's grading sheets. Under the heading "MMR: Faking the Link," Deer reported mismatches between routine (on-duty) pathology reports from the Royal Free Hospital and Dr. Wakefield's summary of Dhillon's and Anthony's results in Table 1 of the *Lancet* article, just as Booth had done in his report four years earlier. Editors at the *BMJ* portrayed Deer as the original source of the analysis of medical records that purportedly unveiled Dr. Wakefield as a fraudster:

> [I]t has taken the diligent skepticism of one man, standing outside medicine and science, to show that the paper was in fact an elaborate fraud. . . . the GMC launched its own proceedings that focused on whether the research was ethical. But while the disciplinary panel was examining the children's medical records in public, Deer compared them with what was published in the Lancet. His focus was now on whether the research was true.[55]

I posted Booth's report on the NWC Research Misconduct Project's website (www.researchmisconduct.org) and pointed out that it raised serious questions about how Deer came to publish, as his own work, the same analysis that the GMC solicitors asked Booth to perform. Outraged over my postings, Deer filed multiple allegations of ethical misconduct against me with the NWC executive director and demanded that my materials be removed from the NWC's website.[56] In an email to the NWC, Deer stated that he had never seen Booth's report prior to its posting on the NWC website in June 2011,[57] but whether or not someone with knowledge of it gave him the idea of comparing the on-duty pathology reports with Table 1 of the *Lancet* article is unknown.

BMJ'S "RAPID RESPONSE"

In September 2011, I submitted evidence from Dr. Wakefield's files to the *BMJ*, along with a commentary outlining my conclusions. Editor in chief Godlee rejected my submission and invited me instead to revise the materials as a "Rapid Response" to Deer's 2010 article on autistic enterocolitis. After I reformatted the information in accordance with the *BMJ*'s instructions to authors, the editors worked with me to make additional minor changes. As soon as these changes were made, the *BMJ* provided me a link to the prepublication version, complete with four attachments: Dhillon's and Anthony's grading sheets, photomicrographs of the children's missing biopsy slides, and my revised commentary. In my commentary, I disputed the GMC's findings over a wide range of issues and addressed the false allegations of ethical misconduct against me, which Deer had submitted to the NWC.

Godlee told me that she was having my Rapid Response, including attachments, externally peer-reviewed—an extraordinary measure for publishing online comments from readers. The *BMJ*'s editors and attorneys completely rewrote it as a brief description of the grading sheets with an explanation of how I obtained them. According to Godlee, the rewrite was based on the peer-reviews; however, in a radical departure from normal procedures for a peer-reviewed, scholarly journal, the editors did not forward the peer-reviews to me and ignored my request to provide me with copies. I also suggested adding a sentence expressing my overall views of the *Lancet* study. Although readers are normally allowed to comment on any part of an article, Godlee stated that I was only permitted to comment on the parts of Deer's articles pertaining to grading sheets. Deputy Editor Tony Delamothe wrote, "We care about getting your opinion on the interpretation of the biopsies into the journal, but nothing more."[58]

On November 9, 2011, the *BMJ* published a Rapid Response, which the editors rewrote and misrepresented as my response to an article Deer published in 2011, titled "How the Case Against the MMR Vaccine was Fixed."[59] The reason that the editors and lawyers prohibited me from commenting on anything but the biopsy pathologies revealed itself when Godlee dismissed my Rapid Response in an interview with *Nature.* "Fiona Godlee, the editor of the *BMJ*, says that the journal's conclusion of fraud was not based on the pathology but on a number of discrepancies between the children's records and the claims in the *Lancet* paper."[60]

Based on the documents I submitted *BMJ*'s editors and attorneys had only two choices: exonerate Dr. Wakefield and get to the bottom of what really went on between Deer, Evan Harris, and the GMC, or publish a new theory in which the *Lancet* authors and virtually everyone they worked with at UCL and Royal Free Hospital were involved in a very elaborate fraud. Unfortunately, they chose the latter and deleted almost all of the evidence in Dr. Wakefield's files that exonerated him and his coauthors. To explain their new theory of research fraud, Godlee accompanied my Rapid Response with an editorial titled "Institutional Research Misconduct." In it, she alleged that UCL administrators, the Royal Free Hospital, and all 13 authors of the *Lancet* study created an MMR scare so that UCL could sell its own safer measles vaccine, diagnostic kits, and "autism products."[61] She called on Parliament to hold an inquiry into the research fraud running rampant in Great Britain's academic institutions.[62] Andrew Miller, who chaired the Science and Technology Committee, responded that this was not the right forum and that he "must be careful not to appear to be vulnerable to public lobbying."[63]

THE UKRIO REPORT

In January 2012, I filed a 167-page report with the UK Research Integrity Office (UKRIO) documenting how the *BMJ*'s editors and lawyers rewrote my Rapid Response to remove important evidence that undermined their allegations against Dr. Wakefield, and how Deer filed false allegations of ethical misconduct against me with the NWC to attempt to prevent me from posting the evidence on the NWC website.[64] Evidence removed from my submission included photomicrographs Dhillon and Anthony took of the *Lancet* children's missing biopsy slides, which Deer called the "ultimate proof" of Dr. Wakefield's innocence or guilt; Anthony's grading sheets where he wrote "colitis" in his marginal notes; and my commentary describing these and other important documents Deer appeared to have concealed from the *BMJ.*

Without checking any of the facts, Deer informed NWC Executive Director Stephen Kohn that I was not qualified to comment on medical records involving colonic biopsies.[65] He also called on the NWC's board of directors to investigate me for ethical misconduct, claiming, for example, that I "improperly exploit[ed] a university email account" when I wrote to Professor Booth, and that I should have used email addresses published on the NWC website and elsewhere.[66] He also claimed that I misled Booth by telling him that I "had been prompted to write to [him] by a question from a *Nature* reporter." In response, I forwarded Deer's emails to the *BMJ* editors with proof that I have been accepted by federal and state courts as an expert regarding the collection of colonic biopsies and that the University of Georgia had approved my use of its computers for my investigations into institutional research misconduct.[67] I had, in fact, written to Professor Booth using my personal AOL email account and given the link to my Research Misconduct Project at the NWC. The *Nature* reporter who suggested I have Professor Booth verify his expert report also contacted the *BMJ*'s editors and disputed Deer's allegations.[68]

Apparently accepting Deer's allegations, Fiona Godlee inserted the following statement in my Rapid Response: "I am not qualified in medicine or histopathology."[69] To counter these misrepresentations, I provided *BMJ*'s editors with an attachment to my Rapid Response that summarized my credentials as an expert in the collection of colonic biopsies. Godlee replaced the false statement, suggesting that I identify myself as "an expert in clinical studies involving the collection and examination of colonic biopsy samples." Then, after deleting my attachment and editing down my Rapid Response, she described me in an editorial only as "a self-employed environmental microbiologist."[70] Likewise, in Deer's feature article addressing my Rapid Response, he described me as a "self-employed American environmental microbiologist working with Dr. Wakefield."[71] I have not been employed as an environmental microbiologist, self or otherwise, since leaving the U.S. Environmental Protection Agency in 2003.

Research involving colonoscopy is one of my main areas of expertise and was my primary area of research at the University of Georgia from 1998–2008, as Deer and Godlee well knew.[72] In 2003, the State of New York accepted me as an expert on colonic biopsy procedures at a hearing in which the State revoked a physician's medical license. I have published research articles and commentaries in leading science and medical journals, including *Nature*, *The Lancet*, and *Nature Medicine*. In 2010, *Annals of Internal Medicine* rated me in the top 10 percent of its peer reviewers.[73]

As is thoroughly documented in my report to the UKRIO, Godlee, Deer, and the *BMJ* removed key evidence exonerating Dr. Wakefield, misrepresented my qualifications, and in Deer's case, created false allegations of scientific misconduct in an effort, I believe, to discredit me. How can anyone, therefore, argue that Godlee, Deer, and the *BMJ* did not do the same in Dr. Wakefield's case?

DR. WAKEFIELD'S LAWSUITS

Dr. Wakefield moved to the United States where, in early 2004, he and others founded the Thoughtful House Center for Children in Austin, Texas. Then, as Deer continued to vilify Dr. Wakefield, Dr. Wakefield filed a defamation lawsuit against Deer and a British television station in 2005.[74] To continue evaluating the safety of pediatric vaccines, Dr. Wakefield collaborated with a primate research group at the University of Pittsburgh.

In 2005, the London court hearing Dr. Wakefield's lawsuit against Deer denied a request from Dr. Wakefield's lawyers to suspend its proceedings temporarily so that they could properly defend him in the GMC hearings.[75] Given legal and financial constraints, Dr. Wakefield voluntarily withdrew his lawsuit against Deer in favor of pursuing the GMC proceedings and, eventually, his appeal against the GMC as well. In 2009, Dr. Wakefield and the primate group in Pittsburgh reported in the journal *Neurotoxicology* that they observed a significant delay in the acquisition of survival reflexes among newborn rhesus macaques vaccinated with thimerosal-containing Hepatitis B vaccine when compared with control animals receiving a saline placebo or no injection.[76] But, when the GMC issued its findings in 2010, editors at *The Lancet* and *Neurotoxicology*, respectively, retracted both the 1998 study of autistic children by Dr. Wakefield et al.,[77] which raised concerns about the MMR vaccine, and the 2009 primate study implicating thimerosal.[78] Finally, in 2012, Dr. Wakefield filed a defamation lawsuit against the *BMJ*, Deer, and Godlee over their allegations of research fraud published in January 2011 and thereafter.[79]

THE HIGH COURT'S RULING

Prof. Walker-Smith contested each and every finding made by the GMC panel in its Fitness to Practice hearing. His lack of confidence in its findings was recently affirmed by the March 2012 High Court ruling of Justice John Mitting, which overturned all of the panel's findings with respect to codefendant

Prof. Walker-Smith.[80] Justice Mitting criticized the GMC panel on many counts. He concluded, for example, that the panel made "fundamental errors,"[81] distorted evidence,[82] and based its findings on an inadequate analysis of the facts.[83] The panel, Justice Mitting ruled, provided inadequate and superficial reasoning and explanation for its conclusions,[84] inappropriately rejected evidence, relied upon "flawed" and "wrong" reasoning, and "numerous and significant inadequacies" in its conclusions, particularly in its findings in the individual cases of the *Lancet* children.[85] Overall, Justice Mitting described the panel's findings as "not legitimate,"[86] "perverse,"[87] "odd" and "unsustainable,"[88] "wrong,"[89] and "untenable."[90] Justice Mitting went on to state that "Universal inadequacies and some errors on the panel's determination accordingly go to the heart of the case. They are not curable."[91] At the hearing before Justice Mitting, the GMC counsel acknowledged "serious weakness in [the GMC panel's] reasoning."[92]

Justice Mitting ruled that the *Lancet* children were, in fact, consecutively referred, and that medical procedures used in the study, including lumbar punctures and endoscopic biopsies, were clinically indicated and, therefore, did not require approval from an ethics committee. These aspects of the High Court's ruling should apply equally to the GMC's findings concerning Dr. Wakefield. Justice Mitting also concluded that the GMC panel failed to recognize that complex studies often contain elements of both medical practice and research, and that the GMC fell short of distinguishing the difference between the two. In the case of Prof. Walker-Smith, the High Court found ample evidence in the record that his primary goal was to diagnose and treat the *Lancet* patients and that some of the patients did, in fact, benefit from his treatment. On the other hand, he opined, "Dr. Wakefield's purpose was undoubtedly research."[93]

The High Court approach was to assess, individually, whether each scientist's primary goal was to benefit the patients in the study versus patients in general. If the scientist's primary aim was to benefit the patients in the *Lancet* study, he did not need an ethics approval for his work on the project. Ethical approvals, however, are granted to *projects*, not *individuals*. Thus, the High Court's assessment yielded no direct insights into whether the primary goal of the *Lancet* study was medical practice or research.

Case studies, such as the *Lancet* study, are by nature "clinical research." They usually come about when one or more physicians recognize that it would be useful to describe the circumstances, treatments, and outcomes of one or more patients under their care in the medical literature. Physicians often have their own ideas as to what may be causing their patients' unusual disorders or diseases, and they experiment with various treatments accordingly. Usually, their primary goal is to diagnose and treat the patients in the study. Publishing a paper in which they describe their patients' circumstances, symptoms, the results of clinical tests, their own hypotheses about underlying causes, and the patients' outcomes is an important, albeit secondary, goal. Another distinguishing characteristic is that case studies, which are typically clinically oriented, usually involve only a few patients. They serve as a starting point for research projects that require large numbers of patients comprising control and experimental groups, or for dose-response analyses. The *Lancet* study clearly fits in this category. It was primarily oriented toward diagnosing and treating the twelve children in the study.

Fortunately, even though the High Court focused on the primary goals of individual scientists rather than the project as a whole, it found that Prof. Walker-Smith's main goal was diagnosing and treating the *Lancet* children. Thus, in the Court's opinion, he did not need ethical approval because his research was clinical in nature; i.e., primarily aimed at diagnosing and treating the *Lancet* patients. But the High Court erred when it concluded that the ethics statement in the *Lancet* paper—"Investigations were approved by the Ethical Practices Committee of the Royal Free Hospital NHS Trust, and parents gave informed consent"—was untrue. It was mistaken because the GMC should have acknowledged that the Ethical Practices Committee approved the research component of the *Lancet* study under Code 162-95, as the evidence showed, and signed consent forms specific for 162-95 were, in fact, included in the children's medical records (Table 1).

Contrary to picturing Dr. Wakefield as a fraudster and child killer, which many in the news media and scientific literature have painted him based on the false allegations published by Deer and Godlee, the High Court's 2012 ruling exonerating Prof. Walker-Smith projects a different image. It points to the transfer factor developed by Dr. Wakefield and others, which Prof. Walker-Smith used in the *Lancet* study and noted that it improved the health of at least one of the immune-compromised *Lancet* children who had evidence of chronic measles infection. Unfortunately, the fraud allegations Harris and Deer launched against Dr. Wakefield and his coauthors in 2004 made it impossible for UCL to further develop this new therapeutic approach after the *Lancet* study was attacked. Such an approach could save countless lives and may be essential to eradicating the measles virus, especially in underdeveloped areas where malnutrition and other factors render patients susceptible to chronic measles infections.[94]

In conclusion, the majority of the GMC panel's findings against Prof. Walker-Smith were based on identical charges leveled against Dr. Wakefield, and upon which the panel found Dr. Wakefield guilty. Therefore, the panel's findings concerning Dr. Wakefield were also "flawed," "wrong," "perverse," "inadequate," "unsustainable," and "superficial." In light of the High Court's ruling, editors at *The Lancet* are now considering whether the retraction of the 1998 *Lancet* article by Dr. Wakefield et al. should be reversed.[95]

"NO RESPECTABLE BODY"

In my experience, the *BMJ*'s and Deer's assaults on Dr. Wakefield lie completely outside the normal realm of reporting and editorial comment for scientific journals. Even Anderson Cooper began CNN's coverage by announcing: "[The *BMJ*] did something extremely rare for a scientific journal. It accused a researcher, Andrew Wakefield, of outright fraud."[96] The explanation may lie in Godlee's editorial responding to my Rapid Response. There, for the first time, she acknowledged that the *BMJ* receives funding from MMR vaccine manufacturers Merck and GSK."[97] Shortly afterward, she corrected one editorial and two related Editor's Choice articles and elaborated on the journal's conflicts of interest: "The BMJ Group receives advertising and sponsorship revenue from vaccine manufacturers, and specifically from Merck and GSK, which both manufacture MMR vaccines."[98]

Although it appears that the *BMJ* is simply using Deer to protect consumer confidence in its sponsors' products, it has no plans to acknowledge the *BMJ*'s conflicts of interests in any of his articles.[99] Still, the *BMJ*'s admissions seriously undermine Brian Deer's credibility. For one thing, they call into question Deer's contract with CNN, which Anderson Cooper used to vouch for Deer's credibility, saying, "He's actually signed a document guaranteeing that he has no financial interest in any of this, or no financial connections to anyone who has an interest in this."[100]

Justice Mitting concluded, "There is now no respectable body of opinion which supports [Dr. Wakefield's] hypothesis, that MMR vaccine and autism/enterocolitis are causally linked."[101] Aside from the fact that the rubella virus, from which the MMR vaccine is manufactured, is indeed causally linked to autism,[102] it seems ironic that the High Court failed to understand that the absence of such a body of science may well be a direct, and intended, result of the GMC's actions against Dr. Wakefield and Prof. Walker-Smith. As mice vacate a field at the sight of an owl devouring even a single mouse, so it is with scientists. It's called the "ecology of fear."[103] If it were not for what was done to Dr. Wakefield and Walker-Smith, many outstanding scientists would likely be willing to conduct objective research on vaccine safety and publish their results regardless of whether or not they may threaten government policies and industry practices.

Over time, science has evolved into a sophisticated marketing tool for supporting government policies and industry practices. As Godlee testified to Parliament, "Even on the peer-reviewed side of things, it has been said that the journals are the marketing arm of the pharmaceutical industry. That is not untrue."[104] Marketing has a lot more to do with hiding the truth than publishing it, as Godlee's convoluted statement makes clear. Virtually everything published in the peer-reviewed scientific literature is funded or carried out by government agencies, corporations, or universities. Governments hire scientists to support their policies. We know this because government agencies fire scientists who don't support their policies. Corporations hire scientists to develop and defend their products and services. Universities hire scientists to bring in grants from government and industry; faculty members who fail at this don't earn tenure. That's not to say that all, or even most, science is untruthful. But what's published is, for the most part, that portion of the truth of which government and industry have little at stake in the outcome. What is quashed is the portion that would pressure government agencies and large corporations to change policies and practices.

CONCLUSIONS

For his investigations uncovering Dr. Wakefield's alleged research fraud, Deer was twice awarded the British Press Award, once in 1999 for articles in *The Sunday Times* "that ranged from vaccine-damaged children to the hidden side-effects of Viagra," and then again in 2011 after the GMC removed Dr. Wakefield and Prof. Walker-Smith from the medical registry. According to Deer's website, the British Press Awards are "the most coveted honors in UK news media—often described as the 'Oscars of British journalism' and compared with US Pulitzer Prizes."[105] When a panel of journalists presented Deer with his second British Press Award, it referred to his accomplishments as a "tremendous righting of a wrong."[106] But it was Dr. Andrew Wakefield and Prof. Walker-Smith, not Brian Deer, who had it right. By clearing Professor John Walker-Smith, England's High Court of Justice has righted a great wrong, but only partly. A full measure of justice requires that Dr. Andrew Wakefield also be exonerated.

The *Lancet* study, and other papers by Dr. Wakefield and his coauthors, documented a relationship between gastrointestinal disease and sudden developmental regression, which is now widely accepted. Researchers at Columbia

University and Harvard Medical School, for example, recently cited papers by Dr. Wakefield and his coauthors and concluded that gastrointestinal disturbances are commonly reported in children with autism and may contribute to behavioral impairment.[107] The only thing that was ever "wrong" with the *Lancet* article was that it documented the fact that some of the patients' parents and doctors linked the onset of their gastrointestinal disease and sudden developmental regression to MMR vaccination. The truth is, large numbers of parents and physicians have linked the two, especially in cases where the children's symptoms become markedly exacerbated as additional shots are administered.

Although Autism Spectrum Disorders (ASD) were exceedingly rare just over two decades ago, the CDC now estimates (based on the year 2000 birth cohort) that 1 in 88 children born in the United States have ASD (2012); and in South Korea, the estimate is 1 in 38 (2011).[108] With numbers rising this fast and reaching these levels, it's clear that environmental factors, not genetics, are mainly driving the epidemic. Scientists have implicated environmental pollutants and pediatric vaccines as possible causes. Unfortunately, government, industry, and the academic institutions they fund are highly motivated to predetermine the outcomes of research in these two areas. They do so by funding scientists who support government policies and industry practices and by targeting those who don't, which is itself a form of institutional research misconduct.[109]

In other words, autism is largely a man-made crisis created by government and industry. Institutional research misconduct is simply perpetuating the calamity. It appears that every nation has been set on a course to inject every human on the planet with increasingly complex mixtures of chemical and biological agents beginning at birth. The science supporting this bold experiment, however, is subject to government and industry manipulation on a scale that few other areas of science have ever experienced. This, combined with silencing reputable scientists who question vaccine safety, is a prescription for turning the hope science offers for future generations into a global disaster.

Disclaimer: The author's views and opinions are his own and do not necessarily represent those of the National Whistleblowers Center.

Chapter Thirty

THE SUPPRESSION OF SCIENCE

Andrew Wakefield, MB, BS, FRCS, FRCPath

I believe that we are coming to the end of a very dark time in the relatively short but dramatic history of autism. That we are in a time of rapid and dramatic change is clearly evident. Earlier denial of the epidemic in the United States and the United Kingdom is giving way to the reality of a pandemic throughout the developed world and the exclusive genetic model of autism is yielding to realization of an environmentally-driven catastrophe. The rejection of a parent's insights by many in the medical system and the prejudice against children through exclusive insurance coverage, medical ignorance, and refusal to honor parental vaccination choices is driving a grassroots revolution.

The medical system—rooted in a profound ignorance of this disorder, yet unwilling to take the clues that parents' narratives and appropriate investigation of affected children have to offer—has failed you, for which I am ashamed. Science has failed parents and children alike, with investment in poor and corrupted population studies reminiscent of the "tobacco is good for you" pseudoscience of the 1950s and 1960s. Inevitably, however, that failure is being exposed: the corruption, the collusion of influence, and the rife and undisclosed conflicts of interest are floating to the surface.

We also have had to deal, inevitably, with a systematic and ruthless attack on the science that is indicative of a vaccine-autism link, in the form of an expensive, deceitful, and ultimately futile public relations campaign intended to protect policy and profit. Unfortunately, this has been at the expense of children.

As explained so clearly by Brian Martin, Professor of Social Sciences in Woolongon, Australia, supporters of dominant scientific theories—in this case, the inviolable status of the vaccine program—sometimes attack competing, less favored theories in ways that conflict with expectations of proper scientific behavior.[1] To reduce concern about their actions, supporters of the dominant theory can—as they do in the case of the vaccine-autism debate—use a variety of techniques:

- Cover up the violation of expectations;
- Devalue the competing theory and its advocates;
- Interpret their own process as proper;
- Use expert panels, meetings, and other formal processes to give a stamp of approval to the dominant view; and
- Intimidate opponents.

All of these methods, and more, are used regularly by perpetrators of actions widely seen as unjust. The ability to exploit these methods is enhanced by seemingly limitless resources, which is vastly disproportionate to the ability and financial resources of those concerned about vaccine safety and vaccination choice.

The media is an obvious and favored conduit through which supporters of orthodoxy have a tactical advantage over challengers. Variously, journalists are "educated" in-house by the Centers for Diseases Control and Prevention (CDC), recruited and paid, lazy, or just plain ignorant of the facts. For those journalists who have been sympathetic to parental and medical concerns, the consequences have been open criticism, intimidation, and dismissal. Mainstream media outlets not only are frequently dependent upon advertising revenues and share board members with Big Pharma companies, but also are very susceptible to the threat that "if you give any credibility to this theory and children die as a consequence, you will be to blame." When these methods fail, however—as they most certainly will—the attack can backfire on the attackers.

Social media, on the other hand, is an Achilles' heel in the ability of orthodoxy to sustain the big, hungry lie. It is also the device through which the backfire is being felt by an increasingly desperate and disconsolate system.

Turning specifically to my own experience in the arena of MMR vaccine and autism, what has happened to me and my colleagues in the United Kingdom is intended as a lesson, as Voltaire observed, "to encourage the others."

When investigated diligently and appropriately, the twelve Lancet children featured in the humble 1998 case series, and thousands who came after them, turned out to have intestinal inflammation, and there are reasons to believe that this inflammation may, in turn, be linked to their neurological disorder. Thus, the paper captured the essential elements of a new disease syndrome—a potentially treatable syndrome—and that should have been cause for some small celebration. Had the children's regression followed natural chickenpox, the response would have been very different. It didn't. For nine children, behavioral changes and subsequent developmental regression followed exposure to the MMR vaccine; and thereby hangs my own tale.

Those potentially responsible for the catastrophe had a problem, however. The 1998 *Lancet* paper that ignited the fuse of the autism-vaccine wars had to be extinguished in florid, repetitious sound bites delivered through a media sometimes conflicted by corporate cross-pollination, compromised by dependence on advertising revenues, and assisted by pharmaceutical special interest groups.

In an act of editorial genocide, there has been a ruthless effort to expunge these children and their disease from the scientific record.

In 1998, I recommended that children continue to be protected against serious infectious diseases by vaccination, but that single vaccines were my preferred option. I urged that parents be given the choice of how to protect their children. Why? Because as part of the process of due diligence, I researched the safety studies of the combination MMR vaccine and came to the conclusion that they were inadequate. Others have since come to this same conclusion.

In the fall of 1998, six months after *The Lancet* paper was published, choice in the United Kingdom was denied to parents when the government removed the license for the importation of the single measles vaccine. Let me state that again. At a time when demand for single vaccines was at its highest and parents were choosing to continue vaccinating against measles, mumps, and rubella with the single vaccines, choice was removed. When I asked a senior member of the United Kingdom's equivalent of the CDC why, she said that allowing parents to choose single vaccines would destroy their established MMR program. Nevertheless, removing the choice between the triple vaccine and single vaccines did not have the effect they intended—parents chose not to vaccinate against

MMR rather than to use the triple vaccine, and MMR vaccination rates fell in the United Kingdom. The regulators chose to protect policy above children.

The more I looked, the more concerned I became. I found a government whistle-blower, evidence of a secret deal between the government and pharma, the licensing of a dangerous MMR vaccine, collusion, lies, and cover-up; all are now revealed in *Callous Disregard: Autism and Vaccines—The Truth Behind a Tragedy*. And this is just the beginning. I am not alone; what befell me also befell my friend and colleague Arthur Krigsman, a U.S. gastroenterologist who has devoted his career to helping heal children with autism. His story also will be told.

Choice involves informed consent. Informed consent is a crucial element of the foundation upon which ethical medical practice rests. Providing parents with an honest assessment of the risks and benefits of any medical procedure allows them to make an informed choice. The process requires the physician to be, to the best of his or her ability, informed. My research revealed that there is no information available on the medium- to long-term health outcomes in vaccinated versus unvaccinated children. As an example of this, the U.K. government introduced and received cursory twenty-one-day follow-up "safety" data on two dangerous MMR vaccines that caused meningitis that, in the majority of cases, didn't even start until after twenty-one days.[2] They knew this. They missed the signal and it was four years later—four years too late—that this dangerous vaccine that should never have been licensed was withdrawn.

Those who are a threat to public confidence, who do not mandate a "safety first" agenda, who deny choice, are the greatest threat to the vaccine program; they are the ones who are ultimately anti-vaccine. Where do the regulators and the vaccine industry stand in 2011 with their coercive vaccination mandates, their costly public relations programs, their ruthless, pragmatic exorcism of dissent, and their public confidence rating? A recent study from University of Michigan shows that one in four parents think vaccines cause autism[3] and fifty-four percent of parents are worried about serious adverse effects caused by vaccines. The Harris Interactive survey[4] conducted just three months later suggests that vaccine safety, parental rights, and vaccination choice are mainstream parental concerns. The regulators and vaccine industry have failed.

The latest initiative: the American Academy of Pediatrics' industry-sponsored Protect Tomorrow campaign is being thrust into our faces and claims to "[bring] to life the memories of the terrible diseases of the past." It is a campaign rooted in fear. The campaign is a measure of the failure of its architects, rather than the success of their policies.

To parents I would say, trust your instincts above all else. When considering how to vaccinate your children, read, get educated, and demand fully informed consent and answers to your questions. When you are stonewalled or these answers are not to your satisfaction, trust your instinct. I say this as someone who has studied and engaged in the science and who has become aware of the limitations of our knowledge and understanding of vaccine safety issues. Maternal instinct, in contrast, has been a steady hand upon the tiller of evolution; we would not be here without it. As the Buddha said, "Believe nothing, no matter where you read it, or who said it, unless your own reason and your own common sense agree."

To the vaccine industry, regulators, public health officials and doctors, pediatricians, and Bill and Melinda Gates, I would say this: the success of vaccination programs requires the willing participation of consumers. The key to any success, therefore, is public confidence in the scientists, doctors, and policy makers—including the pharmaceutical industry—that shape these programs. In turn, the key to that confidence is a safety first vaccine agenda. Those whose priority is safety first are not anti-vaccine. By analogy, those who ordered the recall of multiple Toyota brands for sticking gas pedals are not anti-car.

Callous Disregard might have been the log of a doomed captain written in the cabin of a war-torn frigate, tacking on shredded sails, running from another—perhaps final—broadside; but it is not. It was written from the bridge, the helm secure, the wind at our backs, and the sails full as the aggressor slips back below an uncertain horizon. The day will belong to Reason. Vaccination choice is a human right.

There will be victory of a sort, and it will be a victory from the bottom up. In the true spirit of the U.S. Constitution, the people will have their say—vaccination choice is a human right. The victory will not come from the top down because a phalanx of lobbyists, "experts," and true believers stands between the President and the people he is sworn to serve. In your commitment to achieving this victory, remember: "Every man is guilty of all the good he did not do."

AFTERWORD

MOVING FROM WORDS TO ACTION

Let us not mince words: the right to vaccination choice is at extreme risk in this country. It already exists more in theory than in practice. But now the American Academy of Pediatrics and its spokespeople are suggesting the abolition of philosophical and religious exemption rights. They have a lot of influential support behind them—the pharmaceutical industry, public health, government, and media. State by state, legislation is being introduced that weakens parental rights vis-à-vis mandatory vaccination. In May 2011, Washington Governor Chris Gregoire signed into law ESB5005,[1] which prevents parents from receiving a non-medical exemption to mandatory vaccination without the signature of health care practitioners. In October 2011, California Governor Jerry Brown signed into law AB499,[2] which gives children as young as twelve the right to give their own consent to receive vaccination against sexually-transmitted diseases (i.e., hepatitis B and human papillomavirus vaccines) without parental notification.

We are in a crisis. You can expect more federally recommended vaccines to be added to the schedule; more mandated vaccines for day care, school, and employment; more vaccines of unproven safety to reach the market; more efforts by industry and government to deny adverse events; more efforts to dismiss liability for vaccine injury; more infectious disease scares on the H1N1 swine flu model; more vaccine-induced injuries and death; and more strenuous efforts to weaken and abolish your right to vaccination choice. These are the trends that have been well-established since compulsory vaccination began and that have been accelerating since the 1990s. Unless we change course, these projections will likely become reality.

If you care about your right to make individual vaccination decisions, there is a lot of work to do. The Center for Personal Rights has started a petition, "The Chicago Principles on Vaccination Choice," (appendix B). The petition articulates five basic principles that follow from this book:

(1) The human right to vaccination choice;

(2) The right to non-discrimination based on vaccination status;

(3) The right to complete disclosure of vaccination risks;

(4) A safety first agenda for national vaccine policy; and

(5) Zero tolerance for actual or perceived conflicts of interest in national vaccine policy.

CPR has also called for the following immediate actions in the United States:

(1) The philosophical exemption to vaccination in every state;
(2) Congressional oversight hearings on the national vaccine program;
(3) A prospective study of vaccinated versus unvaccinated populations for long-term health outcomes;
(4) Amendment of the 1986 National Childhood Vaccine Injury Act to reintroduce ordinary tort liability for vaccine manufacturers and medical professionals; and
(5) Dissolution of the Vaccine Injury Compensation Program as currently constituted.

Things You Can Do Now

If you care about this issue, here's what you can do.

1. **Take a Stand.**
   Read the Chicago Principles on Vaccination Choice and the Calls for Immediate Action in the United States. You can find these on the CPR website at www.centerforpersonalrights.org.
2. **Stay Informed.**
   Join our mailing list for news updates relating to personal rights and vaccination choice.
3. **Reach Out.**
   Talk to a friend about this book. Give a copy to your pediatrician, legislator, or colleague. Recommend the book for reading lists in ethics, law, philosophy, history of science, public health, and other classes. Use the Public Discussion Presentation (see appendix) to hold your own meetings. Contact CPR and make arrangements for one of the authors to speak at your church, college, support group, or Parent Teacher Association.
4. **Donate.**
   Give what you can. Make a financial contribution or volunteer your time to raise public awareness about the need for vaccination choice. The majority

of American parents support vaccination choice—they just don't yet know that they are the majority. Learn more at www.centerforpersonalrights.org.

Join with the Center for Personal Rights to protect the human right to vaccination choice. This is an idea whose time has come.

ACKNOWLEDGEMENTS

We are moved by the extraordinary generosity of our contributing authors, who received no compensation and entrusted CPR with the gift of their stories and work. Likewise, there would be no *Vaccine Epidemic* without Tony Lyons of Skyhorse Publishing. We called him for advice on self-publishing a small booklet—his offer to publish our work delighted and inspired us to do more. We acknowledge many who donated their expertise, time, and resources to support CPR's mission. We express special appreciation to Ed and Teri Arranga, Miranda Bailey, Twila Brase, Maureen Drummond, Claire Dwoskin, Laura & Rick Hayes, Jennifer VanDerHorst-Larson & David Larson, Barbara Mullarkey, Barry & Dolly Segal, and F. Edward Yazbak, MD, FAAP.

In addition, we thank the following individuals and organizations for their generous support and contributions:

Laraine Abbey-Katzev, RN
Stacy Senior Allan
Casey Alls
Jo Antonetti
Conne Bard
Kevin Barry, Esq.
Melody Benbow
Chris Birt
Christina Blakey
Victoria Bloch
Jessica Burckhard, DC
Karen Burns
Sandy Caldwell
Hugh Califf
Amy Carson
Gayle Casas
Theresa Cedillo
Laura Cellini
Angela Chiu
Sue Collins
Chris Conley
Teresa Conrick
Joni Cox
Mary Coyle, DI, Hom
Hilary Downing
Jacqueline Dulin
Rebecca Estepp
Malika Felix
Megan Fennelly
Jim Fick, OD
Amy Galarowicz
Kristin Selby Gonzales
Sandy Gottstein
Marian Greene
Ronald Habakus
Janelle Hall
Brian & Carla Hasenkamp
Mary Hirzel
James F. Holland, MD, ScD(hc)
Jimmie Holland, MD
Therese Holliday
Greta Huizenga

Diane Hunter
Laurette Janak
Steve Janak
Max Kane
Jennifer Keefe, Esq.
Penney Kersten
Abdulkadir Khalif
Carly Krakow
Kitty Kurth
Lori Lee, ND, CTN
Sandy Lee, MD
Janelle Lewis
Christina Liberatore
Curt & Kim Linderman
Claudine Liss, Esq.
Sandra Liu
Manette Loudon
Dawn Loughborough
Alison MacNeil
Bobbie Manning
Jennifer Margulies
Karen McDonough
Angela Medlin
Maurine Meleck
Tiffany Montanti
Erik Nanstiel
Mia Nitchun
Gary Null, PhD
Julie Obradovic
Tetyana Obukhanych, PhD
Peggy O'Mara
Michael D. Ostrolenk
Larry Palevsky, MD, FAAP
Alan Phillips, Esq.
Sylvia Pimentel
Kathryne Pirtle
Rebekah Pizana
Nancy Potts
Lisa Purdon
Heidi Roger
Robert Schecter
Kim Scheutte
Harry & Gina Tembenis
Wade Rankin, Esq.
Sym Rankin, RN
Jeanna Reed
Laura Rowley
Lisa Rudley
Lisa Rupe
Linda Smeltzer
Karen Steinberg
M. Kelly Sutton, MD
Eva Vanamee, PhD
Carlos & Branca Veloso
Heather Walker
Angela Warner
Tim Welch
Focus Autism, Inc.
Holistic Moms Network
New Jersey Coalition for Vaccination Choice
ParentalRights.org
Weston A. Price Foundation

From Louise: I am grateful to my brave and devoted parents. I owe a deep debt to my in-laws who saw us through. I acknowledge friends, mentors, and colleagues in the trenches. Thank you, Krista. Above all, this is for R, I, and N, with my love.

From Mary: To A, with love and appreciation. I thank many guides who have led the way and my parents for their steadfast love and support.

From Kim: To H, my inspiration for this work, and C, for all of their love, support, and tolerance. To my family and in-laws for always being there. To all those along this journey who have inspired, shared with, and supported me. I am grateful for each of you.

APPENDIX A: SUPPORTING DOCUMENTS

APPENDIX TO CHAPTER SEVEN: AN URGENT CALL FOR MORE RESEARCH

Abstracts from Underlying Scientific Studies

Account Res. 2012;19(2):65-88.

CONFLICTS OF INTEREST IN VACCINE SAFETY RESEARCH.

DeLong G.
Department of Economics and Finance, Baruch College, New York, New York, 10010, USA. Gayle.delong@baruch.cuny.edu

ABSTRACT

Conflicts of interest (COIs) cloud vaccine safety research. Sponsors of research have competing interests that may impede the objective study of vaccine side effects. Vaccine manufacturers, health officials, and medical journals may have financial and bureaucratic reasons for not wanting to acknowledge the risks of vaccines. Conversely, some advocacy groups may have legislative and financial reasons to sponsor research that finds risks in vaccines. Using the vaccine-autism debate as an illustration, this article details the conflicts of interest each of these groups faces, outlines the current state of vaccine safety research, and suggests remedies to address COIs. Minimizing COIs in vaccine safety research could reduce research bias and restore greater trust in the vaccine program.

PMID: 22375842 [PubMed – indexed for MEDLINE]

Apoptosis, 2012 May;17(5) 516-27. doi: 10.1007/s10495-011-0690-1.

HEPATITIS B VACCINE INDUCES APOPTOTIC DEATH IN HEPA1-6 CELLS.

Hamza H, Cao J, Li X, Li C, Zhu M, Zhao S.

Key Lab of Agricultural Animal Genetics, Breeding, and Reproduction of Ministry of Education, College of Animal Science and Technology, Huazhong Agricultural University, Wuhan, People's Republic of China. Heyam68_hamza@yahoo.com

ABSTRACT

Vaccines can have adverse side-effects, and these are predominantly associated with the inclusion of chemical additives such as aluminum hydroxide adjuvant. The objective of this study was to establish an in vitro model system amenable to mechanistic investigations of cytotoxicity induced by hepatitis B vaccine, and to investigate the mechanisms of vaccine-induced cell death. The mouse liver hepatoma cell line Hepa1-6 was treated with two doses of adjuvanted (aluminium hydroxide) hepatitis B vaccine (0.5 and 1 μg protein per ml) and cell integrity was measured after 24, 48 and 72 h. Hepatitis B vaccine exposure increased cell apoptosis as detected by flow cytometry and TUNEL assay. Vaccine exposure was accompanied by significant increases in the levels of activated caspase 3, a key effector caspase in the apoptosis cascade. Early transcriptional events were detected by qRT-PCR. We report that hepatitis B vaccine exposure resulted in significant upregulation of the key genes encoding caspase 7, caspase 9, Inhibitor caspase-activated DNase (ICAD), Rho-associated coiled-coil containing protein kinase 1 (ROCK-1), and Apoptotic protease activating factor 1 (Apaf-1). Upregulation of cleaved caspase 3,7 were detected by western blot in addition to Apaf-1 and caspase 9 expressions argues that cell death takes place via the intrinsic apoptotic pathway in which release of cytochrome c from the mitochondria triggers the assembly of a caspase activation complex. We conclude that exposure of Hepa1-6 cells to a low dose of adjuvanted hepatitis B vaccine leads to loss of mitochondrial integrity, apoptosis induction, and cell death, apoptosis effect was observed also in C2C12 mouse myoblast cell line after treated with low dose of vaccine (0.3, 0.1, 0.05 μg/ml). In addition In vivo apoptotic effect of hepatitis B vaccine was observed in mouse liver.

PMID: 22249285 [PubMed – in process]

Centers for Disease Control and Prevention Morbidity and Mortality Weekly Report (MMWR); Surveillance Summaries; March 30, 2012 / 61(SS03);1-19.

PREVALENCE OF AUTISM SPECTRUM DISORDERS — AUTISM AND DEVELOPMENTAL DISABILITIES MONITORING NETWORK, 14 SITES, UNITED STATES, 2008.

Autism and Developmental Disabilities Monitoring Network Surveillance Year 2008 Principal Investigators

Corresponding author: Jon Baio, EdS, National Center on Birth Defects and Developmental Disabilities, CDC, 1600 Clifton Road, MS E-86, Atlanta, GA 30333. Telephone: 404-498-3873; Fax: 404-498-3550; Email: jbaio@cdc.gov

ABSTRACT

PROBLEM/CONDITION: Autism spectrum disorders (ASDs) are a group of developmental disabilities characterized by impairments in social interaction and communication and by restricted, repetitive, and stereotyped patterns of behavior. Symptoms typically are apparent before age 3 years. The complex nature of these disorders, coupled with a lack of biologic markers for diagnosis and changes in clinical definitions over time, creates challenges in monitoring the prevalence of ASDs. Accurate reporting of data is essential to understand the prevalence of ASDs in the population and can help direct research.

PERIOD COVERED: 2008.

DESCRIPTION OF SYSTEM: The Autism and Developmental Disabilities Monitoring (ADDM) Network is an active surveillance system that estimates the prevalence of ASDs and describes other characteristics among children aged 8 years whose parents or guardians reside within 14 ADDM sites in the United States. ADDM does not rely on professional or family reporting of an existing ASD diagnosis or classification to ascertain case status. Instead, information is obtained from children's evaluation records to determine the presence of ASD symptoms at any time from birth through the end of the year when the child reaches age 8 years. ADDM focuses on children aged 8 years because a baseline study conducted by CDC demonstrated that this is the age of identified peak prevalence. A child is included as meeting the surveillance case definition for an ASD if he or she displays behaviors (as described on a comprehensive evaluation completed by a qualified professional) consistent with the American Psychiatric

Association's Diagnostic and Statistical Manual-IV, Text Revision (DSM-IV-TR) diagnostic criteria for any of the following conditions: Autistic Disorder; Pervasive Developmental Disorder–Not Otherwise Specified (PDD-NOS, including Atypical Autism); or Asperger Disorder. The first phase of the ADDM methodology involves screening and abstraction of comprehensive evaluations completed by professional providers at multiple data sources in the community. Multiple data sources are included, ranging from general pediatric health clinics to specialized programs for children with developmental disabilities. In addition, many ADDM sites also review and abstract records of children receiving special education services in public schools. In the second phase of the study, all abstracted evaluations are reviewed by trained clinicians to determine ASD case status. Because the case definition and surveillance methods have remained consistent across all ADDM surveillance years to date, comparisons to results for earlier surveillance years can be made. This report provides updated ASD prevalence estimates from the 2008 surveillance year, representing 14 ADDM areas in the United States. In addition to prevalence estimates, characteristics of the population of children with ASDs are described, as well as detailed comparisons of the 2008 surveillance year findings with those for the 2002 and 2006 surveillance years.

RESULTS: For 2008, the overall estimated prevalence of ASDs among the 14 ADDM sites was 11.3 per 1,000 (one in 88) children aged 8 years who were living in these communities during 2008. Overall ASD prevalence estimates varied widely across all sites (range: 4.8–21.2 per 1,000 children aged 8 years). ASD prevalence estimates also varied widely by sex and by racial/ethnic group. Approximately one in 54 boys and one in 252 girls living in the ADDM Network communities were identified as having ASDs. Comparison of 2008 findings with those for earlier surveillance years indicated an increase in estimated ASD prevalence of 23% when the 2008 data were compared with the data for 2006 (from 9.0 per 1,000 children aged 8 years in 2006 to 11.0 in 2008 for the 11 sites that provided data for both surveillance years) and an estimated increase of 78% when the 2008 data were compared with the data for 2002 (from 6.4 per 1,000 children aged 8 years in 2002 to 11.4 in 2008 for the 13 sites that provided data for both surveillance years). Because the ADDM Network sites do not make up a nationally representative sample, these combined prevalence estimates should not be generalized to the United States as a whole.

INTERPRETATION: These data confirm that the estimated prevalence of ASDs identified in the ADDM network surveillance populations continues to increase. The extent to which these increases reflect better case ascertainment as a result of increases in awareness and access to services or true increases in prevalence of ASD symptoms is not known. ASDs continue to be an important public health concern in the United States, underscoring the need for continued resources to identify potential risk factors and to provide essential supports for persons with ASDs and their families.

PUBLIC HEALTH ACTION: Given substantial increases in ASD prevalence estimates over a relatively short period, overall and within various subgroups of the population, continued monitoring is needed to quantify and understand these patterns. With 5 biennial surveillance years completed in the past decade, the ADDM Network continues to monitor prevalence and characteristics of ASDs and other developmental disabilities for the 2010 surveillance year. Further work is needed to evaluate multiple factors contributing to increases in estimated ASD prevalence over time. ADDM Network investigators continue to explore these factors, with a focus on understanding disparities in the identification of ASDs among certain subgroups and on how these disparities have contributed to changes in the estimated prevalence of ASDs. CDC is partnering with other federal and private partners in a coordinated response to identify risk factors for ASDs and to meet the needs of persons with ASDs and their families.

Disabil Health J. 2012 Jan;5(1) 18-25. Epub 2011 Nov 3.

ASSOCIATION BETWEEN PARENTAL NATIVITY AND AUTISM SPECTRUM DISORDER AMONG US-BORN NON-HISPANIC WHITE AND HISPANIC CHILDREN, 2007 NATIONAL SURVEY OF CHILDREN'S HEALTH.

Schieve LA, Boulet SL, Blumberg SJ, Kogan MD, Yeargin-Allsopp M, Boyle CA, Visser SN, Rice C.

National Center on Birth Defects and Developmental Disabilities, Centers for Disease Control and Prevention, Atlanta, GA 30333, USA. LSchieve@cdc.gov

ABSTRACT

BACKGROUND: Limited studies suggest the prevalence of autism spectrum disorders (ASD) varies by whether maternal and child birth country are discordant.

OBJECTIVE/HYPOTHESIS: We explored associations between ASD and maternal and paternal nativity in a sample of US-born non-Hispanic white

(NHW, n = 37,265) and US-born Hispanic (n = 4,690) children in the 2007 National Survey of Children's Health (NSCH).

METHODS: We assessed ASD prevalence within race-ethnicity and parental nativity subgroups. Prevalence ratios (aPR), comparing each group to NHW children with 2 US-born parents, were adjusted for child age, sex, and receipt of care in a medical home. Estimates were weighted to reflect US noninstitutionalized children. Standard errors were adjusted to account for the complex sample design.

RESULTS: In NHW children with 2 US-born parents, ASD prevalence was 1.19%; estimates were similar for NHW children with a foreign-born mother or father. There was a striking heterogeneity between Hispanic children with 2 US-born versus 2 foreign-born parents (ASD prevalence 2.39% versus 0.31%, p = .05). Even after adjustment, PRs comparing ASD prevalence in Hispanic versus NHW children were vastly different for Hispanic subgroups, suggesting a substantially lower prevalence for Hispanic children with both parents foreign-born (aPR 0.2, 95% confidence interval 0.1-0.5) and a higher prevalence for Hispanic children with both parents US-born (aPR 2.0 [0.8-4.6]).

CONCLUSIONS: Previous studies comparing ASD prevalence between NHW and Hispanic children based on a composite Hispanic grouping without consideration of parental nativity likely missed important differences between these racial-ethnic groups. Continuing efforts toward improving early identification in Hispanic children are needed.

PMID: 22226294 [PubMed – indexed for MEDLINE]

J Toxicol Environ Health A. 2011;74(14):903-16.

A POSITIVE ASSOCIATION FOUND BETWEEN AUTISM PREVALENCE AND CHILDHOOD VACCINATION UPTAKE ACROSS THE U.S. POPULATION.

DeLong G.

Department of Economics and Finance, Baruch College, New York, New York, 10010, USA. Gayle.delong@baruch.cuny.edu

ABSTRACT

The reason for the rapid rise of autism in the United States that began in the 1990s is a mystery. Although individuals probably have a genetic predisposition to develop autism, researchers suspect that one or more environmental triggers are also needed. One of those triggers might be the battery of vaccinations that young children receive. Using regression analysis and controlling for family income and ethnicity, the relationship between the proportion of children who received the recommended vaccines by age 2 years and the prevalence of autism (AUT) or speech or language impairment (SLI) in each U.S. state from 2001 and 2007 was determined. A positive and statistically significant relationship was found: The higher the proportion of children receiving recommended vaccinations, the higher was the prevalence of AUT or SLI. A 1% increase in vaccination was associated with an additional 680 children having AUT or SLI. Neither parental behavior nor access to care affected the results, since vaccination proportions were not significantly related (statistically) to any other disability or to the number of pediatricians in a U.S. state. The results suggest that although mercury has been removed from many vaccines, other culprits may link vaccines to autism. Further study into the relationship between vaccines and autism is warranted.

PMID: 21623535 [PubMed – indexed for MEDLINE]

J Inorg Biochem. 2011 Nov;105(11):1489-99. Epub 2011 Aug 23.

DO ALUMINUM VACCINE ADJUVANTS CONTRIBUTE TO THE RISING PREVALENCE OF AUTISM?

Tomljenovic L, Shaw CA.
Neural Dynamics Research Group, Department of Ophthalmology and Visual Sciences, University of British Columbia, 828 W. 10th Ave, Vancouver, BC, Canada V5Z 1L8. lucijat77@gmail.com

ABSTRACT

Autism spectrum disorders (ASD) are serious multisystem developmental disorders and an urgent global public health concern. Dysfunctional immunity and impaired brain function are core deficits in ASD. Aluminum (Al), the most commonly used vaccine adjuvant, is a demonstrated neurotoxin and a strong immune stimulator. Hence, adjuvant Al has the potential to induce neuroimmune disorders. When assessing adjuvant toxicity in children, two key points ought to be considered: (i) children should not be viewed as "small adults" as their unique physiology makes them much more vulnerable to toxic insults; and (ii) if exposure to Al from only few vaccines can lead to cognitive impairment and autoimmunity in adults, is it unreasonable to question whether the current pediatric schedules, often containing 18 Al adjuvanted vaccines, are safe for children? By applying Hill's criteria for establishing causality between exposure and outcome we investigated whether exposure to Al from vaccines could be contributing to the rise in ASD prevalence in the Western world. Our results show that: (i) children from countries with the highest ASD prevalence appear to have the highest exposure to Al from vaccines; (ii) the increase in exposure to Al adjuvants significantly correlates with the increase in ASD prevalence in the United States observed over the last two decades (Pearson $r=0.92$, $p<0.0001$); and (iii) a significant correlation exists between the amounts of Al administered to preschool children and the current prevalence of ASD in seven Western countries, particularly at 3-4 months of age (Pearson r=0.89-0.94, p=0.0018-0.0248). The application of the Hill's criteria to these data indicates that the correlation between Al in vaccines and ASD may be causal. Because children represent a fraction of the population most at risk for complications following exposure to Al, a more rigorous evaluation of Al adjuvant safety seems warranted.

PMID: 22099159 [PubMed – indexed for MEDLINE]

Arch Gen Psychiatry. 2011 Nov;68(11):1095-102. doi: 10.1001/archgenpsychiatry.2011.76. Epub 2011 Jul 4.

GENETIC HERITABILITY AND SHARED ENVIRONMENTAL FACTORS AMONG TWIN PAIRS WITH AUTISM.

Hallmayer J, Cleveland S, Torres A, Phillips J, Cohen B, Torigoe T, Miller J, Fedele A, Collins J, Smith K, Lotspeich L, Croen LA, Ozonoff S, Lajonchere C, Grether JK, Risch N.

Department of Psychiatry, Stanford University School of Medicine, USA. joachimh@stanford.edu

ABSTRACT

CONTEXT: Autism is considered the most heritable of neurodevelopmental disorders, mainly because of the large difference in concordance rates between monozygotic and dizygotic twins.

OBJECTIVE: To provide rigorous quantitative estimates of genetic heritability of autism and the effects of shared environment. Design, Setting, and

PARTICIPANTS: Twin pairs with at least 1 twin with an autism spectrum disorder (ASD) born between 1987 and 2004 were identified through the California Department of Developmental Services.

MAIN OUTCOME MEASURES: Structured diagnostic assessments (Autism Diagnostic Interview-Revised and Autism Diagnostic Observation Schedule) were completed on 192 twin pairs. Concordance rates were calculated and parametric models were fitted for 2 definitions, 1 narrow (strict autism) and 1 broad (ASD).

RESULTS: For strict autism, probandwise concordance for male twins was 0.58 for 40 monozygotic pairs (95% confidence interval [CI], 0.42-0.74) and 0.21 for 31 dizygotic pairs (95% CI, 0.09-0.43); for female twins, the concordance was 0.60 for 7 monozygotic pairs (95% CI, 0.28-0.90) and 0.27 for 10 dizygotic pairs (95% CI, 0.09-0.69). For ASD, the probandwise concordance for male twins was 0.77 for 45 monozygotic pairs (95% CI, 0.65-0.86) and 0.31 for 45 dizygotic pairs (95% CI, 0.16-0.46); for female twins, the concordance was 0.50 for 9 monozygotic pairs (95% CI, 0.16-0.84) and 0.36 for 13 dizygotic pairs (95% CI, 0.11-0.60). A large proportion of the variance in liability can be explained by shared environmental factors (55%; 95% CI, 9%-81% for autism and 58%; 95% CI, 30%-80% for ASD) in addition to moderate genetic heritability (37%; 95% CI, 8%-84% for autism and 38%; 95% CI, 14%-67% for ASD)..

CONCLUSION: Susceptibility to ASD has moderate genetic heritability and a substantial shared twin environmental component.

PMID: 21727249 [PubMed – indexed for MEDLINE]

Hum Exp Toxicol. 2011 Sep:30(9):1420-8. Epub 2011 May 4.

INFANT MORTALITY RATES REGRESSED AGAINST NUMBER OF VACCINE DOSES ROUTINELY GIVEN: IS THERE A BIOCHEMICAL OR SYNERGISTIC TOXICITY?

Miller NZ, Goldman GS.
Think Twice Global Vaccine Institute, USA. neilzmiller@gmail.com [corrected]

ABSTRACT

The infant mortality rate (IMR) is one of the most important indicators of the socio-economic well-being and public health conditions of a country. The US childhood immunization schedule specifies 26 vaccine doses for infants aged less than 1 year—the most in the world—yet 33 nations have lower IMRs. Using linear regression, the immunization schedules of these 34 nations were examined and a correlation coefficient of $r = 0.70$ ($p < 0.0001$) was found between IMRs and the number of vaccine doses routinely given to infants. Nations were also grouped into five different vaccine dose ranges: 12–14, 15–17, 18–20, 21–23, and 24–26. The mean IMRs of all nations within each group were then calculated. Linear regression analysis of unweighted mean IMRs showed a high statistically significant correlation between increasing number of vaccine doses and increasing infant mortality rates, with $r = 0.992$ ($p = 0.0009$). Using the Tukey-Kramer test, statistically significant differences in mean IMRs were found between nations giving 12–14 vaccine doses and those giving 21–23, and 24–26 doses. A closer inspection of correlations between vaccine doses, biochemical or synergistic toxicity, and IMRs is essential.

PMID: 21543527 [PubMed – indexed for MEDLINE] PMCID: PMC3170075

Acta Neurobiol Exp (Wars). 2010;70(2):165-76.

SORTING OUT THE SPINNING OF AUTISM: HEAVY METALS AND THE QUESTION OF INCIDENCE.

DeSoto MC, Hitlan RT.
Department of Psychology, University of Northern Iowa, Cedar Falls, Iowa, USA. cathy.desoto@uni.edu

ABSTRACT

The reasons for the rise in autism prevalence are a subject of heated professional debate. Featuring a critical appraisal of some research used to question whether rising levels of autism are related to environmental exposure to toxins (Soden et al. 2007, Barbaresi et al. 2009, Thompson et al. 2007) we aim to evaluate the actual state of scientific knowledge. In addition, we surveyed the empirical research on the topic of autism and heavy metal toxins. In our opinion empirical investigations are finding support for a link with heavy metal toxins. The various causes that have led to the increase in autism diagnosis are likely multi-faceted, and understanding the causes is one of the most important health topics today. We argue that scientific research does not support rejecting the link between the neurodevelopmental disorder of autism and toxic exposures.

PMID: 20628440 [PubMed – indexed for MEDLINE]

Environ Sci Technol. 2010 Mar 15;44(6):2112-8.

TIMING OF INCREASED AUTISTIC DISORDER CUMULATIVE INCIDENCE.

McDonald ME, Paul JF.

National Health and Environmental Effects Research Laboratory, U.S. Environmental Protection Agency, MD-B343-06, Research Triangle Park, North Carolina 27711, USA. mcdonald.michael@epa.gov

ABSTRACT

Autistic disorder (AD) is a severe neurodevelopmental disorder typically identified in early childhood. Both genetic and environmental factors are implicated in its etiology. The number of individuals identified as having autism has increased dramatically in recent years, but whether some proportion of this increase is real is unknown. If real, susceptible populations may have exposure to controllable exogenous stressors. Using literature AD data from long-term (approximately 10-year) studies, we determined cumulative incidence of AD for each cohort within each study. These data for each study were examined for a changepoint year in which the AD cumulative incidence first increased. We used data sets from Denmark, California, Japan, and a worldwide composite of studies. In the Danish, California, and worldwide data sets, we found that an increase in AD cumulative incidence began about 1988-1989. The Japanese study (1988-1996) had AD cumulative incidence increasing continuously, and no changepoint year could be calculated. Although the debate about the nature of increasing autism continues, the potential for this increase to be real and involve exogenous environmental stressors exists. The timing of an increase in autism incidence may help in screening for potential candidate environmental stressors.

PMID: 20158232 [PubMed – indexed for MEDLINE]

J Toxicol Environ Health A. 2010;73(24):1665-77.

HEPATITIS B VACCINATION OF MALE NEONATES AND AUTISM DIAGNOSIS, NHIS 1997-2002.

Gallagher CM, Goodman MS.
PhD Program in Population Health and Clinical Outcomes Research, Stony Brook University Medical Center, State University of New York at Stony Brook, Stony Brook, New York, USA. cmgallagher@notes.cc.sunysb.edu

ABSTRACT

Universal hepatitis B vaccination was recommended for U.S. newborns in 1991; however, safety findings are mixed. The association between hepatitis B vaccination of male neonates and parental report of autism diagnosis was determined. This cross-sectional study used weighted probability samples obtained from National Health Interview Survey 1997-2002 data sets. Vaccination status was determined from the vaccination record. Logistic regression was used to estimate the odds for autism diagnosis associated with neonatal hepatitis B vaccination among boys age 3-17 years, born before 1999, adjusted for race, maternal education, and two-parent household. Boys vaccinated as neonates had threefold greater odds for autism diagnosis compared to boys never vaccinated or vaccinated after the first month of life. Non-Hispanic white boys were 64% less likely to have autism diagnosis relative to non white boys. Findings suggest that U.S. male neonates vaccinated with the hepatitis B vaccine prior to 1999 (from vaccination record) had a threefold higher risk for parental report of autism diagnosis compared to boys not vaccinated as neonates during that same time period. Nonwhite boys bore a greater risk.

PMID: 21058170 [PubMed – indexed for MEDLINE]

Acta Neurobiol Exp (Wars). 2010;70(2):147-64.

INFLUENCE OF PEDIATRIC VACCINES ON AMYGDALA GROWTH AND OPIOID LIGAND BINDING IN RHESUS MACAQUE INFANTS: A PILOT STUDY.

Hewitson L, Lopresti BJ, Stott C, Mason NS, Tomko J.

Department of Obstetrics and Gynecology, University of Pittsburgh School of Medicine, Pittsburgh, PA, USA. lch1@pitt.edu

ABSTRACT

This longitudinal, case-control pilot study examined amygdala growth in rhesus macaque infants receiving the complete US childhood vaccine schedule (1994-1999). Longitudinal structural and functional neuroimaging was undertaken to examine central effects of the vaccine regimen on the developing brain. Vaccine-exposed and saline-injected control infants underwent MRI and PET imaging at approximately 4 and 6 months of age, representing two specific timeframes within the vaccination schedule. Volumetric analyses showed that exposed animals did not undergo the maturational changes over time in amygdala volume that was observed in unexposed animals. After controlling for left amygdala volume, the binding of the opioid antagonist [(11)C]diprenorphine (DPN) in exposed animals remained relatively constant over time, compared with unexposed animals, in which a significant decrease in [(11)C]DPN binding occurred. These results suggest that maturational changes in amygdala volume and the binding capacity of [(11)C]DPN in the amygdala was significantly altered in infant macaques receiving the vaccine schedule. The macaque infant is a relevant animal model in which to investigate specific environmental exposures and structural/functional neuroimaging during neurodevelopment.

PMID: 20628439 [PubMed – indexed for MEDLINE]

J Toxicol Environ Health A. 2010 Jan;73(19):1298-313.

DELAYED ACQUISITION OF NEONATAL REFLEXES IN NEWBORN PRIMATES RECEIVING A THIMEROSAL-CONTAINING HEPATITIS B VACCINE: INFLUENCE OF GESTATIONAL AGE AND BIRTH WEIGHT.

Hewitson L, Houser LA, Stott C, Sackett G, Tomko JL, Atwood D, Blue L, White ER.

Department of Obstetrics and Gynecology, University of Pittsburgh School of Medicine, Pittsburgh, Pennsylvania, USA. lch1@pitt.edu

ABSTRACT

This study examined whether acquisition of neonatal reflexes in newborn rhesus macaques was influenced by receipt of a single neonatal dose of hepatitis B vaccine containing the preservative thimerosal (Th). Hepatitis B vaccine containing a weight-adjusted Th dose was administered to male macaques within 24 h of birth (n = 13). Unexposed animals received saline placebo (n = 4) or no injection (n = 3). Infants were tested daily for acquisition of nine survival, motor, and sensorimotor reflexes. In exposed animals there was a significant delay in the acquisition of root, snout, and suck reflexes, compared with unexposed animals. No neonatal responses were significantly delayed in unexposed animals. Gestational age (GA) and birth weight (BW) were not significantly correlated. Cox regression models were used to evaluate main effects and interactions of exposure with BW and GA as independent predictors and time-invariant covariates. Significant main effects remained for exposure on root and suck when controlling for GA and BW, such that exposed animals were relatively delayed in time-to-criterion. Interaction models indicated there were various interactions between exposure, GA, and BW and that inclusion of the relevant interaction terms significantly improved model fit. This, in turn, indicated that lower BW and/or lower GA exacerbated the adverse effects following vaccine exposure. This primate model provides a possible means of assessing adverse neurodevelopmental outcomes from neonatal Th-containing hepatitis B vaccine exposure, particularly in infants of lower GA or BW. The mechanisms underlying these effects and the requirements for Th requires further study.

PMID: 20711932 [PubMed – indexed for MEDLINE]

Neurotoxicology. 2009 May;30(3):331-7. Epub 2009 Mar 21.

OCKHAM'S RAZOR AND AUTISM: THE CASE FOR DEVELOPMENTAL NEUROTOXINS CONTRIBUTING TO A DISEASE OF NEURODEVELOPMENT.

DeSoto MC, Hitlan RT.

Department of Psychology, University of Northern Iowa, Baker Hall, Cedar Falls, IA 50614-0505, United States. cathy.desoto@uni.edu

ABSTRACT

Much professional awareness regarding environmental triggers for ASD has been narrowly focused on a single possible exposure pathway (vaccines). Meanwhile, empirical support for environmental toxins as a broad class has been quietly accumulating. Recent research has shown that persons with ASD have comparatively higher levels of various toxins and are more likely to have reduced detoxifying ability, and, that rates of ASD may be higher in areas with greater pollution. This report documents that within the state with the highest rate of ASD, the rate is higher for schools near EPA Superfund sites, $t(332)=3.84$, $p=.0001$. The reasons for the rise in diagnoses likely involve genetically predisposed individuals being exposed to various environmental triggers at higher rates than in past generations.

PMID: 19442816 [PubMed – indexed for MEDLINE]

Health Place, 2009 Mar;15(1):18-24. Epub 2008 Feb 12.

PROXIMITY TO POINT SOURCES OF ENVIRONMENTAL MERCURY RELEASE AS A PREDICTOR OF AUTISM PREVALENCE.

Palmer RF, Blanchard S, Wood R.

University of Texas Health Science Center, San Antonio Department of Family and Community Medicine, 7703 Floyd Curl Drive, San Antonio, Texas Mail Code 7794, TX 78229-3900, USA. palmerr@uthscsa.edu

ABSTRACT

The objective of this study was to determine if proximity to sources of mercury pollution in 1998 were related to autism prevalence in 2002. Autism count data from the Texas Educational Agency and environmental mercury release data from the Environmental Protection Agency were used. We found that for every 1000 pounds of industrial release, there was a corresponding 2.6% increase in autism rates ($p<.05$) and a 3.7% increase associated with power plant emissions($P<.05$). Distances to these sources were independent predictors after adjustment for relevant covariates. For every 10 miles from industrial or power plant sources, there was an associated decreased autism Incident Risk of 2.0% and 1.4%, respectively ($p<.05$). While design limitations preclude interpretation of individual risk, further investigations of environmental risks to child development issues are warranted.

PMID: 18353703 [PubMed – indexed for MEDLINE]

Epidemiology. 2009 Jan;20(1):84-90.

THE RISE IN AUTISM AND THE ROLE OF AGE AT DIAGNOSIS.

Hertz-Picciotto I, Delwiche L.

Department of Public Health Sciences, University of California, Davis, California 95616, USA. ihp@ucdavis.edu

ABSTRACT

BACKGROUND: Autism prevalence in California, based on individuals eligible for state-funded services, rose throughout the 1990s. The extent to which this trend is explained by changes in age at diagnosis or inclusion of milder cases has not been previously evaluated.

METHODS: Autism cases were identified from 1990 through 2006 in databases of the California Department of Developmental Services, which coordinates services for individuals with specific developmental disorders. The main outcomes were population incident cases younger than age 10 years for each quarter, cumulative incidence by age and birth year, age-specific incidence rates stratified by birth year, and proportions of diagnoses by age across birth years.

RESULTS: Autism incidence in children rose throughout the period. Cumulative incidence to 5 years of age per 10,000 births rose consistently from 6.2 for 1990 births to 42.5 for 2001 births. Age-specific incidence rates increased most steeply for 2- and 3-year olds. The proportion diagnosed by age 5 years increased only slightly, from 54% for 1990 births to 61% for 1996 births. Changing age at diagnosis can explain a 12% increase, and inclusion of milder cases, a 56% increase.

CONCLUSIONS: Autism incidence in California shows no sign yet of plateauing. Younger ages at diagnosis, differential migration, changes in diagnostic criteria, and inclusion of milder cases do not fully explain the observed increases. Other artifacts have yet to be quantified, and as a result, the extent to which the continued rise represents a true increase in the occurrence of autism remains unclear.

PMID: 19234401 [PubMed – indexed for MEDLINE]

Toxicological & Environmental Chemistry, 1029-0486, Volume 90, Issue 5, 2008, Pages 997-1008.

HEPATITIS B TRIPLE SERIES VACCINE AND DEVELOPMENTAL DISABILITY IN U.S. CHILDREN AGED 1–9 YEARS.

Gallagher CM, Goodman MS.

Graduate Program in Public Health, Stony Brook University Medical Center, Health Sciences Center, State University of New York at Stony Brook, Stony Brook, New York, USA

This study investigated the association between vaccination with the Hepatitis B triple series vaccine prior to 2000 and developmental disability in children aged 1–9 years (n¼1824), proxied by parental report that their child receives early intervention or special education services (EIS). National Health and Nutrition Examination Survey 1999–2000 data were analyzed and adjusted for survey design by Taylor Linearization using SAS version 9.1 software, with SAS callable SUDAAN version 9.0.1. The odds of receiving EIS were approximately nine times as great for vaccinated boys (n¼46) as for unvaccinated boys (n¼7), after adjustment for confounders. This study found statistically significant evidence to suggest that boys in United States who were vaccinated with the triple series Hepatitis B vaccine, during the time period in which vaccines were manufactured with thimerosal, were more susceptible to developmental disability than were unvaccinated boys.

J Allergy Clin Immunol. 2008 Mar;121(3):626-31. Epub 2008 Jan 18.

DELAY IN DIPHTHERIA, PERTUSSIS, TETANUS VACCINATION IS ASSOCIATED WITH A REDUCED RISK OF CHILDHOOD ASTHMA.

McDonald KL, Huq SI, Lix LM, Becker AB, Kozyrskyj AL.
Faculty of Medicine, Department of Community Health Sciences, University of Manitoba, Winnipeg, Manitoba, Canada.

ABSTRACT

BACKGROUND: Early childhood immunizations have been viewed as promoters of asthma development by stimulating a T(H)2-type immune response or decreasing microbial pressure, which shifts the balance between T(H)1 and T(H)2 immunity.

OBJECTIVE: Differing time schedules for childhood immunizations may explain the discrepant findings of an association with asthma reported in observational studies. This research was undertaken to determine whether timing of diphtheria, pertussis, tetanus (DPT) immunization has an effect on the development of childhood asthma by age 7 years.

METHODS: This was a retrospective longitudinal study of a cohort of children born in Manitoba in 1995. The complete immunization and healthcare records of cohort children from birth until age 7 years were available for analysis. The adjusted odds ratio for asthma at age 7 years according to timing of DPT immunization was computed from multivariable logistic regression.

RESULTS: Among 11,531 children who received at least 4 doses of DPT, the risk of asthma was reduced to (1/2) in children whose first dose of DPT was delayed by more than 2 months. The likelihood of asthma in children with delays in all 3 doses was 0.39 (95% CI, 0.18-0.86).

CONCLUSION: We found a negative association between delay in administration of the first dose of whole-cell DPT immunization in childhood and the development of asthma; the association was greater with delays in all of the first 3 doses. The mechanism for this phenomenon requires further research.

PMID: 18207561 [PubMed – indexed for MEDLINE]

Pediatrics. 2008 Jul;122(1):149-53.

PROTECTING PUBLIC TRUST IN IMMUNIZATION.

Cooper LZ, Larson HJ, Katz SL.
Department of Pediatrics, College of Physicians & Surgeons, Columbia University, New York, New York, USA. loucooper@att.net

ABSTRACT

Public trust in the safety and efficacy of vaccines is one key to the remarkable success of immunization programs within the United States and globally. Allegations of harm from vaccination have raised parental, political, and clinical anxiety to a level that now threatens the ability of children to receive timely, full immunization. Multiple factors have contributed to current concerns, including the interdependent issues of an evolving communications environment and shortfalls in structure and resources that constrain research on immunization safety (immunization-safety science). Prompt attention by public health leadership to spreading concern about the safety of immunization is essential for protecting deserved public trust in immunization.

PMID: 18595998 [PubMed – indexed for MEDLINE]

Environ Health Perspect. 2007 May;115(5):792-8. Epub 2007 Jan 4.

ORGANOPHOSPHATE PESTICIDE EXPOSURE AND NEURODEVELOPMENT IN YOUNG MEXICAN-AMERICAN CHILDREN.

Eskenazi B, Marks AR, Bradman A, Harley K, Barr DB, Johnson C, Morga N, Jewell NP.

Center for Children's Environmental Health Research, School of Public Health, University of California, Berkeley, California 94720-7380, USA. eskenazi@ berkeley.edu

ABSTRACT

BACKGROUND: Organophosphate (OP) pesticides are widely used in agriculture and homes. Animal studies suggest that even moderate doses are neurodevelopmental toxicants, but there are few studies in humans.

OBJECTIVES: We investigated the relationship of prenatal and child OP urinary metabolite levels with children's neurodevelopment.

METHODS: Participating children were from a longitudinal birth cohort of primarily Latino farm-worker families in California. We measured six nonspecific dialkylphosphate (DAP) metabolites in maternal and child urine as well as metabolites specific to malathion (MDA) and chlorpyrifos (TCPy) in maternal urine. We examined their association with children's performance at 6 (n = 396), 12 (n = 395), and 24 (n = 372) months of age on the Bayley Scales of Infant Development [Mental Development (MDI) and Psychomotor Development (PDI) Indices] and mother's report on the Child Behavior Checklist (CBCL) (n = 356).

RESULTS: Generally, pregnancy DAP levels were negatively associated with MDI, but child measures were positively associated. At 24 months of age, these associations reached statistical significance [per 10-fold increase in prenatal DAPs: beta = -3.5 points; 95% confidence interval (CI), -6.6 to -0.5; child DAPs: beta = 2.4 points; 95% CI, 0.5 to 4.2]. Neither prenatal nor child DAPs were associated with PDI or CBCL attention problems, but both prenatal and postnatal DAPs were associated with risk of pervasive developmental disorder [per 10-fold increase in prenatal DAPs: odds ratio (OR) = 2.3, p = 0.05; child DAPs OR = 1.7, p = 0.04]. MDA and TCPy were not associated with any outcome.

CONCLUSIONS: We report adverse associations of prenatal DAPs with mental development and pervasive developmental problems at 24 months of age. Results should be interpreted with caution given the observed positive relationship with postnatal DAPs.

PMID: 17520070 [PubMed – indexed for MEDLINE] PMCID: PMC1867968

J Neurochem. 2006 Apr;97(1):69-78. Epub 2006 Mar 8.

HIGH SUSCEPTIBILITY OF NEURAL STEM CELLS TO METHYLMERCURY TOXICITY: EFFECTS ON CELL SURVIVAL AND NEURONAL DIFFERENTIATION.

Tamm C, Duckworth J, Hermanson O, Ceccatelli S.

Institute of Environmental Medicine, Division of Toxicology and Neurotoxicology, Karolinska Institutet, Stockholm, Sweden.

ABSTRACT

Neural stem cells (NSCs) play an essential role in both the developing embryonic nervous system through to adulthood where the capacity for self-renewal may be important for normal function of the CNS, such as in learning, memory and response to injury. There has been much excitement about the possibility of transplantation of NSCs to replace damaged or lost neurones, or by recruitment of endogenous precursors. However, before the full potential of NSCs can be realized, it is essential to understand the physiological pathways that control their proliferation and differentiation, as well as the influence of extrinsic factors on these processes. In the present study we used the NSC line C17.2 and primary embryonic cortical NSCs (cNSCs) to investigate the effects of the environmental contaminant methylmercury (MeHg) on survival and differentiation of NSCs. The results show that NSCs, in particular cNSCs, are highly sensitive to MeHg. MeHg induced apoptosis in both models via Bax activation, cytochrome c translocation, and caspase and calpain activation. Remarkably, exposure to MeHg at concentrations comparable to the current developmental exposure (via cord blood) of the general population in many countries inhibited spontaneous neuronal differentiation of NSCs. Our studies also identified the intracellular pathway leading to MeHg-induced apoptosis, and indicate that NSCs are more sensitive than differentiated neurones or glia to MeHg-induced cytotoxicity. The observed effects of MeHg on NSC differentiation offer new perspectives for evaluating the biological significance of MeHg exposure at low levels.

PMID: 16524380 [PubMed – indexed for MEDLINE]

Environ Res. 2004 Jul;95(3):363-74.

TEMPORAL VARIATION OF BLOOD AND HAIR MERCURY LEVELS IN PREGNANCY IN RELATION TO FISH CONSUMPTION HISTORY IN A POPULATION LIVING ALONG THE ST. LAWRENCE RIVER.

Morrissette J, Takser L, St-Amour G, Smargiassi A, Lafond J, Mergler D.
Centre des interactions entre la santé et l'environnement (CINBIOSE), Université du Québec à Montréal, C.P. 8888, Succ. Centre-ville, Montreal, Canada H3P 3P8

ABSTRACT

Fish consumption from the Great Lakes and the St. Lawrence River has been decreasing over the last years due to advisories and increased awareness of the presence of several contaminants. Methylmercury (MeHg), a well-established neurotoxicant even at low levels of exposure, bioaccumulates to differing degrees in various fish species and can have serious adverse effects on the development and functioning of the human central nervous system, especially during prenatal exposure. Most studies on MeHg exposure have focussed on high-level consumers from local fish sources, although mercury (Hg) is also present in fresh, frozen, and canned market fish. Moreover, little information exists on the temporal variation of blood and hair Hg in pregnant women, particularly in populations with low levels of Hg. The aim of the present study was to characterize the temporal variation of Hg during pregnancy and to investigate the relation between fish consumption from various sources prior to and during pregnancy and maternal cord blood and mother's hair Hg levels. We recruited 159 pregnant women from Southwest Quebec through two prenatal clinics of the Quebec Public Health System. All women completed two detailed questionnaires concerning their fish consumption (species and frequency) prior to and during pregnancy. The women also provided blood samples for all three trimesters of pregnancy and hair samples after delivery of up to 9 cm in length. Blood and hair Hg levels were analyzed by cold-vapor atomic-absorption and -fluorescence spectrometry methods, respectively. Results showed that maternal blood and hair Hg levels decreased significantly between the second and third trimesters of pregnancy. However, cord blood Hg was significantly higher than maternal blood at birth. Maternal hair was correlated with Hg blood concentration and was highly predictive of the organic fraction in cord blood. A strong dose relation was observed between the frequency of fish consumption before and during pregnancy and Hg exposure in mothers and newborns. Fish consumption prior to and during pregnancy explained 26% and 20% of cord blood Hg variance, respectively. For this population, detailed multivariate analyses showed that during pregnancy market fish (fresh, canned, and frozen) were more important sources of Hg exposure than were fish from the St. Lawrence River. These results should be taken into account for future advisories and intervention strategies, which should consider Hg levels in different species from all sources in order to maximize the nutritional input from fish and minimize the toxic risk.

PMID: 15220070 [PubMed – indexed for MEDLINE]

Environ Health Perspect. 2003 Sep;111(12):1465-70.

AN ASSESSMENT OF THE CORD BLOOD: MATERNAL BLOOD METHYLMERCURY RATIO: IMPLICATIONS FOR RISK ASSESSMENT.

Stern AH, Smith AE.
Division of Science, Research and Technology, New Jersey Department of Environmental Protection, PO Box 409, 401 E. State Street, Trenton, NJ 08625, USA. astern@dep.state.nj.us

ABSTRACT

In the current U.S. Environmental Protection Agency reference dose (RfD) for methylmercury, the one-compartment pharmacokinetic model is used to convert fetal cord blood mercury (Hg) concentration to a maternal intake dose. This requires a ratio relating cord blood Hg concentration to maternal blood Hg concentration. No formal analysis of either the central tendency or variability of this ratio has been done. This variability contributes to the overall variability in the dose estimate. A ratio of 1.0 is implicitly used in the model, but an uncertainty factor adjustment is applied to the central tendency estimate of dose to address variability in that estimate. Thus, incorporation of the cord:maternal ratio and its variability into the estimate of intake dose could result in a significant change in the value of the RfD. We analyzed studies providing data on the cord:maternal blood Hg ratio and conducted a Monte Carlo-based meta-analysis of 10 studies meeting all inclusion criteria to generate a comprehensive estimate of the central tendency and variability of the ratio. This analysis results in a recommended central tendency estimate of 1.7, a coefficient of variation of 0.56, and a 95th percentile of 3.4. By analogy to the impact of the similar hair:blood Hg ratio on the overall variability in the dose estimate, incorporation of the cord:maternal ratio may support a 3-fold uncertainty factor adjustment to the central tendency estimate of dose to account for pharmacokinetic variability. Whether the information generated in this analysis is sufficient to warrant a revision to the RfD will depend on the outcome of a comprehensive reanalysis of the entire one-compartment model. We are currently engaged in such an analysis.

PMID: 12948885 [PubMed – indexed for MEDLINE] PMCID: PMC1241648

Vaccine. 1997 Jun;15(8):903-8.

NONHUMAN PRIMATE MODELS TO EVALUATE VACCINE SAFETY AND IMMUNOGENICITY.

Kennedy RC, Shearer MH, Hildebrand W.

Department of Microbiology and Immunology, University of Oklahoma Health Sciences Center, Oklahoma City 73190, USA

ABSTRACT

When considering preclinical studies to evaluate the safety and immunogenicity of putative vaccine candidates, such as nucleic acid vaccines, species most closely related to humans should be considered. Phylogenetically, the great apes (chimpanzees, orangutans, gorillas, and gibbons) are most closely related to humans. However, the great apes, which diverged from humans over 5 million years ago, represent endangered or threatened species that limits their utility in preclinical studies. In addition, cost considerations for using great apes in biomedical studies represents another serious limitation. The Old World monkeys, (macaques, baboons, mandrills, and mangabeys), diverged from humans over 15 million years ago. A number of the Old World monkey species including rhesus, cynomolgus, and African green monkeys, have also been employed in biomedical research to evaluate vaccine safety and immunogenicity. New World monkeys (aotus, owl, cebus monkeys, and marmosets) are the most phylogenetically divergent from humans, yet they have also been utilized to develop nonhuman primate models for a number of human infectious diseases and tumors. The advantages and disadvantages in selecting a particular nonhuman primate species for studies to evaluate DNA vaccine safety and immunogenicity are briefly discussed. Comparative immunology, reproductive physiology, endogenous infectious agents, and cost considerations are briefly described.

PMID: 9234544 [PubMed – indexed for MEDLINE]

APPENDIX TO CHAPTER THIRTEEN: PEDIATRICS: SICK IS THE NEW HEALTHY

HARVARD MEDICAL SCHOOL
DEPARTMENT OF SURGERY

FRANCIS D. MOORE, M. D.
Moseley Professor of Surgery, Emeritus
Harvard Medical School
Surgeon-in-Chief, Emeritus
Peter Bent Brigham Hospital

The Countway Library
10 Shattuck Street
Boston, Massachusetts 02115
(617) 734-0420

December 2, 1999

Judy Converse, MPH, RD
P.O. Box 992
West Falmouth, MA 02574

Dear Judy:

Thanks for your very full letter and also for your card with the lovely picture of Ben.

I know you had a wonderful time in Honolulu, and I hope that Ben "weathered" it very well.

As regards neonatal active immunization ("vaccination") as a possible cause of autism, maybe the matter can rest temporarily in its present state.

The point that I am so anxious to emphasize is that the "window of vulnerability" occurs right around the time of birth, and is probably of longer duration in babies that are "slightly premature."

Very few women make notes of their monthly periods with sufficient accuracy to date the exact time of insemination. Therefore, the exact pregnancy duration of many so-called "full-term" infants may vary over a length of time as much as a month. Stated otherwise, many full-term infants are actually somewhat premature.

The spotty, random, and unpredictable nature of this "random prematurity of full-term infants" may well be involved in the spotty, random, and unpredictable incidence of immunological consequences of neonatal active immunization.

Judy Converse, MPH, RD
December 2, 1999
Page 2.

I am sending a copy of this letter to Dr. Yazbak, because I am very impressed with his knowledge of this field and his interest in the problem.

The only solution that I know for this is the absolute avoidance of neonatal immunization, particularly against hepatitis B, which is a disease of very young children that is of the utmost rarity. To put any child at hazard of autism with the excuse of avoiding this extremely rare disease, is of course preposterous.

Let's keep in touch.

Very truly yours,

Francis D. Moore, M.D.

FDM:sml

cc: F. Edward Yazbak, M.D., F.A.A.P.

APPENDIX TO CHAPTER SIXTEEN: WHO WILL DEFEND THE DEFENDERS?

Excerpts from Richard Rovet's report on anthrax vaccine injury to his superiors while serving at Dover Air Force Base from 1997 to 1999:

DOVER AFB 1997-1999

"It was clearly delineated in my job description to collect, review, and coordinate information with various healthcare providers and case manage all patients who were on our long-term ill list. This included those who displayed a systemic illness to the anthrax vaccine. I often coordinated care with Walter Reed Army Hospital's Gulf War Illness Clinic called the Medical Evaluation and Treatment Unit (METU) for our patients who had fallen ill in so called "temporal relation" to anthrax inoculation. I find it disturbing that my medical leadership, at the time, counseled me for inquiring and reporting to them the number of individuals who had systemic reaction to the anthrax vaccine and subsequently seroconverted to display a documented blood marker for autoimmunity."

• • •

"For instance, Patient TSGT D.B. (active duty aircrew member) became ill following anthrax vaccination. He developed myocarditis, (possible autoimmune etiology) cardiac arrhythmia and inflammation around the heart sac. A short time later, patient D.B. had to have his heart shocked out of an arrhythmia. Several days later a clot dislodged from his heart and he suffered a mini stroke.

To the best of my knowledge this vital aircrew member was medically boarded never to fly again. This was not an isolated incident."

• • •

"In fact Dover's Chief of Safety, Colonel J, a C-5 Galaxy pilot, became so ill following vaccination that he reportedly had to turn the flight controls over to the co-pilot and slept on the aircraft's bunk during a flight over the Atlantic. Another pilot who also was in transient flight over the Atlantic reported a similar problem. His name was Captain ▬▬. He developed unexplained vertigo (GRAY-OUTS), severe joint aches and pain and rare lesions throughout his internal organs."

• • •

"Another detailed fact concerning a rapid systemic reaction was from an African American Female Staff Sergeant. SSGT W. became extremely ill following anthrax vaccination . . . Several hours later the patient reportedly developed a fever of 104-105 degrees and a large injection site nodule with severe arm soreness. On 25 FEB 1999 the patient deployed to Saudi Arabia. SSGT W. stated that while in transit she experienced severe vertigo (GRAY-OUT) and almost passed out. She reported being very ill for several days. On 6 MAR 1999 SSGT W. reported to the deployment hospital to receive her second vaccination. She related her persistent medical problems since she initially received her first anthrax vaccination to the hospital's medical officer. According to SSGT W. this individual stated that her illness was in no way related to the anthrax vaccine and the Dover personnel administered the shot improperly. Despite her concern and illness the patient was given her second anthrax vaccination from Lot # FAV 036. SSGT W.'s health steadily declined. Her headaches became worse she developed ringing in her ears, frequent bouts of dizziness, severe joint aches and pain, chronic fatigue and unexplained rashes.

According to the anthrax vaccine package insert SSGT W. was having a purported "rare" systemic reaction to the anthrax vaccine. This precludes further vaccination. Nevertheless after explaining this to the USAF medical professional in Saudi Arabia she was given her third vaccination on 20 MAR 1999 from Lot # FAV 033.When she arrived back at Dover she was once again ordered to continue the series of vaccinations . . . After months of suffering, the patient was sent to Walter Reed Army Hospital where she was diagnosed with an AUTOIMMUNE ARTHRITIC condition, chronic fatigue and was reportedly told by Colonel ▬▬ that she had signs and symptoms consistent with Gulf War Illness."

• • •

"On 21 SEP 1999, at approximately 1700 a reserve aircrew member walked into my office in physical distress. He was pale and gaunt and his gross motor ability appeared to be impaired. He was drooling. He appeared drunk, yet was intellectually lucid. He comprehended what I was saying. He stated, "I want someone here to see one of my spells when it's happening." I sat him down and began to take his vital signs. I was then about to go to get the flight surgeon when Major ███████ came into the room and demanded that the patient, in acute physical distress, stand to attention when she walked unannounced into my office. The patient, in civilian clothes, whom she never met before, attempted to get up and fell back and hit his head on the wall."

• • •

"Former Senior Airman M.D. developed severe bone and joint pain following vaccination. In addition she was diagnosed with autoimmune optic neuritis which is a condition seen in Multiple Sclerosis (MS)."

• • •

"Captain J.R. a versatile and combat tested pilot found it difficult to get out of bed due to arthritic bone and joint pain. His experience with the anthrax vaccine is well documented in congressional record. His post vaccination blood work showed seroconversion to the autoimmune marker, ANA."

• • •

"Lieutenant Colonel Jay Lacklen is a decorated combat pilot who had no pre-existing arthritic symptoms. Post vaccination he developed an autoimmune arthritic condition and displayed a positive blood marker for autoimmunity."

• • •

"Captain Cheryl Angerer, another highly skilled pilot developed numbness and tingling (paresthesias) on one side of her body following anthrax vaccination from the same lot number as others who had fallen ill. Captain Angerer was so incensed at her medical treatment and anthrax vaccine induced illness that she resigned publicly after developing an autoimmune condition. Her post vaccination blood work showed the blood marker for autoimmunity, ANA."

• • •

"TSGT J.M. is a former enlisted aircrew member who is permanently grounded due to a severe autoimmune condition involving severe bone and joint pain, memory loss and unexplained "gray-outs." His pre-vaccination blood work showed no evidence of autoimmunity while his post vaccination blood work showed positive evidence of autoimmunity."

• • •

"Master Sergeant M.M. developed life threatening autoimmune condition, Gullain-Barré syndrome following vaccination and was placed on a respirator, in Intensive Care for weeks."

• • •

"Master Sergeant (RET) J.K. developed autoimmune diabetes following anthrax vaccination in addition to severely elevated liver and pancreatic enzymes."

• • •

"Master Sergeant J. B. developed severe bone and joint pain following vaccination. His blood work showed positive evidence of autoimmunity."

• • •

"Major D.D. developed autoimmune optic neuritis following vaccination. He also developed numbness and tingling on one half of his body. He has signs and symptoms consistent with Multiple Sclerosis (MS). To the best of my knowledge he no longer flies."

• • •

"Technical Sergeant Earl Stauffer developed severe dizziness, gray-outs with continued ringing in his ears following vaccination."

• • •

"Former Staff Sergeant James Picconi suffered dizziness, gray-outs and severe heart arrhythmia following vaccination."

• • •

"Technical Sergeant M.M. suffered unexplained seizures and gray-outs, ringing in the ears (tinnitus), dizziness, bone and joint pain following vaccination."

• • •

"Colonel J.M. developed bone and joint pain, chronic fatigue, ringing in the ears and memory loss following vaccination."

• • •

"Staff Sergeant N. P. lost over sixty pounds and developed lesions on his throat and lungs following vaccination."

• • •

"Technical Sergeant B.B. developed severe bone and joint pain following vaccination. He also lost a great deal of weight and developed lesions on his pelvis and ribs."

• • •

"Lieutenant Jamie Martin developed severe bone and joint pain following vaccination. This highly trained pilot refused his third vaccination due to previously encountered illness associated with the anthrax vaccine and was discharged from the service."

• • •

"As stated in her testimony (under oath) Captain Michelle Piel experienced a severe autoimmune reaction following vaccination from lot # FAV 030. A disturbing fact that was not related to Captain Piel prior to her testimony was that her pre-vaccination blood work showed NO evidence of autoimmunity and her post vaccination blood work showed POSITIVE evidence of autoimmunity.

Unbeknownst to this USAF Academy graduate and pilot, Lieutenant Colonel (Dr.) Thomas ██████, on several occasions called her a "malingerer, liar and whiner" in front of his medical staff. Please note that Doctor ███ was a field grade officer, a commander and the head of the anthrax program at Dover."

• • •

"Another highly qualified pilot from the USAFR at Dover also testified that day (under oath), Captain Jon Richter. He developed autoimmune arthritis following vaccination."

• • •

"There were more individuals within the Reserve Airlift Wing at Dover AFB who received the same lot number of vaccines that also reported severe and sub-clinical unexplained illness and autoimmunity. There were also many more individuals with sub-clinical ailments who were afraid to come forward for fear of reprisal. In the end the problems at Dover were ascribed to: whining pilots, malingerers, children, troublemakers, liars, rumormongers, group-think, stress, psychosomatic maladies and artifact in reporting. Unfortunately this view is the accepted assessment of the anthrax vaccine problems at Dover by our military and civilian leadership as well as a corporate media that has abandoned its obligation to roll up its sleeves and ferret out the truth. In the end something very wrong occurred at Dover."

• • •

"Lt. Gen. Roadman assured everyone the vaccine was completely safe and that only a minute percentage of those military personnel inoculated had had a negative reaction. Meanwhile, I was encountering more of my squadron mates who were vaccinated that said they too had experienced various reactions including tinnitus, dizziness, muscle and joint pain and in one case black-outs. However, most were attempting to keep a low profile and did not readily discuss these matters for fear of reprisal."

• • •

"The National Academy of Sciences' Institute of Medicine in 2002 conducted a comprehensive review of anthrax vaccine safety and effectiveness. Personnel from Dover Air Force Base contributed to the information gathered for their review. This prestigious group of independent experts concluded that trace amounts of squalene are not associated with an increase in the rates of adverse events and that further investigation of possible contamination is not warranted.'

William Winkenwerder, Jr., MD, MBA
Former Assistant Secretary of Defense Health Affairs
Letter to reporter Lester Holt of MSNBC"

• • •

"**In Memory**

Specialist Rachel Lacy, USAFR, combat medic, died on 4 April 2003 from an autoimmune "lupus-like" illness. She also had diffuse alveolar lung tissue damage and inflammation of the heart sac.

Technical Sergeant Clarence Glover
NCOIC Flight Medicine Clinic Dover AFB, Delaware
TSGT Glover passed away in February 2000 from a "heart condition" following anthrax vaccination.

To the best of my knowledge the autopsy results were not made public."

Richard Rovet's letter of complaint concerning his treatment following his reports on the anthrax vaccine adverse reactions:

█████ RICHARD J. ROVET
████████████████
████████████████
████████████████
████████████████
████████████████
████████████
████████████
████████████
██████████████████████

Dear Sir or Ma'am:

On 21 July 1999 I delivered protected communication before the now National Security Subcommittee that was investigating anthrax vaccine adverse reactions and the challenges of reporting them. I accepted this obligation freely and without reservation after exhausting all means within my chain-of-command. My protected communication was delivered in accordance with my duty as an officer in the USAF, a medical professional and most importantly as a Christian. It is within that spirit that I will pursue the truth in regards to this important issue. For over five years now I have faced overt and subtle reprisal for delivering my protected communication to the subcommittee. My intentions were to bring to light the illnesses associated with the vaccine and the challenges related to the reporting of these illnesses.

On 3 April 2003, I brought these aforementioned actions, which violate United States Code Military Justice (UCMJ), to the attention of HQ ASC/IG, Wright Patterson Air Force Base Ohio. I was told by the Air Systems Command Inspector General (IG), ████████ that "I shouldn't hang my hat on any resolution to my case" and that I needed to "move on." After months of transferring my complaint through Air Force channels I was finally given an action officer to conduct an investigation addressing my allegations. To my dismay and protestations my case was to be handled by the IG from my former base, the 81st Training Wing, Keesler Air Force Base, Mississippi. This was the very same office that I approached in the spring of 2001 that in turn disclosed my protected communication to them and threatened me to "back off" my complaint and allegations concerning the problems and illnesses associated with the anthrax vaccine. I was told in May of 2001 that Master Sergeant [redacted] of 81st TW/IG conveyed this threat through my former Non Commissioned Officer In Charge (NCOIC) of KAFB ER. This occurred during the same time frame that public scrutiny was heightened, once again, concerning the anthrax vaccine: In the spring of 2001 the trial of the first military physician to refuse the anthrax vaccine was underway at Keesler AFB. Please review the statement provided to me by Master Sergeant [redacted].

My interactions with the USAF Inspector General system since 1999 to present indicate an unwillingness to ferret out the truth and protect whistleblowers in accordance with Title 10 United States Code Section 1034, Department of Defense Directive 7050.6 and Air Force Instruction 90-301.

The information contained herein will not focus on the willful reprisal and damage to my career that has occurred since I testified before the subcommittee on 21 July 1999. That evidence will be delivered in a separate report. However, in some instances I will include excerpts from the USAF Inspector General's Reprisal Complaint Analysis (RCA, completed by the 81st TW KAFB) to provide the necessary framework that will illustrate the less-than-forthcoming behavior by some who continue to defend the indefensible.

To this day the light of truth has not penetrated the misinformation and continued patterns of deception that have become the hallmark of the anthrax vaccination program. Unfortunately public, congressional and media scrutiny have yet been able to breach the public affairs barbican that surrounds the program since the first Gulf War. Program protection seems to have replaced force protection.

Before proceeding I must make perfectly clear that I do not have an anti-vaccine agenda (my own shot record is a testimony). I also fully support biological countermeasures in this age of asymmetrical combat. As a medic to think otherwise would be akin to negligence. I also fully agree that the anthrax vaccine seems to be well tolerated by the majority of individuals despite the lack of long-term health data and a transparent active surveillance system. Yet, there are those individuals who have been injured by the anthrax vaccine beyond the purported "rare" systemic event. It is either through ignorance or as the General Accounting Office stated "a continued pattern of deception" that this mythology is propagated; propagated by an active Public Affairs (PA) apparatus worthy of a "Bronze Anvil Award."

I fully understand the nature of my allegations and will provide evidence to verify my charges. I further understand that a considerable amount of time has passed. Yet, due to the nature of my allegations and information that I am providing you, I sincerely hope that any investigation will, at a minimum, be unbiased in its approach. I will also provide evidence that senior officials at Dover in 1999 may have provided false information to federal investigators concerning the extent of anthrax vaccine adverse reactions at Dover AFB, Delaware.

I am a veteran of over nineteen years' service to our great country. During the first Gulf War I served stateside as an enlisted aircraft mechanic. After the War I noticed many individuals that were deployed to theater of operation, and some of those who were not in theater, developed unexplained illnesses collectively known as Gulf War Illness. I also worked briefly at the Veterans Administration following the war and further witnessed more "unexplained illnesses" from service during that time frame. In 1997 I was stationed at Dover AFB, Delaware. In the course my tenure at Dover I helped follow-up on ill Gulf War era veterans for the Comprehensive Clinical Evaluation Program (CCEP) in the Tri-State region. Most reported autoimmune illnesses such as: severe arthritic pain, rashes, nervous system damage, numbness and tingling in extremities, excessive fatigue, Multiple Sclerosis, vertigo and "gray-outs," cardiac and gastrointestinal problems. They also complained of problems receiving care, often citing active duty medical providers who told them that their illnesses were all in their head. Unfortunately this message is still being propagated in military medicine and being reportedly conveyed to new medical providers who enter active service.

In the fall of 1998 I began noticing similar clusters of unexplained illnesses within our healthy and young aircrew population. I had witnessed first hand bizarre symptomatology and illnesses atypical for the population and age group. Strikingly the illness reported were many of the same as were previously reported by Gulf War era veterans seven years earlier. Unfortunately the common denominator was the anthrax vaccine. Equally unfortunate, the patients were treated the same as veterans who had served seven years earlier. It is interesting to note that a recent Veterans Affairs report in fall of 2004 by the esteemed Research Advisory Counsel For Gulf War Veteran Illnesses recognizes sick veterans from the first Gulf War as no longer having a syndrome but rather an illness connected with service. It further implies that the anthrax vaccine may be a precipitous factor in some of the illnesses being reported. I also would like to share an excerpt from After Action Report (AAR) from the first Gulf War, which states that, the anthrax vaccine was associated with "high rate of side effects." It is also interesting to note that according to Jeffery Smith of the *Washington Post* stated in his January 1991 article that,

> Military officials told Glenn's (Senator John Glenn) committee in August 1989 that the only available anthrax vaccine was not suited to mass troop vaccinations because of a 'higher than normal rate of reactogenicity', or adverse medical effects and its relative lack of effectiveness once troops were exposed to the germ.

Over the past decade congressional hearing after hearing has been devoted to the problems associated with the anthrax vaccine. The General Accounting Office (GAO) also clearly demonstrated that there are significant problems with the anthrax vaccine and the vaccine injury surveillance system. The most damning report on the anthrax vaccine occurred on 27 SEP 2000 before the National Security, Veterans Affairs, and International Relations congressional subcommittee delivered by former Congressman Jack Metcalf. His report was the culmination of a three-year investigation into the use of an experimental substance called squalene that has been found in certain lot numbers of the anthrax vaccine. Many of those contaminated lots of anthrax vaccine were used at Dover AFB, Delaware. Many reportedly were used in the first Gulf War. Metcalf's report also extensively documents, at length, "stonewalling" on the squalene issue and a consistent "pattern of deception" by Defense officials involved with the anthrax vaccination program.

On 5 May 1999 the first of two "town hall" meetings arrived at Dover. Lieutenant Colonel introduced himself as an action officer from the USAF Surgeon General's office. When posed a question by a concerned aircrew member about squalene in the anthrax vaccine ██████ stated, "Not in the licensed human vaccine. There is a difference there. Squalene has been used in a human experimental vaccine, but not in the licensed vaccine. The licensed vaccine is what we all have been given."

On 11 May 1999 during the next "town hall meeting" at Dover AFB, Delaware, the former Surgeon General USAF, General Roadman gave his word as an officer that squalene was never used in the anthrax vaccine and the issue was a "red herring." In a video of that meeting one can see Colonel Alving, of USAMRID, behind General Roadman. It has been reported that Colonel Alving apparently conducted extensive research on squalene adjuvants and reproduced similar autoimmune illnesses in laboratory animals as those seen in individuals who were harmed by the anthrax vaccine.

Why is it that the DOD denied the use of squalene as an adjuvant in an experimental vaccine when they had indeed used it in clinical trials?

If squalene is naturally occurring in the vaccine, then where is the peer reviewed science to prove this?

In September 2000 the FDA found squalene in several of the lots of anthrax vaccine that were used at Dover. The majority of illnesses that I had personally witnessed were in people who were injected with these lots of vaccine. Several statements made by DOD officials attributed the squalene found in the vaccine to the oil from someone's fingers in the manufacturing process. Does this mean that America's sons and daughters were injected with a contaminated, unsterile product?

Why is it that several ill individuals at Dover developed antibodies to squalene after receiving the anthrax vaccine?

To understand how this dangerous and controversial pharmaceutical has been able to endure public, congressional and legal scrutiny since the first Gulf War one must look at the shield it has hidden behind. The public relation shield was originally named Office of Special Assistant for Gulf War Illness (OSAGWI). Currently it operates under the name Deployment Health Support Directorate. A tactic used by this organization is outlined clearly in their public relations strategy," Bronze Anvil Communication Plan." This information was previously presented (under oath) to the Senate Veterans Affairs Committee on July 10, 2002 by Richard Weidman and Dr. Linda Spoonster-Schwartz of the Vietnam Veterans of America (VVA).

I strongly believe that on 11 May 1999 the "Bronze Anvil" strike team arrived at Dover AFB, DE to provide damage control over the "problems" that had arisen there. The team was comprised of researchers, adjuvant specialists, scientists, the USAF Surgeon General and members from Office of Special Assistant for OSAGWI. If the anthrax vaccination is not related to Gulf War Illness, then why was their presence necessary?

At the very heart of this travesty are the men, women and families who have suffered because of a dangerous pharmaceutical. Why is it that America's sons and daughters (of an all-volunteer force) can be entrusted with the defense of this great nation, often serving in austere conditions while operating millions of dollars worth of sophisticated equipment? Why is it that they can be trusted with this, yet when they raise legitimate concerns about their health and with a vaccine that has a conspicuous past they are deemed malingerers, liars, whiners, malcontents, misinformed and sent to mental health for so called psychosomatic illnesses? We will never, ever, get a handle on this problem as long as civilian and military leaders and those injured from the vaccine are afraid to come forward because they fear reprisal. From my own personal experience I now understand why.

Sincerely,

RICHARD J. ROVET, [redacted]

APPENDIX TO CHAPTER TWENTY: THE ROLE OF GOVERNMENT AND MEDIA

Research supporting the vaccine-autism causation theory:

MERCURY–AUTISM RESEARCH

1. Gallagher, Carolyn, and Melody Goodman. "Hepatitis B triple series vaccine and developmental disability in US children aged 1-9 years," *Journal of Toxicological & Environmental Chemistry* 90, no. 5 (Sep 2008): 997-1008.
   Boys who received the triple series of mercury-containing hepatitis B vaccine were more susceptible to developmental disability than were boys who were unvaccinated.
2. Gallagher, C.M., and M.S. Goodman. "Hepatitis B Vaccination of Male Neonates and Autism," *Annals of Epidemiology* 19 no. 9 (Sep 2009): 651-80.
   The birth dose of the thimerosal-containing hepatitis B vaccine triples a boy's chance of developing autism.
3. Windham, Gayle C., Lixia Zhang, Robert Gunier, Lisa A. Croen, and Judith K. Grether. "Autism Spectrum Disorders in Relation to Distribution of Hazardous Air Pollutants in the San Francisco Bay Area," *Environmental Health Perspectives.* 114 no. 9 (Sep 2006).
   Results suggest a potential association between autism and estimated metal concentrations, and possibly solvents, in ambient air around the birth residence.
4. Palmer, Raymond F., Stephen Blanchard, and Robert Wood. "Proximity to point sources of environmental mercury release as a predictor of autism prevalence," *Health & Place* 15 no. 1 (2008): 18-24.

The closer a child lives to a mercury emitting, coal-burning power plant, the greater is his chance of developing autism.

5. Palmer, Raymond F., Stephen Blanchard, Zachary Stein, David Mandell, and Claudia Miller. "Environmental mercury release, special education rates, and autism disorder: an ecological study of Texas," *Health & Place* 12 no. 2 (2006): 203-9.
   For every 1000 pounds of environmentally released mercury, there was a forty-three percent increase in the rate of special education services and a sixty-one percent increase in the rate of autism. The association between environmentally released mercury and special education rates was fully mediated by increased autism rates.
6. Rose, Shannon, Stepan Melnyk, Alena Savenka, Amanda Hubanks, Stefanie Jernigan, Mario Cleves, and S. Jill James. "The Frequency of Polymorphisms affecting Lead and Mercury Toxicity among Children with Autism," *American Journal of Biochemistry and Biotechnology* 4 no 2. (2008): 85-94.
   Children with autism, as compared to neurologically typical children, more frequently display a gene variant associated with lower glutathione levels. Glutathione is an amino acid that is one of the body's primary detoxification mechanisms. These children may be at increased risk for toxicity during prenatal and postnatal development when the brain and immune system are most sensitive to low levels of lead and mercury.
7. Goth, Samuel R., Ruth A. Chu, Jeffrey P. Gregg, Gennady Cherednichenko, and Isaac Pessah. "Uncoupling of ATP-mediated Calcium Signaling and Dysregulated IL-6 Secretion in Dendritic Cells by Nanomolar Thimerosal," *Environmental Health Perspectives* 114 no. 7 (2006).
   Thimerosal, at a tiny fraction of what is in a full-dose mercury-containing vaccine such as the current flu vaccine, may impair the immune system: "Our findings that DCs primarily express the RyR1 channel complex and that this complex is uncoupled by very low levels of THI with dysregulated IL-6 secretion raise intriguing questions about a molecular basis for immune dysregulation and the possible role of the RyR1 complex in genetic susceptibility of the immune system to mercury."
8. Nataf, Robert, Corrinne Skorupka, Lorene Amet, Alain Lam, Anthea Springbett, and Richard Lathe. "Porphyrinuria in childhood autistic disorder: Implications for environmental toxicity," *Toxicology and Applied Pharmacology*, 214 (2006): 99–108.
   Porphyrin levels are a marker for mercury toxicity. This study found that the children with autism studied had elevated porphyrin levels; those with Asperger's were also elevated, but not as high as those with autism, and when a subgroup of children with autism was treated with the chelator DMSA, the result was a significant drop in urinary porphyrin excretion. These data implicate environmental toxicity in childhood autistic disorder.

9. Burbacher, Thomas M., Danny D. Shen, Noelle Liberato, Kimberly S. Grant, Elsa Cerniciari, and Thomas Clarkson. "Comparison of Blood and Brain Mercury Levels in Infant Monkeys Exposed to Methylmercury or Vaccines Containing Thimerosal," *Environmental Health Perspectives* 113 no. 8 (Aug 2005): 1015-21.
   The CDC has used the ingested methylmercury safety standard to claim that injected ethylmercury is safe. This study demonstrates that methylmercury standards are not a suitable measure for determining risk from injected ethylmercury exposure, as injected ethylmercury becomes trapped in the brain at a much higher rate than ingested methylmercury.
10. Waly, M., H. Olteanu, R. Banerjee, S.W. Choi, J.B. Mason, B.S. Parker, S. Sukumar, S. Shim, A. Sharma, J.M. Benzecry, V.A. Power-Charnitsky, and R.C. Deth. "Activation of Methionine Synthase by Insulin-like Growth Factor-1 and Dopamine: a Target for Neurodevelopmental Toxins and Thimerosal," *Molecular Psychiatry* 9 no. 4 (July 2004): 358-70.
    Thimerosal impairs methylation. Methylation produces glutathione, which is the body's primary means of removing toxic metals from the body.
11. James, S.J., William Slikker, III, Stepan Melnyk, Elizabeth New, Marta Pogribna, and Stefanie Jernigan. "Thimerosal Neurotoxicity is Associated with Glutathione Depletion: Protection with Glutathione Precursors," *Neurotoxicology* 26 (Jan 2005): 1-8.
    Lower levels of glutathione are associated with increased neuron damage from exposure to thimerosal.
12. James, S. Jill, Shannon Rose, Stepan Melnyk, Stefanie Jernigan, Sarah Blossom, Oleksandra Pavliv, and David W. Gaylor. "Cellular and mitochondrial glutathione redox imbalance in lymphoblastoid cells derived from children with autism," *FASEB Journal* 23 (Mar 2009): 2374-83.
    Autism is associated with reduced glutathione, mitochondrial dysfunction, impaired antioxidant defense, and impaired detoxification.
13. Yel, L., L.E. Brown, K. Su, S. Gollapudi, and S. Gupta. "Thimerosal induces neuronal cell apoptosis by causing cytochrome c and apoptosis-inducing factor release from mitochondria," *International Journal of Molecular Medicine*, 16 no. 6 (Dec 2005): 971-7.
    Low levels of exposure to thimerosal, tiny fractions of what is in a flu vaccine, can cause mitochondrial dysfunction so severe that the mitochondria cause neuroblastoma cells to self-destruct.
14. Humphreya, Michelle L., Marsha P. Coleb, James C. Pendergrassc, and Kinsley K. Kiningham. "Mitochondrial mediated thimerosal-induced apoptosis in a human neuroblastoma cell line (SK-N-SH)," *Neurotoxicology* 26 no. 3 (Jun 2005): 407-16.
    Low levels of exposure to thimerosal can cause mitochondrial dysfunction so severe that the mitochondria causes the cell to self-destruct.

15. Minami, T., E. Miyata, Y. Sakamoto, H. Yamazaki, and S. Ichida. "Induction of metallothionein in mouse cerebellum and cerebrum with low-dose thimerosal injection," *Cell Biology and Toxicology* 26 no. 2 (Apr 2009): 143-52.
    Shows how low-dose thimerosal exposure may be associated with autism, by demonstrating its effects on the cerebellum of mice, in comparison with the brain pathology observed in patients diagnosed with autism.
16. Kempuraj, Duraisamy, Shahrzad Asadi, Bodi Zhang, Akrivi Manola, Jennifer Hogan, Erika Peterson, and C. Theoharides. "Mercury induces inflammatory mediator release from human mast cells," *Journal of Neuroinflammation* 7 no. 20 (2010).
    Mercury chloride stimulates VEGF and IL-6 release from human mast cells. This phenomenon could disrupt the blood-brain-barrier and permit brain inflammation. As a result, the findings of the present study provide a biological mechanism for how low levels of mercury may contribute to ASD pathogenesis.
17. Agrawal, A., P. Kaushal, S. Agrawal, S. Gollapudi, S. Gupta, and J. Leukoc. "Thimerosal induces the responses via influencing cytokine secretion by human dendritic cells," *Journal of Leukocyte Biology* 81 no. 2 (Feb 2007): 474-82.
    Thimerosal depletes glutathione and causes inflammation.
18. Vargas, D.L., C. Nascimbene, C. Krishnan, A.W. Zimmerman, and C.A. Pardo. "Neuroglial Activation and Neuroinflammation in the Brain of Patients with Autism," *Annals of Neurology* 57 no.1 (Feb 2005): 67-81.
    Brain inflammation and autoimmunity are present in those with autism.
19. Charleston. J.S., R.P. Bolender, N.K. Mottet, R.L Body, M.E. Vahter, and T.M. Burbacher. "Increases in the number of reactive glia in the visual cortex of Macaca fascicularis following subclinical long-term methyl mercury exposure," *Toxicology and Applied Pharmacology* 129 no. 2 (Dec 1994): 196-206.
    Autopsies on people with autism found chronic brain inflammation, apparently linked to the brain's immune system and produced by activation of its glial cells.
20. Hewitson, Laura, Brian J. Lopresti, Carol Stott, N. Scott Mason, and Jaime Tomko. "Influence of pediatric vaccines on amygdala growth and opioid ligand binding in rhesus macaque infants: A pilot study," *Acta Neurobiol Experimentalis* 70 (2010): 147–164.
    Infant primates given the full vaccination schedule displayed altered brain chemistry and structures as compared to non exposed animals. The vaccinated group displayed larger brain size, as is seen in those with autism.
21. Sajdel-Sulkowska, Elizabeth M., Boguslaw Lipinski, Herb Windom, Tapan Audhya, and Woody McGinnis. "Oxidative Stress in Autism: Elevated Cerebellar 3-nitrotyrosine Levels," *American Journal of Biochemistry and Biotechnology* 4 no. 2 (2008): 73-84.
    Oxidative stress is associated with autism.

22. Kern, Janet K., and Anne M. Jones. "Evidence of Toxicity, Oxidative Stress, and Neuronal Insult in Autism," *Journal of Toxicology and Environmental Health* 9 (2006): 485–99.
Associations between neuronal death, toxicity, oxidative stress, increased brain size, glutathione involvement, and autism.
23. Chauhan, A., and V. Chauhan. "Oxidative Stress in Autism," *Pathophysiology* 13 no. 3 (2006): 717-81.
Oxidative stress may contribute to the development of autism.
24. Cheuk, D.K.L., and V. Wong. "Attention-deficit hyperactivity disorder and blood mercury level: A case-control study in Chinese children," *Neuropediatrics* 37 (2006): 234-40.
High blood mercury level is associated with ADHD.
25. DeSoto, M. Catherine, and Robert T. Hilan. "Blood Levels of Mercury Are Related to Diagnosis of Autism: A Reanalysis of an Important Data Set," *Journal of Child Neurolog.* 22 no. 11 (2007): 1308-11.
A significant relation does exist between the blood levels of mercury and diagnosis of an autism spectrum disorder.

ADDITIONAL RESEARCH SUPPORTING VACCINE-AUTISM THEORY

26. Herbert, Martha R. "Autism: A Brain Disorder, or A Disorder That Affects the Brain?" *Clinical Neuropsychiatry* 2 no. 6 (2005): 354-79.
27. Werner, E., and G. Dawson. "Validation of the Phenomenon of Autistic Regression Using Home Videotapes," *Archives of General Psychiatry* 62 no. 8 (2005): 889-95.
Confirms the reliability of parent reports of neurological regression.
28. Poling J.S., R.E. Frye, J. Shoffner, A.W. Zimmerman. "Developmental Regression and Mitochondrial Dysfunction in a Child with Autism," *Journal of Child Neurology* 21 no. 2 (Feb 2006): 170-2.
A case study of Hannah Poling, who was paid by the Vaccine Injury Compensation Program, and her mitochondrial associated autism.
29. Herbert, Martha R. "Large Brains in Autism: The Challenge of Pervasive Abnormality," *The Neuroscientist* 11 no. 5 (2005): 417-40.
Increased brain size and toxic injury are associated with autism.
30. Petrik, Michael S., Margaret C. Wong, Rena C. Tabata, Robert F. Garry, and Christopher A. Shaw. "Aluminum adjuvant linked to gulf war illness induces motor neuron death in mice." *Neuromolecular Medicine* 9 no. 1 (2007): 83-100
Aluminum adjuvants used in childhood vaccines cause neuron death.

31. Gargus, J. Jay, and Faiqa Imtiaz. "Mitochondrial Energy-Deficient Endophenotype in Autism." *American Journal of Biochemistry and Biotechnology* 4 no. 2 (2008): 198-207.
    Mitochondrial disorders are associated with autism.
32. Anderson, Matthew P., Brian S. Hooker, and Martha R. Herbert. "Bridging from Cells to Cognition in Autism Pathophysiology: Biological Pathways to Defective Brain Function and Plasticity." *American Journal of Biochemistry and Biotechnology* 4 no. 2 (2008): 167-76.
    Evidence to support a model where the disease process underlying autism may begin when an in utero or early postnatal environmental, infectious, seizure, or autoimmune insult triggers an immune response that increases reactive oxygen species (ROS) production in the brain that leads to DNA damage (nuclear and mitochondrial) and metabolic enzyme blockade and that these inflammatory and oxidative stressors persist beyond early development (with potential further exacerbations), producing ongoing functional consequences.
33. Oliveira, G., A. Ataíde, C. Marques, T.S. Miguel, A.M. Coutinho, L. Mota-Vieira, E. Gonçalves, N.M. Lopes, V. Rodrigues, H. Carmona da Mota, and A.M. Vicente. "Epidemiology of autism spectrum disorder in Portugal: prevalence, clinical characterization, and medical conditions," *Developmental Medicine & Child Neurology* 49 no. 10 (Oct 2007): 726-33.
    Autism is associated with a high rate of mitochondrial disorder.
34. Kawashti, M.I., O.R. Amin, and N.G. Rowehy. "Possible Immunological Disorders in Autism: Concomitant Autoimmunity and Immune Tolerance," *The Egyptian Journal of Immunology* 13 no. 1 (2006): 99-104.
    Autism is associated with autoimmunity, and some autoimmune features may be linked to the MMR vaccine.
35. Singh, V.K., and R.L. Jensen. "Elevated levels of measles antibodies in children with autism." *Pediatric Neurology*, 28 no. 4 (2003 Apr): 292-4.
    Autistic children have a hyperimmune response to measles virus, which in the absence of a wild type of measles infection might be a sign of an abnormal immune reaction to the vaccine strain or virus reactivation.
36. Hertz-Picciotto, I., L. Delwiche. "The rise in autism and the role of age at diagnosis." *Epidemiology* 20 no. 1 (Jan 2009): 84-90.
    The rise in autism rates is not a statistical artifact explained away by better diagnosing or diagnostic substitution. The increase in diagnosis represents a true rise in cases of autism. As there is no such thing as a genetic epidemic, the cause of the autism epidemic cannot be genetic, but must be change in the environment in which children are being raised.

APPENDIX TO CHAPTER TWENTY-SEVEN: WHAT SHOULD PARENTS DO?

COUNTING VACCINE DOSES AND COMBINATION VACCINES

The Centers for Disease Control and Prevention (CDC) recommends seventy doses of sixteen different vaccines. Certain childhood vaccines are only available in combination with others (DTaP and MMR). Combination vaccines are "generally preferred" by the CDC and the American Academy of Pediatrics.[1] For the purpose of counting doses, however, their component parts represent separate vaccines against discrete diseases and should be tallied individually. See the color photo insert in the center of the book for the CDC schedules for children.

COUNTING DOSES

Children Ages 0 Through 6 Years

The CDC recommends 50 doses of fourteen vaccines for all children ages 0 through six years, except certain high-risk groups:

- 5 Diphtheria, Tetanus, Pertussis or "DTaP" = 15 total doses
  - 5 Diphtheria
  - 5 Tetanus
  - 5 Pertussis
- 2 Hepatitis A or "HepA"
- 3 Hepatitis B or "HepB"
- 4 *Haemophilus influenzae* type b or "Hib"

2 Measles, mumps, rubella or "MMR" = 6 total doses
- 2 Measles
- 2 Mumps
- 2 Rubella

7 Influenza[2]
4 Pneumococcal or "PCV" (*Prevnar*)[3]
4 Polio or "IPV"
3 Rotavirus or "RV"
2 Chickenpox or "varicella"

DTaP and MMR are combination vaccines; one syringe contains one dose each of three vaccines. When DTaP or MMR is administered, the child receives one injection, but the body experiences the impacts of three vaccines against three different diseases.

Children Ages 7 Through 18 Years

The CDC recommends 20 doses of six vaccines for all children ages 7 through 18, except certain high-risk groups:

2 Meningococcal or "MCV" (*Menactra*)[4]
1 Tetanus, diphtheria, pertussis or "Tdap" = 3 total doses
- 1 Tetanus
- 1 Diphtheria
- 1 Pertussis

12 Influenza
3 Human papillomavirus or "HPV" (*Gardasil, Cervarix*)

COMBINATION VACCINES

In 2010, the CDC and the American Academy of Pediatrics announced that "the use of a combination vaccine series is now considered generally preferred over separate injections of the equivalent component vaccines."[5] This is a stronger statement of support than the previous recommendation which deferred to the clinician to make the final decision.[6] The reasons for the preference of combination vaccines include cost, storage considerations, product availability, fewer injections and visits, provider assessment, improved recordkeeping and tracking, and enhanced patient compliance.[7]

The development of combination vaccines is not a simple matter of mixing vaccine A, vaccine B, and vaccine C into the same syringe and rolling up a child's sleeve. In his textbook *Combination Vaccines: Development, Clinical Research, and Approval*, editor Ronald Ellis and his contributing authors acknowledge some important challenges: "Vaccine antigens, which have been developed and licensed separately as vaccines, are brought together in physical mixtures which are not always stable and potent."[8] To justify the use of a combination vaccine relative to its individual component vaccines, it is important to demonstrate "equivalence" with respect to safety (rates of injury) and efficacy (antibody levels).[9]

Safety

In their chapter on pediatric combination vaccines, rubella vaccine inventor and industry adviser Stanley Plotkin and colleagues admit that "little is known about the immunological interactions among vaccine components."[10] When multiple antigens are placed into one injection, the combination may cause more *reactogenicity*, or an "undesirable immunological reaction."[11] Therefore, when it comes to prioritizing combination vaccines over individual component vaccines, "safety is the single most important issue."[12] The science of vaccine development describes different mechanisms or routes by which combination vaccines can cause a greater degree of injury.[13]

When evaluating the safety of combination vaccines, the trials "should be prospective, randomized, double blinded and have appropriate control (comparison) groups."[14] Safety must be evaluated in absolute terms, in comparison to the separately-administered components, based on standard practice, and when administered with other vaccines typically given at the same time.[15]

Merck's MMRV or *ProQuad* is a recent example of a combination vaccine's greater risk of causing harm. In 2008, the CDC announced that it no longer preferred the combination measles, mumps, rubella, and varicella vaccine (MMRV) relative to separate injections of equivalent component vaccines (MMR-V) because MMRV was associated with a higher incidence of febrile seizures in children aged 12 to 23 months.[16] In 2009, the CDC said that the MMRV vaccine was acceptable but "physicians should discuss the risks and benefits of both options with parents."[17] In 2010, the CDC resumed its preference for MMRV in children aged 4 years and older. Children aged 12 to 47 months receiving the first dose could receive either vaccine. Children receiving a second dose, and any child with personal or family history of seizures, should not receive MMRV.[18]

Efficacy

In the past, most vaccines were generic. Today, vaccine development uses new technologies that result in a greater diversity of final products.[19] When scientists talk about the immune response a vaccine induces, they use the term *immunogenicity.* Vaccines contain antigens, which are substances that produce antibodies when injected. A variety of issues can cause quantitative and qualitative differences in overall effectiveness of a combination vaccine. Antigens that are combined in a single injection can "interfere" with each other. When live viruses are combined, for example, one may inhibit the replication of another. Another example involves engineering to boost weak antigens by attaching them to a different carrier to get the job done ("conjugation"). When the efficacy of vaccine components is diminished in combination, the phenomenon is known as "interference."[20]

On the pharmaceutical side, some vaccines are less effective when paired with certain ingredients, for example:[21]

- Certain vaccines—measles, mumps, rubella, varicella, and inactivated polio—are less effective when combined with the mercury-based preservative thimerosal
- *Haemophilus influenzae* type b vaccine (Hib) is less effective when combined with an adjuvant derived from aluminum salts

When conducting efficacy trials,[22] the combination vaccine must have an effectiveness that is equivalent to or greater than the separately-administered components. There must also be consistency or reliability of the immune response. This means that different vaccine lots must be tested and their effectiveness cannot vary statistically from lot to lot.

Public health officials prefer combination vaccines because they make it easier to vaccinate larger numbers of children, thereby facilitating compliance. Vaccine manufacturers prefer combination vaccines because they are more profitable. Parents want childhood vaccines that are safe and effective. While clinicians play a key role in supporting public health, they are required to serve the best interests of their patients. Vaccine package inserts provide detailed information to help parents and physicians make their decisions.

Combination Vaccines for Use in Children[23]

Two

Comvax® = Haemophilus b Conjugate (Meningococcal Protein Conjugate) and Hepatitis B (Recombinant) Vaccine

DECAVAC® = Tetanus and Diphtheria Toxoids Adsorbed

TENIVAC® = Tetanus and Diphtheria Toxoids Adsorbed

Three

Adacel® = Tetanus Toxoid, Reduced Diphtheria Toxoid and Acellular Pertussis Vaccine, Adsorbed

Boostrix® = Tetanus Toxoid, Reduced Diphtheria Toxoid and Acellular Pertussis Vaccine, Adsorbed

DAPTACEL® = Diphtheria and Tetanus Toxoids and Acellular Pertussis Vaccine Adsorbed

Infanrix® = Diphtheria and Tetanus Toxoids and Acellular Pertussis Vaccine Adsorbed

Tripedia® = Diphtheria and Tetanus Toxoids and Acellular Pertussis Vaccine Adsorbed

MMR II® = Measles, Mumps, and Rubella Virus Vaccine, Live

Four

KINRIX® = Diphtheria and Tetanus Toxoids and Acellular Pertussis Adsorbed and Inactivated Poliovirus Vaccine

ProQuad® = Measles, Mumps, Rubella and Varicella Virus Vaccine Live

Five

Pediarix® = Diphtheria and Tetanus Toxoids and Acellular Pertussis Adsorbed, Hepatitis B (Recombinant) and Inactivated Poliovirus Vaccine Combined

Pentacel® = Diphtheria and Tetanus Toxoids and Acellular Pertussis Adsorbed, Inactivated Poliovirus and Haemophilus b Conjugate (Tetanus Toxoid Conjugate) Vaccine

APPENDIX TO CHAPTER TWENTY-EIGHT: WHO IS DR. ANDREW WAKEFIELD?

Statement from Dr. Andrew Wakefield
Regarding the GMC Hearing Sanctions
April 5, 2010

On Wednesday April 7th, General Medical Council (GMC) lawyers will demand that I and likely two other doctors involved in the MMR-autism case should be erased from the U.K.'s medical register, removing our license to practice medicine. Doctors' regulators have found the three of us—Professor John Walker-Smith, Professor Simon Murch and me—guilty of undertaking research on children with autism without approval from an ethics committee. We can prove, with extensive documentary evidence, that this conclusion is false.

Let me make it absolutely clear that, at its heart, the GMC hearing has been about the protection of MMR vaccination policy. The case has been driven by an agenda to crush dissent that in my opinion serves the government and the pharmaceutical industry—not the welfare of children. It's important to note that there has never been a complaint against any of the doctors by any parent involved in this case—only universal parental support and gratitude.

My colleagues, Professors Walker-Smith and Murch, are outstanding pediatricians and pediatric gastroenterologists. They have led the field of pediatric gastroenterology for decades, devoting their lives to caring for sick children. Our only "crime" in this matter has been to listen to the concerns of parents, act according to the demands of our professional training, and provide appropriate care to this neglected population of children. It is unthinkable that at the end of an unimpeachable career, Professor Walker-Smith would even consider unethical experimentation on children under his care.

In the course of our work, we discovered and treated a new intestinal disease syndrome in children with autism, alleviating suffering in affected children around the world. This should be cause for celebration. Instead, we have been vilified in the press, and demonized by a wasteful PR campaign by the Department of Health. The aim of this negative publicity was to discredit my criticism of vaccine safety research.

Sadly, my colleagues have suffered severe collateral damage in this effort to prevent valid scientific enquiry. They should be exonerated, and left alone with their reputations intact, in the certain knowledge that they have done only what is right.

The loss of my own medical license is, unfortunately, the cost of doing business. Although I do not take this loss lightly, the suffering—so much of it unnecessary—that I have seen among those affected by this devastating disease makes the professional consequences for me a small price to pay by comparison.

As long as a question mark remains over vaccine safety; as long as a safety-first vaccine policy is subordinate to profit and self-interest; as long as the benefits of vaccines are threatened by those who have compromised public confidence by denial of vaccine damage, and as long as these children need help; I will continue my work.

*Dr. Wakefield Statement on GMC Hearing Sanctions available at www.ageofautism.com/2010/04/statement-from-dr-andrew-wakefield-regarding-gmc-hearing-sanctions.html

APPENDIX TO CHAPTER TWENTY-NINE: THE EXONERATION OF PROFESSOR JOHN WALKER-SMITH

David L. Lewis, PhD

Below, I discuss two of my research projects, dental infection control and land application of biosolids, that changed government policies. In these and other projects that I carried out at the U.S. Environmental Protection Agency and University of Georgia, private corporations, industry-supported organizations, and government agencies used false allegations of ethical and research misconduct as part of an effort to discredit the researchers and stop the research.

I. DENTAL INFECTION CONTROL

Like the 1998 *Lancet* article by Wakefield et al, my research published in *The Lancet*[1] and *Nature Medicine*[2] linking HIV transmission to dental drills was highly controversial. When dental visits precipitously dropped, we were blamed for causing a dental scare worldwide and causing illnesses and deaths from a lack of proper dental care. One dentist who coauthored the research articles, who also took a controversial stand against mercury amalgams in fillings, lost his license to practice. Dentists balked at the solution we proposed, which was to heat-sterilize high-speed dental handpieces and other devices when the Centers for Disease Control and Prevention (CDC) considered high-level disinfection acceptable.

In an attempt to discredit the work, the American Dental Association (ADA) published an editorial claiming that I was a dentist with a patent in which I stood to profit if heat-sterilization became the standard of practice. I have never attended dental school, and the only patent I ever owned was for a home water filter that I never sold. Fortunately, the U.S. Departments of Navy and Air Force confirmed our results concerning the potential for certain common dental devices to transmit infections and recommended heat-sterilization.[3] As a result of our research, the CDC, FDA, ADA, and other public health organizations worldwide adopted the current heat-sterilization standard for all reused dental devices entering the oral cavity.

Every institution involved with dental health policies from the CDC to the ADA believed that all we would accomplish would be generating preventable illnesses and deaths by creating irrational public fears. Their approach took a different course in 1992, however, when ABC's *Primetime Live* aired a segment called "Under the Gum." It showed CDC Director Harold Jaffee watching a simple demonstration of our research for the first time. To illustrate how much blood and saliva were being retracted and ejected by the devices, I operated dental drills and prophy angles (used to polish teeth) in a laboratory model of the oral cavity that was contaminated with small amounts of blood. Then I handed them to a dentist to prepare for reuse following CDC guidelines.

Dr. Jaffee watched visible amounts of blood spit back out of the patient-ready devices when they were subsequently operated in a white porcelain container. ABC medical correspondent Sylvia Chase asked, "Is it not the same thing—this kind of blood transfer—as sharing a needle?" Jaffee opened his mouth to speak, but no words came out. After a long pause, he said: "Clearly, we don't want one patient to be exposed to another's blood."

That night, common sense prevailed over peer-reviewed scientific articles published over the previous three decades supporting government infection-control guidelines. Virtually everyone in the dental and medical community thought that science had established beyond any doubt that dentistry's high-level disinfection standard fully protected patients from infection. But now all could see with their own eyes how six patients in a Florida dental practice could have contracted HIV from their infected dentist by sharing reused prophy angles. We demonstrated that lubricants contaminated with blood from AIDS patients could efficiently infect human lymphocyte cultures even after disinfection with FDA-approved germicides.

With no system for tracking sporadic infections in dentistry and with leaky dental devices designed with no thought of infection control, dentistry had a big problem. Patients' fears, in this case, turned out to be justified. Government agencies adopted a heat-sterilization standard, and patient visits to the dental office soon returned to normal.

II. LINKING ILLNESSES TO BIOSOLIDS

My duties as a research microbiologist in the EPA's Office of Research and Development (ORD) included investigating illnesses and deaths linked to land application of processed sewage sludge (biosolids). My team included a medical microbiologist, an environmental engineer, and a pediatrician treating children exposed to biosolids. Together, we published case studies linking biosolids to skin, gastrointestinal, and respiratory problems.[4]

To stop our research, employees in EPA's Office of Water, including the director of the Office of Wastewater Management, requested help from corporate executives in the biosolids business. Several weeks later, one of the executives sent the EPA officials an anonymous white paper falsely accusing me of research misconduct and criminal fraud. One of the EPA officials distributed the allegations at public meetings in Georgia while the corporate executives brought them to the attention of EPA Administrator Christie Whitman.

EPA's efforts to silence me and other scientists documenting problems with biosolids prompted science committee hearings in the U.S. House of Representatives in 2000.[5,6] The hearings focused on actions taken by former EPA Administrator Carol Browner and Robert Perciasepe, who was assistant administrator of EPA's Office of Water. Browner's assistant administrators cut off my research funding and distributed allegations that I had violated the Hatch Act and ethics rules by writing a commentary in *Nature*[7] critical of EPA's science. My research included investigating a potential deep-sea oil spill in the Gulf of Mexico.[8]

Perciasepe, who currently serves as EPA Deputy Administrator under President Obama, authorized funding for a Biosolids Incident Response Team (BIRT) to investigate cattle deaths on two dairy farms in Georgia linked to biosolids produced by the city of Augusta. Robert Brobst, an EPA employee who headed BIRT, arranged for the EPA to fund the University of Georgia (UGA) to determine whether or not Augusta's biosolids posed a risk to cattle. On the dairy farms, the cattle developed AIDS-like symptoms after consuming forage grown on Augusta's biosolids and died of various infections. Local veterinarians and environmental experts hired by the dairy farmers discovered high levels of cadmium, molybdenum, and other heavy metals in soil and forage samples from areas treated with the biosolids. They also found potentially toxic levels of the heavy metals in the cattle's liver, kidney, and milk samples. Augusta's biosolids also contained potentially toxic levels of organic chemical wastes, including PCBs and chlordane, which were also detected in environmental samples collected from the dairy farms.

UGA provided me an office in its Department of Marine Sciences and promised to pursue hiring me for a tenured faculty position. Corporate executives who provided EPA and the NRC with the anonymous white paper, however, filed the allegations of research misconduct at UGA. Some of the allegations closely paralleled Wakefield's case. The company alleged that my research done as an EPA employee had not been properly approved and that my private work as an expert witness on behalf of patients constituted a serious conflict of interest and ethics violation.

By this time, the EPA's Office of General Counsel had determined that none of the allegations were based in any facts.[11] Serving as an expert witness was the only way I could have access to patient records tied up in lawsuits involving illnesses and deaths linked to biosolids. EPA informed UGA that my research and private expert witness work had been approved by EPA ethics officials, and UGA informed the company that it did not intend to pursue its misconduct petition. To overcome this obstacle, the company hired Georgia Senator Kasim Reed, now mayor of Atlanta, to pressure UGA not to dismiss its petition.[12] As a result, UGA left the allegations in a state of limbo and never ruled on them.

Additional allegations paralleling those against Wakefield were filed against me by a city being sued over biosolids. The city alleged that it was a conflict of interest for me to donate my expert witness fees to UGA to fund my research on biosolids. UGA cleared me of any wrongdoing. My department head later testified, however, that UGA administrators no longer supported hiring me because UGA depended on "money either from possible future EPA grants or connections there might be between the waste-disposal community [and] members of faculty."

In a further attempt to discredit our research, the EPA funded UGA to publish data fabricated by the city of Augusta, Georgia, to cover up cattle deaths that I was investigating on two dairy farms treated with Augusta's biosolids. Predictably, the study concluded that Augusta's biosolids did not pose a risk to cattle. UGA issued a press release quoting lead author Julia Gaskin: "Some individuals have questioned whether the 503 regulations are protective of the public and the environment. This study puts some of those fears to rest."

The McElmurray family, which owned one of the dairy farms, sued the U.S. Department of Agriculture to recover damages from crops they lost as a result of Augusta's biosolids contaminating their land with heavy metals and other hazardous wastes. Robert Brobst, Gaskin's EPA coauthor, used the UGA study to argue that the biosolids did not contaminate the land. Judge Anthony Alaimo in the U.S. District Court in the Southern District of Georgia, however, ruled in favor of plaintiffs: "Brobst opined in a letter that the McElmurrays' land was not contaminated. [But] Brobst concedes that his conclusion is based on Augusta's unreliable, and to some extent invented, data."[13]

Alaimo's ruling and a multiuniversity study in Ohio confirming our findings were covered in an editorial and news article published in *Nature.*[14] Editors described the EPA's biosolids program as "a failure of three presidential administrations." The dairy farmers and I filed a separate lawsuit on behalf of the United States to compel UGA to return the misused EPA funds and withdraw the fabricated data. When deposed under oath, Gaskin admitted she knew there were problems with the data when she submitted the paper, and she believed that Augusta's biosolids harmed the dairy farms.[15] EPA and UGA, nevertheless, refused to withdraw the fabricated data. In 2012, Judge Clay Land of the Athens Division of the Middle District of Georgia ordered the dairy farmers and me to pay more than $61,000 in court costs, which the EPA and the University of Georgia had promised to cover if we would stop commenting on the case publicly.[16]

APPENDIX B: TOOLS FOR VACCINATION CHOICE

i. Chicago Principles on Vaccination Choice

ii. Calls for Immediate Action in the United States

iii. Harris Interactive Poll—May 2010

iv. Public Discussion Presentation

v. Frequently Asked Questions about Vaccines

THE CHICAGO PRINCIPLES ON VACCINATION CHOICE

We, the people who affirm our belief in personal rights, in order to promote the general health and welfare for ourselves and our children and to establish justice, advocate the following principles:

1. The right to free and informed consent to all medical interventions, including vaccination, is a fundamental human right. This includes the right to refuse any medical intervention, including vaccination.
2. Laws that make education, employment, day care and public benefits contingent on vaccination status, except in extreme public health emergencies, violate the fundamental human right to vaccination choice.
3. Laws that sanction nondisclosure of known medical risks, including risks of vaccination, violate the fundamental human right to free and informed consent for all medical interventions.
4. A safety first agenda must apply to vaccination policy. No government should have the right to recommend or mandate vaccines until impartial scientific research has documented their relative safety.
5. Individuals who evaluate, recommend, and mandate vaccines must be free of all actual and perceived conflicts of interest.

For more information:
www.centerforpersonalrights.org

CALLS FOR IMMEDIATE ACTION IN THE UNITED STATES

1. Every state should permit philosophical exemption to vaccination mandates, upholding the human right to free and informed consent for all medical interventions. Without the real right to say no, current policy is coercive and lacks legitimacy.
2. Congress should conduct oversight hearings on the national vaccine program, including mandates for the military and immigrants, vaccine safety, conflicts of interest, suppression of science, evidence of vaccine injury, and comparative empirical data from countries with differing vaccination schedules.

Harris Interactive ParentQuery
Fielding Period: May 5-11, 2010
Center For Personal Rights, Inc.
Weighted To The U.S. Parent Population - Propensity

Page 1

Q1605 Please indicate the extent to which you agree or disagree with each of the following statements about vaccination.

12 May 2010

Summary Of Strongly/Somewhat Agree

Base: Qualified Parents

		Parent Gender		Parent Age			Region				Parent Education			# of Parents in HH		HH Income			
	Total	Male	Female	18-34	35-44	45+	North-east	Mid-west	South	West	H.S. or Less	Some College	College Grad. +	Single Parent	2 Parents	<$35K	$35K - $49.9K	$50K - $74.9K	$75K+
	(A)	(B)	(C)	(D)	(E)	(F)	(G)	(H)	(I)	(J)	(K)	(L)	(M)	(N)	(O)	(P)	(Q)	(R)	(S)
Unweighted Base	1144	494	650	380	482	282	258	320	363	203	207	433	504	226	918	275	152	254	353
Weighted Base	1144	473	671	361	474	309	245	272	378	249	417	342	385	216	928	254	129*	216	433
I am concerned that the pharmaceutical industry has undue influence over government vaccine mandates.	618 54%	268 57%	349 52%	168 47%	273 58% D	176 57% d	137 56%	153 56%	212 56%	116 46%	220 53%	190 56%	208 54%	125 58%	493 53%	131 52%	78 60%	121 56%	225 52%
The government should fund an independent scientific study of fully vaccinated vs. unvaccinated individuals to assess long-term health outcomes.	613 54%	251 53%	362 54%	200 55%	248 52%	166 54%	128 52%	150 55%	207 55%	128 51%	212 51%	186 54%	215 56%	129 60%	484 52%	140 55%	84 65% S	118 55%	220 51%
Parents should have the right to decide which vaccines their children receive without government mandates.	592 52%	254 54%	338 50%	179 50%	255 54%	158 51%	125 51%	155 57%	191 50%	121 48%	211 51%	195 57% m	186 48%	126 58%	465 50%	137 54%	74 58%	114 53%	206 48%
I am concerned about serious adverse effects of vaccines.	544 48%	208 44%	336 50%	170 47%	229 48%	144 47%	113 46%	130 48%	195 51%	106 43%	197 47%	171 50%	175 46%	119 55% o	425 46%	131 52%	81 63% RS	95 44%	187 43%
All children should receive 69 doses of 16 vaccines before age 18, as recommended by the federal government.	477 42%	204 43%	273 41%	148 41%	211 45%	118 38%	100 41%	105 39%	164 43%	108 43%	148 36%	126 37%	203 53% KL	81 37%	397 43%	99 39%	49 38%	78 36%	214 49% pR

Proportions/Means: Columns Tested (5%, 10% risk level) - B/C - D/E/F - G/H/I/J - K/L/M - N/O - P/Q/R/S
Overlap formulae used. * small base

3. Congress should immediately initiate a prospective study of vaccinated versus unvaccinated populations for long-term health outcomes. Such a baseline study has never been done, suggesting that the vaccination schedule as a whole is an experiment on human subjects. This bill is called the "Comprehensive Comparative Study of Vaccinated and Unvaccinated Populations Act" and was originally cosponsored by Rep. Maloney (D-NY) and Rep. Osborne (R-NE).
4. Congress should amend the 1986 National Childhood Vaccine Injury Act to reinstate ordinary tort liability for vaccine manufacturers and medical professionals. Liability protections in the 1986 Law have fueled the expansion of the vaccine schedule and have not made vaccines safer, as the Law required.
5. Congress should abolish the Vaccine Injury Compensation Program or make it optional. The Program has failed in its purpose to quickly, generously, and administratively compensate families for vaccine injury.

HARRIS INTERACTIVE POLL—MAY 2010

In May 2010, global marketing research firm Harris Interactive conducted an online survey of 1,144 parents with children age seventeen or younger within the United States. This survey was conducted on behalf of the Center for Personal Rights, Inc. via its ParentQuery omnibus product.

Figures for age, sex, race and ethnicity, education, region, household income, and the age of children in the household were weighted where necessary to bring them into line with their actual proportions in the population. Propensity score weighting was used to adjust for respondents' propensity to be online.

Respondents for this survey were selected from among those who have agreed to participate in Harris Interactive surveys. The data have been weighted to reflect the composition of the parent population. Because the sample is based on those who agreed to participate in the Harris Interactive panel, no estimates of theoretical sampling error can be calculated.

According to the final report:

- 52% believe that "parents should have the right to decide which vaccines their children receive without government mandates."
- 54% of parents are "concerned that the pharmaceutical industry has undue influence over government vaccine mandates."

- 54% of parents agree that "the government should fund an independent scientific study of fully vaccinated vs. unvaccinated individuals to assess long-term health outcomes."
- 48% of parents are "concerned about serious adverse effects of vaccines."
- 42% of parents, a minority, agree that "all children should receive 69 doses of 16 vaccines before age 18, as recommended by the federal government."*

For full poll results and complete survey methodology, including weighting variables, see the seventy-eight-page full statistical report available at www.centerforpersonalrights.org.

CENTER FOR PERSONAL RIGHTS

Public Discussion Presentations

The Center for Personal Rights (CPR) is a non-profit organization that advocates for the rights to life, liberty, and personal security for ourselves and our children. CPR demands the universal human rights standard of informed consent to all medical interventions. Compulsory vaccination is morally and legally unjustifiable.

As part of its mission, CPR has compiled information and data on vaccination choice for public presentations. Below are a few slides from the presentation, "The Conversation: Vaccination Is A Human Right." You can download the full PowerPoint presentation at www.centerforpersonalrights.org.

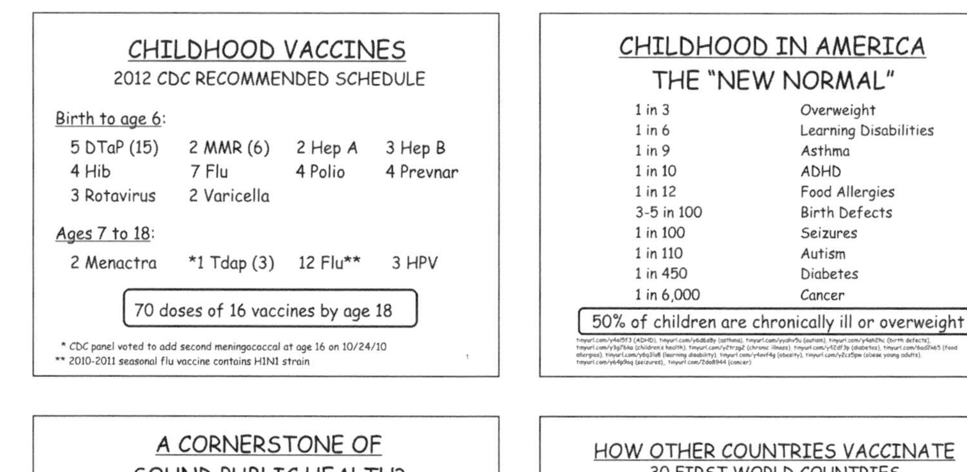

FREQUENTLY ASKED QUESTIONS ABOUT VACCINES

"Vaccines are Effective"

1. **Vaccines are modern medicine's greatest achievement. Haven't they saved the developed world from the scourge of infectious disease?**

   Response: Better nutrition, clean water, modern sanitation, hygiene, improved living and working conditions, and antibiotics have played essential and under-appreciated roles in public health. Many infectious diseases were almost gone by the time vaccination mandates began. Vaccines have played a role in decreasing infectious disease, but they've also injured and taken lives. Vaccine package inserts document a long list of possible side effects and death that can occur after vaccination. The government administers a special program to compensate for vaccine injury or death. Because every vaccination decision is potentially a life-and-death decision or may affect future quality of life, free and informed consent is imperative. Individuals must have the right to decide for themselves and their children if and when to vaccinate. If vaccines are truly safe and effective, the vast majority will want them, with no need for mandates.

2. **Vaccines save lives. Don't vaccines overwhelmingly benefit the individual and society?**

   Response: Look at the health of children in the United States today. One in six has a learning disability; one in nine has asthma; one in twelve has ADD; one in eighty-eight has autism. The United States ranks forty-ninth in infant mortality in the world, below every industrialized country except Poland (www.tinyurl.com/24s4d7). Vaccines have helped to diminish infectious diseases, but they have not ensured good health. We are experiencing an alarming increase in a wide range of chronic, neurological, and autoimmune childhood diseases and disorders. The extent to which vaccines play a role in these childhood diseases and disorders is largely unstudied, but some evidence suggests strong associations.

3. **Vaccines work. Don't vaccines protect us against deadly diseases?**

   Response: Immunity from vaccines is temporary; it wears off over time. Only natural immunity acquired through actual infection and boosted through exposure to disease in the community gives lifelong immunity. Vaccines rely on "booster" shots to maintain antibodies. This is the reason it is recommended that children get six doses of each vaccine for diphtheria, pertussis, and tetanus, four doses of each vaccine for polio, *haemophilus influenzae* type b and pneumococcal disease, three doses of each vaccine for hepatitis B, rotavirus and human papillomavirus, and two doses of each vaccine for measles, mumps, rubella, chickenpox and hepatitis A. Scientists initially believed that vaccines would afford lifelong immunity. As disease outbreaks increasingly occurred among the vaccinated, mandates for vaccine boosters became commonplace. Even in communities where 98+% of the population is "fully vaccinated" for a particular disease, creating so-called herd immunity, outbreaks among the vaccinated still occur (www.tinyurl.com/d5l8vm, www.tinyurl.com/27sexxd).

"Vaccines Are Safe"

4. **Adverse reactions to vaccines are exceptionally rare. Aren't death and injuries caused by vaccines necessary risks that all members of the community must accept as the price of citizenship?**
   Response: All human beings are entitled to life, liberty, and security of person. A society that protects fundamental human rights does not discriminate on the basis of age or genetic predisposition or mandate a practice in peacetime or in the absence of a true public health emergency that is known to cause injury and death. Doctors often fail to recognize injuries and underreport them because we have a voluntary Vaccine Adverse Events Reporting System (VAERS). There are no penalties for failure to report injuries and it is estimated that one to ten percent of adverse events are reported. Inclusion in VAERS does not prove causation, and little research has investigated alleged vaccine injuries. Most doctors are uninformed about vaccine adverse events and are unaware of how to best help their patients when these adverse events happen. By law, doctors bear no liability for vaccine injury. Furthermore, our understanding of the nature and mechanism of vaccine-induced injury and death is limited, in part because we have failed to study the long-term health outcomes of fully vaccinated compared with unvaccinated individuals.

5. **Vaccines are scientifically sound. Aren't vaccines extremely well-researched for optimal safety by independent government scientists with children's and society's best interests at heart?**
   Response: Like all prescription medications, vaccines are considered "unavoidably unsafe" by law. Doctors give infants over thirty doses of "unavoidably unsafe" medications by fifteen months of age, usually without full and detailed informed consent. Nonetheless, the Centers for Disease Control and Prevention (CDC) states on its website that "vaccines are held to the highest standard of safety" (www.tinyurl.com/2dwf8x9). In reality, vaccines have not been well-researched. There has never been a study comparing the long-term health outcomes of vaccinated versus unvaccinated populations. Vaccines have not been studied cumulatively, in the way that they are administered. They are tested individually, in isolation, for very short periods of

time by the vaccine manufacturers themselves, who are protected from almost all legal liability. Furthermore, many of the decision makers for vaccination recommendations and mandates have serious conflicts of interest—some are even vaccine patent holders. In 2007, a majority of outside scientific expert advisors to the CDC were found to have potential violations of the agency's conflict of interest regulations (www.tinyurl.com/24u4jen).

6. **Side effects are slight. To the extent that there are any side effects from vaccines, aren't they are mild, such as soreness at the site of vaccination and short-term fever?**
Response: All vaccines can result in injury or death; legally, they are "unavoidably unsafe" products. This is the reason that Congress created the Vaccine Injury Compensation Program—because it recognized that vaccines injure and cause death in some individuals. Although screening mechanisms and risk factors could be identified to protect those at risk, they do not exist today. Leading voices, such as Dr. Bernadine Healy, former head of the National Institutes of Health (www.tinyurl.com/59wns7) have called for studies of those who may have been vaccine-injured. Nonetheless, all fifty U.S. states impose "one-size-fits-all" vaccination mandates as a requirement for day care and school admission, regardless of family medical history and infant weight.

7. **The book is closed. Every major study looking for connections between vaccines and autism has failed to find any link. Isn't this book closed?**
Response: Dr. Robert Chen, Chief of Vaccine Safety and Development at the CDC, said in 1995, "the only line item for vaccine safety research is I think on the order of a little less than $2 million per year. That basically covers operation of VAERS (Vaccine Adverse Events Reporting System) period, and nothing else." (Advisory Commission on Childhood Vaccines (ACCV) and National Vaccine Advisory Committee (NVAC) Subcommittees on Vaccine Safety, May 31, 1995.) More recently, Dr. Louis Cooper, vaccine inventor and former President of the American Academy of Pediatrics admitted that the vaccine safety science research budget was twenty million dollars or a mere 0.5 percent of the four billion dollar total vaccine budget for purchase, promotion, and delivery in 2008. (www.pediatrics.org/cgi/content/full/122/1/149). This is especially significant if you consider the dramatic rise in autism incidence in the past two decades, from four to five in 10,000 in the early 1990s, to one in 110 today.

The studies that the U.S. government uses to disprove causation of harm are limited and flawed (see www.fourteenstudies.org). Several rigorous studies have found links between vaccines, developmental disability, and chronic illness. These include the Gallagher-Goodman Stony Brook University Medical Center study tying hepatitis B triple series vaccine and developmental disability (www.tinyurl.com/2ejdrd8) and the University of Manitoba study funded by the Canadian Institutes of Health Research that found an association between delayed diphtheria, pertussis, and tetanus vaccination and a reduced risk of childhood asthma (www.tinyurl.com/2edrl7c). There have been few studies on children with regressive autism. The 2004 Institute of Medicine Report did not rule out the possible existence of a vulnerable subset of children in whom vaccines could cause autism. And, in late 2009, the government spokesman on autism, Dr. Thomas Insel, acknowledged that the real rise in autism prevalence must be attributed at least in part to environmental causes.

8. **Thimerosal is no smoking gun. Isn't it true that we know that vaccines don't cause autism because thimerosal, the mercury-based preservative, has been removed from all pediatric vaccines and autism rates have continued to rise?**

   Response: It is misleading to assume a direct and causal relationship between those two facts; the inference fails to account for confounding factors. In other words, there were other potentially significant events occurring at the same time that must be considered. For example, as we decreased the amount of thimerosal in pediatric vaccines, we increased the total number of vaccines as well as the cumulative exposure to other neurotoxic ingredients, including aluminum. It is not possible to ascertain whether vaccines are causing the dramatic increase in autism incidence until we dedicate the funds to pursue a long-term controlled study. Furthermore, thimerosal remains in some childhood vaccines today. Flu shots are recommended for all children and most seasonal flu shots, including those routinely given to children and pregnant women, contain ten to twenty-five micrograms of thimerosal or approximately 10,000 to 25,000 parts per billion (ppb). Thimerosal is nearly fifty percent ethylmercury by weight (www.tinyurl.com/y7pdm5m) and continues to be used in the manufacturing process of many vaccines. After filtration, vaccines still have "trace" amounts of thimerosal equivalent to 0.5 microgram or 2,000 ppb. Two hundred ppb meets the EPA classification for hazardous waste and two ppb is the EPA's limit for safe drinking water (www.tinyurl.com/y2zfe58). Mercury is the most neurotoxic nonradioactive substance on the planet. It has been removed from over-the-counter medicines and veterinary vaccines. It should not be injected in any amount into humans of any age.

9. **Pediatricians are knowledgeable. Pediatricians are well-educated about the risks and benefits of vaccines. Can't they tell if a vaccine is contraindicated for a particular child?**
Response: The full ramifications of the vaccination schedule are unknown because researchers have not studied vaccines cumulatively in the way that they are administered. Pediatricians and family practice doctors receive little education about the possible adverse consequences that can occur after the administration of vaccines. The information they do receive is typically limited to a description of the vaccines on the CDC schedule, how to administer them, the importance of full compliance, and how to catch children up who have missed doses. State departments of health apply pressure on physicians who write "too many" medical exemptions. Many doctors are not eager to discuss vaccination decisions with parents and rarely inform them of their rights and the legal exemptions available to them. Pediatricians are even supported by the American Academy of Pediatrics in their decision to discharge parents who are not willing to vaccinate their children according to CDC-recommended vaccination schedules (www.tinyurl.com/4mdze2).

"Vaccination Compulsion is Ethical"

10. **Public health trumps individual choice. If you don't vaccinate yourself and your children, aren't you endangering others?**
Response: If you have been vaccinated against an infectious disease and your blood antibody titers indicate that you have levels providing specific immunity, you cannot contract that disease from someone, whether that person was vaccinated or not. People who are unvaccinated, and those who are vaccinated but whose immunity has worn off, can be disease carriers if they do not stay at home when they are sick. In the event of an infectious disease outbreak, schools require that unvaccinated students remain at home. Due to viral shedding from live virus vaccines, a recently vaccinated person can infect the unvaccinated and those who are vaccinated but whose immunity has worn off. Vaccine package inserts do not rule out the possibility of disease transmission via viral shedding and close personal contact (www.tinyurl.com/24ldtn).

Admittedly, some people are unable to be vaccinated due to age or health status. There have been vitriolic discussions about the selfishness of parents who do not vaccinate their children. In the absence of a true public health emergency, the morality of requiring someone to vaccinate to protect others must be carefully weighed against the morality of mandating a medical intervention for healthy individuals that is known to injure and cause death to some. The government has committed or paid out over two billion dollars to about 2,500 families to compensate for vaccine injury and death. A society that protects fundamental human rights does not discriminate on the basis of genetic predisposition or mandate a practice that inevitably causes injury and death, regardless of the number of victims. This is especially true when the scientific foundation supporting long-term safety and efficacy is unknown due to government failure to provide rigorous unbiased science on vaccine safety. The decision to be vaccinated should rest with the individual or the parent, not the state.

11. **Mandatory vaccination is legal: Aren't vaccination laws constitutional and for the public good?**

    Response: Jacobson v. Massachusetts, a 1905 U.S. Supreme Court case, established the precedent that states may impose vaccination mandates during public health emergencies. This decision, arising from a deadly smallpox epidemic in a different era, predates the doctrines of free and informed consent and medical autonomy and most of the scientific advances of the twentieth century. State-mandated vaccination during a smallpox epidemic in 1905 reflects a radically different reality than ours today. In 1905, there was a single vaccination mandate at issue during a widespread and contagious smallpox epidemic, and Jacobson's noncompliance resulted in a monetary fine. At no point was Jacobson denied employment, or his child denied school admission, due to the failure to vaccinate. Jacobson and his son did not face mandates with up to seventy doses of sixteen vaccines that included foreign proteins and neurotoxic chemicals. And even *Jacobson v. Massachusetts*, which upheld state mandates, noted the risks of "regulations so arbitrary and oppressive" that they would be "cruel and inhuman in the last degree." Jacobson also provided for the medical exemption to mandatory vaccination which is upheld in all fifty states. Current vaccination mandates, required in non-emergency situations, based on inadequate science, rest on a questionable constitutional foundation.

12. **Justified liability protection. Congress rightly limited the liability of doctors and vaccine manufacturers in 1986 to safeguard the nation's vaccine supply. Don't we need insulation of doctors and industry from liability to protect the vaccine supply?**

    Response: In 1986, Congress granted tort liability protection to doctors and the vaccine industry against vaccine injuries and death. This means that people harmed by vaccines cannot sue vaccine manufacturers or the physicians who administered the vaccines. In 2011, the Supreme Court ruled in *Bruesewitz v. Wyeth* that families of the vaccine-injured do not have the right to sue for a defective product design claim in civil court before a jury. Justice Sotomayor wrote in her dissent that the majority's decision leaves a regulatory vacuum in which no one—no federal agency, no jury—ensures that vaccine manufacturers incorporate scientific and technological advancements. There is no incentive for corporations to improve the design of products that are already generating significant profits. Blanket liability protection and the lack of free and informed consent, have led to the large increase in compulsory childhood vaccination mandates. A responsible universal vaccination program must include rigorous, impartial science on vaccine safety, free and informed consent of individuals to vaccination, and corporate accountability for vaccines. We must change the laws that underpin today's childhood mass vaccination program to create real choice and real accountability.

13. **Schools as the gatekeeper: It makes sense to have schools police kids' vaccination status. Shouldn't schools require immunization?**
    Response: Education is a fundamental right for children to receive and a responsibility for parents to provide. The right to education should not be contingent on vaccination status. It is profoundly unwise to turn schools into vaccination police. In the event of infectious disease outbreaks, unvaccinated children may be required to stay at home. This does not interfere with their general right to participate in school.

"Vaccines Promise Ever-Increasing Benefits"

14. **Unlimited benefits: There appear to be no limits to the diseases we can conquer through vaccines. Scientists are working on vaccines for cancer and HIV. Isn't this welcome?**
    Response: As is the case with all medical interventions, people should have the right to free and informed consent for vaccination. This means there should be complete disclosure regarding the risks and benefits and all vaccines must be rigorously studied for safety and effectiveness before use. For a variety of reasons, including family history, prior disease exposure, and a varying assessment of the dangers and likelihood of contracting the disease, all people will not make the same decisions about the risk-reward tradeoffs. Individuals must make those choices, not government officials or scientists.

15. **More is better: Aren't relatively new compulsory vaccines, such as vaccines for flu, hepatitis B, and HPV, welcome because they protect us from deadly diseases?**
    Response: The legal justification for compulsory vaccination is to protect people in the event of contagious, widespread infectious disease epidemics or true public health emergencies. The U.S. Supreme Court's 1905 decision, *Jacobson v. Massachusetts*, cre-

ated this precedent during a smallpox outbreak. Citizens have the right today to obtain exemption from vaccination for medical, religious and philosophical reasons, depending on the state. The scope of vaccination mandates must be proportionate to the public health emergency. While there are no epidemics of hepatitis B in the United States, for example, this vaccine for a sexually-transmitted disease is compulsory for babies and children in forty-seven states, based on the premise that government may impose vaccines for the "greater good." More vaccination mandates are not necessarily better, and state requirements must respect fundamental human rights.

NOTES

CHAPTER 1: VACCINATION CHOICE IS A HUMAN RIGHT —MARY HOLLAND, JD

1. The U.N. Charter states in its Preamble, "We the peoples of the United Nations determined . . . to reaffirm faith in fundamental human rights, in the dignity and worth of the human person, in the equal rights of men and women." www.un.org/en/documents/charter/index.shtml.
2 Preamble, the Universal Declaration of Human Rights, www.un.org/en/documents/udhr/index.shtml.
3 The Nuremberg Code states as the first principle that "[t]he voluntary consent of the human subject is absolutely essential," www.hhs.gov/ohrp/irb/irb_appendices.htm#j5.
4 The Council of Europe's 1997 Convention on Human Rights and Biomedicine Article 5 states: "An intervention in the health field may only be carried out after the person concerned has given free and informed consent to it," http://conventions.coe.int/Treaty/en/Treaties/Html/164.htm. UNESCO's 2005 Universal Declaration on Bioethics and Human Rights Article 6 states: "Any preventive, diagnostic and therapeutic medical intervention is only to be carried out with the prior, free and informed consent of the person concerned, based on adequate information," www.unesco.org/new/en/social-and-human-sciences/themes/bioethics/bioethics-and-human rights. For more on contemporary international human rights norms on bioethics and medicine, see chapter 2.
5 The 1986 National Childhood Vaccine Injury Act acknowledges that vaccine injury or death may be "unavoidable even though the vaccine was properly prepared and accompanied by proper directions and warnings." 42 U.S.C. 300aa-22(b)(1). The "unavoidable" language in the Act is from the Restatement (Second) of Torts that applies to "products which, in the present state of human knowledge, are quite incapable of being made safe." Restatement (Second) of Torts Section 402A, comment k (1965).
6 Vaccine Information Statements (VISs) are information sheets produced by the CDC. Federal law requires that VISs be handed out before certain vaccinations are given. See the Center for Disease Control and Prevention's website, www.cdc.gov/vaccines/pubs/vis/default.htm.
7 The 1986 National Childhood Vaccine Injury Act states that vaccine manufacturers are free from tort liability "if the injury or death resulted from side effects that were unavoidable even though the vaccine was properly prepared and was accompanied by proper directions and warnings." 42 U.S.C. 300aa-22(b)(1).
8 *Jacobson v. Massachusetts*, 197 U.S. 11 (1905).
9 Recording of Martin Luther King, www.entertonement.com/clips/cbrtmmrrvb—Arc-of-the-moral-universeMartin-Luther-King-Jr-MLK-Civil-Rights-How-long-. For discussion of the origin of the Theodore Parker quotation, see National Public Radio's website, www.npr.org/templates/story/story.php?storyId=129609461.
10 For a detailed account of the legal problems with compulsory vaccination, *Jacobson v. Massachusetts* and the 1986 National Childhood Vaccine Injury Act, see Mary Holland, *Reconsidering Compulsory Childhood Vaccination*, http://papers.ssrn.com/sol3/papers.cfm?abstract_id=1677565.

CHAPTER 2: THE INTERNATIONAL HUMAN RIGHTS STANDARD —SOOKYUNG SONG, JD

1 The trovafloxacin experiment on 100 Nigerian children took a heavy toll, causing death to five and inflicting severe injuries to many others. The survivors subsequently brought an action against Pfizer, Inc., which orchestrated the clinical trials, claiming that the doctors working on Pfizer's behalf failed to acquire informed consent prior to the experiments and thus had violated customary international law. *Abdullahi v. Pfizer, Inc.*, 562 F. 3d 163 (2009). In a landmark decision, the Second Circuit upheld the court's jurisdiction and found a binding customary international law norm prohibiting nonconsensual medical experimentation. The parties eventually settled.

2 See Trials of War Criminals before the Nuremberg Military Tribunal, Vol. I and II, The Medical Case, U.S. Government Printing Office (1949) at the National Institutes of Health website, http://ohsr.od.nih.gov/guidelines/nuremberg.html.

3 General Assembly resolution 217A (III), U.N. Doc A/810 at 71 (1948) at the United Nations website, www.un.org/en/documents/udhr/.

4 Ibid, art. 5.

5 G.A. res. 2200A (XXI), Dec. 16, 1966, 993 U.N.T.S. 3 at the United Nations website, http://treaties.un.org/doc/Publication/UNTS/Volume%20999/volume-999-I-14668-English.pdf.

6 Jonathan Todres, "Can Research Subjects of Clinical Trials in Developing Countries Sue Physician-Investigators for Human Rights Violations?" 16 N.Y.L. Sch. J. Hum. Rts. 737, 745 (2000).

7 Convention on Human Rights and Biomedicine, Apr. 4, 1997, 2137 U.N.T.S. 171 at the Council of Europe website, http://conventions.coe.int/Treaty/en/Treaties/Html/164.htm.

8 Explanatory Report on the Convention on Human Rights and Biomedicine, at para. 34, http://conventions.coe.int/Treaty/en/Reports/Html/164.htm.

9 Ibid

10 UNESCO website, http://portal.unesco.org/en/ev.php-URL_ID=31058& URL_DO=DO_TOPIC&URL_SECTION=201.html.

11 Ibid, art. 3.

CHAPTER 3: A HUMAN RIGHTS ASSESSMENT —LOUISE KUO HABAKUS, MA

1 Jonathan Mann et al, *Health and Human Rights, A Reader*, (New York: Routledge, 1999), 3.

2 Ibid, 9.

3 Ibid, 47-48.

4 Preamble, Charter of the United Nations, signed June 26, 1945, www.un.org/en/documents/charter/preamble.shtml.

5 *Jacobson v. Massachusetts*, 197 U.S. 11 (1905), 31.

6 Ibid, 28.

7 "FACA: Conflicts of Interest and Vaccine Development—Preserving the Integrity of the Process," Hearing before the Committee on Government Reform, House of Representatives, One Hundred Sixth Congress, Second Session, June 15, 2000, http://frwebgate.access.gpo.gov/cgi-bin/getdoc.cgi?dbname=106_house_hearings&docid=f:73042.wais.

8 Mary Holland, "Reconsidering Compulsory Childhood Vaccination," http://papers.ssrn.com/sol3/papers.cfm?abstract_id=1677565.

9 Bernard Guyer et al., "Annual Summary of Vital Statistics: Trends in Health of Americans During the 20th Century," *Pediatrics*, 106 (2000):1315, www.pediatrics.org/cgi/content/full/106/6/1307.

10 Ibid.

11 Takis Panagiotopoulos et al., "Increase in congenital rubella occurrence after immunization in Greece: retrospective survey and systematic review," *BMJ* (1999), December 4, 319(7223):1462–1467, www.ncbi.nlm.nih.gov/pmc/articles/PMC28289/?tool=pubmed.

12 "Exemptions from School Immunization Requirements," Table of the National Conference of State Legislatures June 2010, www.ncsl.org/IssuesResearch/Health/SchoolImmunizationExemptionLaws/tabid/14376/Default.aspx.

13 James Colgrove, *State of Immunity: The Politics of Vaccination in Twentieth-Century America* (Berkeley: University of California Press, 2006), 158.

14 Ibid.

15 Ibid, 81.

16 Ibid, 113.

17 Ibid, 158.

18 Benjamin M. Nkowane et al., "Measles Outbreak in a Vaccinated School Population: Epidemiology, Chains of Transmission and the Role of Vaccine Failures," *American Journal of Public Health*, 1987 April; 77(4): 434–438, www.ncbi.nlm.nih.gov/pmc/articles/PMC1646939/; TL Gustafson et al., "Measles outbreak in a fully immunized secondary-school population," *New England Journal of Medicine*, 1987 March 26, 316(13):771-4, www.ncbi.nlm.nih.gov/pubmed/3821823; HJ Brockoff et al., "Mumps Outbreak in a Highly Vaccinated Student Population, The Netherlands, 2004," *Vaccine*, 2010 April 9, 28(17):2932-6, E-pub 2010 Feb 25, www.ncbi.nlm.nih.gov/pubmed/20188683; and BD Tugwell et al., "Chickenpox outbreak in a highly vaccinated school population," *Pediatrics*, 2004 March, 113(3 Pt 1):455–9, www.ncbi.nlm.nih.gov/pubmed/14993534.

19 "CDC: Most U.S. children receiving recommended vaccinations," *Pediatric Supersite*, September 16, 2010, www.pediatricsupersite.com/view.aspx?rid=70382.

20 GM Lee et al., "Pertussis in adolescents and adults: should we vaccinate?" *Pediatrics*, 2005 June, 115(6):1675–84, www.ncbi.nlm.nih.gov/pubmed/ 15930232; and "Whooping cough on the rise in New Jersey," *The Record*, July 6, 2012, http://www.northjersey.com/news/Whooping_cough_on_the_rise_in_North_Jersey.html.

21 "Pertussis (Whooping Cough)—What You Need to Know," Centers for Disease Control and Prevention, www.cdc.gov/features/pertussis/.

CHAPTER 4: DUE PROCESS AND THE AMERICAN CONSTITUTION —JAMES TURNER, JD

1 National Vaccine Injury Compensation Program, www.hrsa.gov/vaccinecompensation/statistics_report.htm.

2 The Center for Biologics Evaluation and Research (CBER) is a center within the Food and Drug Administration, an agency within the United States Government's Department of Health and Human Services. CBER's mission is to protect and enhance the public health through the regulation of biological and related products including blood, vaccines, allergenics, tissues, and cellular and gene therapies.

3 Joseph Edward Smadel was a U.S. physician and virologist. He was the first recipient of the Albert Lasker Award for Clinical Medical Research. In the 1950s, under Smadel's leadership, Walter Reed Army Institute of Research established itself as one of the premier institutes for the study of infectious diseases, http://en.wikipedia.org/wiki/Joseph_Edward_Smadel.

4 Statement of Barbara Loe Fisher, President, National Vaccine Information Center, to California State Senate Committee on Health and Human Services, January 23, 2002, www.whale.to/v/fisher7.html.

5 Testimony of Barbara Loe Fisher, Co-Founder and President, National Vaccine Information Center, U.S. House Government Reform Committee, August 3, 1999, "Vaccines: Finding a Balance Between Public Safety and Personal Choice," www.whale.to/vaccines/fisher.html.

6 www.exemptmychild.com/media//DIR_16101/Arkansas$20Vaccine$20Exemption$20Laws.pdf section d(4).

7 All fifty states have legislation mandating specified vaccines for students. Although exemptions vary from state to state, all school immunization laws grant exemptions to children for medical reasons. Almost all states, except Mississippi and West Virginia, grant religious exemptions for people who have religious beliefs against immunizations. Twenty states allow philosophical exemptions for those who object to immunizations because of personal, moral or other beliefs. The National Conference of State Legislatures June 2010 Table "Exemptions from School Immunization Requirements" show individual state exemptions, www.ncsl.org/IssuesResearch/Health/SchoolImmunization ExemptionLaws/tabid/14376/Default.aspx.

8 *Furman v. Georgia*, 408 U.S. 238 (1972).

9 In rare situations when the voluntary incidence rate is low and the population effect of the vaccine so well established that the government can prove the requirement for a vaccine, a mandate might be required for certain vaccines. However this should be a rare occurrence.

10 *Cruzan v. Dir., Mo. Dep't of Health*, 497 U.S. 261, 278 (1990).

11 *Parham v. J.R.*, 422 U.S. 584, 602; see also, *Troxel v. Granville*, 530 U.S. 57 (2000).

12 See, e.g. *Dela Rosa v. Sec'y of HHS*, 2001 U.S. Claims LEXIS 173, No. 93-433V at *3, 4 (encephalitis, intractable seizures, pancreatitis, and parotitis. Dela Rosa also suffered from dementia with a decline in IQ from 115 to 77, agitation and aggressive behaviors, and medically intractable localization-related epilepsy, according to medical records submitted in the case by the Mayo Clinic.); *Rice v. Sec'y of HHS*, 2001 U.S. Claims LEXIS 68, No. 98-438V at *39 (encephalopathy and seizures); *Miller v. Sec'y of HHS*, 2000 U.S. Claims Lexis 277, No. 95-0196 V, reissued 2002, at *1, 33 (acute encephalopathy and its sequelae, namely, permanent neurological damage including seizures and developmental delays).

13 See generally National Childhood Vaccine Injury Act, 42 U.S.C. 300aa-1 et seq.

14 *Jacobson v. Massachusetts*, 197 U.S. 11 (1905).

15 *McCarthy v. Boozman*, 212 F.Supp.2d 945 (W.D. Ark. 2002); and *Boone v. Boozman*, 217 F. Supp. 2d 938 (E.D. Ark. 2002).

16 *Jacobson*, note 14, 30 (emphasis added).

17 Ibid, 33, citing *Wisconsin & R.R. Co. v. Jacobson*, 179 U.S. 287, 301; 1 Dillon Mun. Corp., 4th ed., §§319 to 325 and Freund's *Police Power*, §63 et seq.

18 *State v. Martin & Lipe*, 134 Ark. 420 (1918) and *Brazil v. State*, 134 Ark. 420 (1918).

19 Ibid, *Brazil v. State*, 427.

20 Brazil, 428.

21 Brockman, Leslie N., "Enforcing The Right To a Public Education For Children Afflicted With AIDS," 36 *Emory Law Journal*, 603, 615 n.70 (1987). The cases include *Burroughs v. Mortensen*, 312 Ill. 163, (1924); *Blue v. Beach*, 155 Ind. 121 (1900); *Osborn v. Russell*, 64 Kan. 507 (1902); *People ex rel. Hill v. Board of Educ.*, 224 Mich. 388 (1923); *State ex rel. Freeman v. Zimmerman*, 86 Minn. 353 (1902); *State ex rel. O'Bannon v. Cole*, 220 Mo. 697 (1909); *Glover v. Board of Educ.*, 14 S.D. 139 (1900); *McSween v. Board of School Trustees*, 60 Tex. Civ. App. 270 (1910); *State ex rel. Adams v. Burdge*, 95 Wis. 390 (1897).

22 Ibid, 615 n. 70. The cases include *People ex rel. Jenkins v. Board of Educ.*, 234 Ill. 422 (1908); *People ex rel. Lawbaugh v. Board of Educ.*, 177 Ill. 572 (1899); *Potts v. Breen*, 167 Ill. 67 (1897); *Rhea v. Board of Educ.*, 41 N.D. 449 (1919).

23 *United States v. Quinones*, 205 F. Supp. 2d 256 (S.D.N.Y.), rev'd, 313 F.3d 49 (2d Cir. 2002), cert. denied, 540 U.S. 1051 (2003).

24 *United States v. Quinones*, 196 F. Supp. 2d 416, 420 (S.D.N.Y.), rev'd, 313 F.3d 49 (2d Cir. 2002), cert. denied, 540 U.S. 1051 (2003).

25 *Quinones*, 205 F. Supp. 2d, 256.

26 *Quinones*, 196 F. Supp. 2d, 420.

27 *United States v. Quinones*, 313 F.3d 49 (2d Cir. 2002), cert. denied, 540 U.S. 1051 (2003).

28 *Herrera v. Collins*, 506 U.S. 390 (1993), concurring opinion.

CHAPTER 5: THE RIGHT TO LEGAL REDRESS —MARY HOLLAND, JD, AND ROBERT J. KRAKOW, JD

1 The 1986 National Childhood Vaccine Injury Act acknowledges that vaccine injury or death may be "unavoidable even though the vaccine was properly prepared and accompanied by proper directions and warnings." 42 U.S.C. 300aa-22(b)(1). The "unavoidable" language in the Act is from the Restatement (Second) of Torts that applies to "products which, in the present state of human knowledge, are quite incapable of being made safe." Restatement (Second) of Torts Section 402A, comment k (1965).

2 According to the Centers for Disease Control and Prevention, "[y]ears of testing are required by law before a vaccine can be licensed. Once licensed and in use, vaccines are continuously monitored for safety and efficacy," www.cdc.gov/vaccinesafety/Vaccines/Index1.html.

3 The National Childhood Vaccine Injury Act of 1986, 42 U.S.C. 300aa-1 et seq.

4 42 U.S.C. § 300aa- 14; see also www.hrsa.gov/vaccinecompensation/table.htm.

5 HHS information on vaccine injury compensation, www.hrsa.gov/vaccinecompensation/statistics_report.htm.

6 For example, Dr. Martin Smith, the President of the American Academy of Pediatrics, wrote in an article about the 1986 act to his members that "[t]his was the best compromise settlement . . . that could be reached." Martin H. Smith, National Childhood Vaccine Injury Act, *Pediatrics* 82 (1988):264. From a different perspective, Barbara Loe Fisher, President of the National Vaccine Information Center, said in an address to the Advisory Commission on Childhood Vaccines about the work of parents in crafting the 1986 Act, "[w]e believed we were participating in the development of a law which would give—in the words of the then AAP Chairman—'simple justice to children,'" www.nvic.org/injury-compensation/vaccineinjury.aspx.

7 Remarks of Robert Krakow at American Rally for Personal Rights, Chicago, IL, May 26, 2010, www.americanpersonalrights.org.

8 See note 5 for HHS information.

9 The Omnibus Autism Proceeding was a set of six test cases chosen from approximately 5,000 petitions to determine whether vaccines caused autism in those cases. The cases lasted approximately eight years, presenting thousands of pages of science and testimony and tens of witnesses on two theories, that the mercury-containing preservative thimerosal causes autism alone and that thimerosal together with the measles-mumps-rubella vaccine causes autism, www.uscfc.uscourts.gov/sites/default/files/autism.background.2010_0.pdf.

10 To be sure, there is evidence that individual Special Masters exercise appropriate independence. Our critique is directed to the VICP as an institution and not to the performance of individual Special Masters.

11 Information on the Vaccine Safety Datalink at the CDC website, www.cdc.gov/vaccinesafety/Activities/VSD.html.

12 Institute of Medicine, "Vaccine Safety Research, Data Access, and Public Trust," www.iom.edu/Reports/2005/Vaccine-Safety-Research-Data-Access-and-Public-Trust.aspx, noting that the CDC's handling of the VSD thimerosal data suffered from a "lack of transparency," that datasets "were not archived in a standard manner," and "are not available for reanalysis," and that CDC officials should "seek legal

advice" regarding possible violations of the Federal Data Quality Act. The critical VSD information is allegedly available offshore in the archives of a private healthcare provider. The Special Masters of the Court of Federal Claims presumably could have issued subpoenas for this information but did not. For a detailed discussion of the VSD and the possible thimerosal-autism link, see David Kirby, *Evidence of Harm: Mercury in Vaccines and the Autism Epidemic: A Medical Controversy* (New York: St. Martin's Press, 2005).

13 *Althen v. Secretary of HHS*, 418 F.3d 1274 (Fed. Cir. 2005).

14 Concession agreement linked to David Kirby, "Government Concedes Vaccine-Autism Case in Federal Court—Now What?" *Huffington Post*, February 25, 2008, www.huffingtonpost.com/david-kirby/government-concedes-vacci_b_88323.html.

15 Compensation decisions at the Court of Federal Claims website, *Child Doe/77 v. Secretary of HHS*, 2010, http://www.uscfc.uscourts.gov/sites/default/files/CAMPBELLSMITH.%20DOE77082710.pdf.

16 *Hazlehurst v. HHS*, Court of Federal Claims, 2009, ftp://autism.uscfc.uscourts.gov/autism/vaccine/Campbell-Smith%20Hazlehurst%20Decision.pdf; and *Hazlehurst v. HHS*, U.S. Court of Appeals for the Federal Circuit, 2010, www.cafc.uscourts.gov/images/stories/opinions-orders/09-5128.pdf.

17 Mary Holland, Louis Conte, Robert Krakow, and Lisa Colin, *Unanswered Questions from the Vaccine Injury Compensation Program: A Review of Compensated Cases of Vaccine-Induced Brain Injury*, 28 Pace Envtl. L. Rev. 480 (2011) . Available at: http://digitalcommons.pace.edu/pelr/vol28/iss2/6. This summary of Unanswered Questions is adapted from the Executive Summary found at http://www.ebcala.org/unanswered-questions.

18 *See, e.g., The National Vaccine Injury Compensation Program (VICP) Division of Vaccine Injury Compensation Update*, Advisory Commission on Childhood Vaccines (March 8, 2012) at http://www.hrsa.gov/vaccinecompensation/divisionupdate030812.pdf (at slide 4).

19 The Special Master in the VICP admitted DOJ's expert report of Dr. George Bustin without admitting the underlying data verifying his expert report. This data was crucial to the final decision that Michelle Cedillo's autism was not caused by the measles virus. The judges of the U.S. Court of Appeals for the Federal Circuit acknowledged that the lack of underlying data for the Bustin report was "troubling," but did not require that the data be produced in part because the formal rules of civil procedure do not apply. *Cedillo v. HHS*, 2010-5004, (Fed. Cir. August 27, 2010), www.cafc.uscourts.gov/images/stories/opinions-orders/10-5004.pdf.

20 In the Omnibus Autism Proceeding, DOJ frequently referred to Dr. Andrew Wakefield's character and laboratory, although he was not a party to the litigation and had no direct relevance to it. DOJ had Dr. Chadwick testify because Dr. Chadwick had worked in Dr. Wakefield's laboratory, although he had no knowledge about the laboratory where Michelle Cedillo's measles biopsy laboratory analyses were actually done. *Cedillo v. HHS*, 2009 WL 331968 at 56. Similarly, in closing arguments DOJ's lead attorney Vincent Matanoski stated that "all the strands through these cases come back to him [Andrew Wakefield]. He presented bad science." *Cedillo* Transcript, June 26, 2007, 2898-99 at A599–A600.

21 *Daubert v. Merrell Dow Pharmaceuticals, Inc.*, 43 F.3d 1311 (9th Cir. 1995).

22 *Althen v. Sec'y of HHS*, 418 F.3d 1274, 1280.

23 Rebecca Estepp, parent of child in the Omnibus Autism Proceeding and vaccine safety advocate, as quoted in Coalition for Vaccine Safety press release, www.coalitionforvaccinesafety.org/press.htm.

24 See, e.g., Carolyn Gallagher & Melody Goodman, "Hepatitis B Triple Series Vaccine and Developmental Disability in U.S. Children Aged 1-9 Years," *Toxicology and Environmental Chemistry* 997 (2008); Kara L. McDonald et al., "Delay in Diphtheria, Pertussis, Tetanus Vaccination Is Associated With a Reduced Risk of Childhood Asthma," *Journal of Allergy and Clinical Immunology* 121 (2008):626.

25 131 S. Ct. 1068, 1086, 179 L. Ed. 2d 1 (2011).

26 *Bruesewitz v. Wyeth* at 1086.

27 Congressional record at 1986 Report 26, U.S.Code Cong. & Admin.News, 1986, at p. 6367.

28 *Bruesewitz v. Wyeth* at 1090-91.

29 *Bruesewitz v. Wyeth* at 1091-92.

30 *Bruesewitz v. Wyeth* at 1097-99.

31 Ibid.

31 *Bruesewitz v. Wyeth* at 1100-01.

32 *Ante*, at 1080.

33 Transcript of oral argument, *Bruesewitz v. Wyeth*, No. 09-152, October 12, 2010, 30, www.supremecourt.gov/oral_arguments/argument_transcripts/09-152.pdf.

CHAPTER 6: GOD AND THE GOVERNMENT —WILLIAM WAGNER, JD

1 Early in history, God, through Moses, introduces the divine law and mandates that parents are to talk about these commandments at home and impress them upon their children. Children are to obey and honor their parents. Deuteronomy 5:16; Ephesians 6:1–3. Here God also promises a long life and that it will go well for those who follow the commandments. Later, in Proverbs, God provides divine wisdom, guiding parents to "train" their children in the way they should go, so that when they grow old they "will not depart from it." Proverbs 22:6. In the raising of children, God provides that parents should exercise appropriate corrective discipline, that is to say, discipline grounded in love. Proverbs 13:24. In Paul's letter to the Ephesians, God confirms that parents are to bring up their children "in the training and instruction of the Lord." Ephesians 6:4.

2 See Joshua 24:15.

3 The writings of Pufendorf, Grotius, and Locke, for example, all expressly recognized the right of parents to direct and control decisions concerning the upbringing of their children. See, e.g., Pufendorf, *The Two Books on the duty of man and citizen according to the natural law*, Chapter 3, sections 1 and 2, www.constitution.org/puf/puf-dut_203.txt. Pufendorf, when discussing parents and children, writes:

> From marriage spring children, over whom paternal authority has been established—the most ancient and at the same time the most sacred kind of rule, under which children are bound to respect the commands and recognize the superiority of parents.
>
> The authority of parents over their children arises from two main causes: first, because the natural law itself, in commanding man to be social, enjoined upon parents the care of their children; and that this might not be neglected, Nature at the same time implanted in them the tenderest affection for their offspring. For the exercise of that care there is needed the power to direct the actions of children for their own welfare, which they do not yet understand themselves, owing to their lack of judgment. And then that authority rests upon the tacit consent also of the offspring. For it is rightly presumed that, if an infant had had the use of reason at the time of its birth, and had seen that it could not save its life without the parents' care and the authority therewith connected, it would gladly have consented to it, and would in turn have made an agreement with them for a suitable bringing-up. Actually, however, the parents' authority over their offspring is established when they take up the child and nurture it, and undertake to form it, to the best of their ability, into a fit member of human society.

Thus, for Pufendorf, it is for the parent to "support their children in comfort, and so shape their body and mind by skillful and wise education, that they become fit and useful members of human and civil society." Ibid., section 11. Such parental direction results in "good" and "wise" adult children of character.

Grotius, in his writings on the natural law likewise recognized the right of parents to direct and control the upbringing of their children. HUGO GROTIUS, *The Rights of War and Peace*, Book 1, The Preliminary Discourse XV.

Locke, too, understood that parents have the natural right to direct and control the upbringing of their children for their good, including their instruction. JOHN LOCKE, *The Second Treatise of Government*, 96 (PRENTICE HALL 1952) (1690).

4 See JOHN LOCKE, *The Second Treatise of Government*, 96-99 (Prentice Hall 1952) (1690). Locke writes: Section 169. "THOUGH I have had occasion to speak of these separately before, yet the great mistakes of late about government, having, as I suppose, arisen from confounding these distinct powers one with another, it may not, perhaps, be amiss to consider them here together." Locke states:

> First, then, Paternal or parental power is nothing but that which parents have over their children, to govern them for the children's good. . . . The affection and tenderness which God hath planted in the breast of parents toward their children, makes it evident, that this is not intended to be a severe arbitrary government, but only for the help, instruction, and preservation of their offspring. Section 170, 96–97.
>
> And thus, 'tis true, the paternal is a natural government, but not at all extending itself to the ends and jurisdictions of that which is political. The power of the father doth not reach at all to the property of the child, which is only in his own disposing. Section 170, 97.
>
> Secondly, Political power is that power, which every man having in the state of nature, has given up into the hands of the society, and therein to the governors, whom the society hath set over itself, with this express or tacit trust, that it shall be employed for their good, and the preservation of their property. Section 171.
>
> Nature gives the first of these, viz. paternal power to parents for the benefit of their children during their minority, to supply their want of ability, and understanding how to manage their property. . . . Voluntary agreement gives the second, viz. political power to governors for the benefit of their subjects, to secure them in the possession and use of their properties. And forfeiture gives the third despotical power to lords for their own benefit, over those who are stripped of all property. Section 173, 98–99.

Thus, the natural law thinkers also recognized that a parent may delegate, while retaining oversight of the delegation, aspects of the upbringing of their children.

Pufendorf writes:

> [A]lthough the obligation to educate their children has been imposed upon parents by nature, this does not prevent the direction of the same from being entrusted to another, if the advantage or need of the child require, with the understanding, however, that the parent reserves to himself the oversight of the person so delegated. (emphasis added) PUFENDORF, *The Two Books on the duty of man and citizen according to the natural law*, Chapter 3, section 9, www.constitution.org/puf/pufdut_203.txt.

5 In particular, parents hold the authority and responsibility over the main-tenance, protection, and education of their children. For example, Blackstone writes in *The Rights of Persons*, Book I. Ch. 16, 435-36; 439–42, Chapter 16, "Parent and Child")

> THE duty of parents to provide for the maintenance of their children is a principle of natural law; an obligation, says Pufendorf, laid on them not only by nature herself, but by their own proper act, in bringing them into the world: for they would be in the highest manner injurious to their issue, if they only gave the children life, that they might afterward see them perish. By begetting them therefore they have entered into a voluntary obligation, to endeavour, as far as in them lies, that the life which they have bestowed shall be supported and preserved. And thus the children will have a perfect right of receiving maintenance from their parents. . . .

> THE last duty of parents to their children is that of giving them an education suitable to their station in life: a duty pointed out by reason, and of far the greatest importance of any. For, as Pufendorf very well observes, it is not easy to imagine or allow, that a parent has conferred any considerable benefit upon his child, by bringing him into the world; if he afterward entirely neglects his culture and education, and suffers him to grow up like a mere beast, to lead a life useless to others, and shameful to himself.

6 *Pierce v. Society of Sisters*, 268 U.S. 510 (1925).

7 *Wisconsin v. Yoder*, 406 U.S. 205 (1972).

8 See e.g., the Declaration of Independence.

9 The state-mandated vaccination progeny of the Supreme Court decision *Jacobson v Massachusetts*, 197 U.S. 11 (1905), is a case in point. In the midst of a national smallpox epidemic, the Supreme Court upheld a government mandated adult vaccination program. The *Jacobson* Court opined that "the rights of the individual in respect of his liberty may at times, under the pressure of great dangers, be subjected to [government] restraint . . . as the safety of the general public may demand." 197 U. S. at 26. Chapter 1 explains how governments subsequently relied on this decision improperly to justify mandated vaccinations for exponentially milder public safety interests.

CHAPTER 7: AN URGENT CALL FOR MORE RESEARCH —CAROL STOTT, PHD (CANTAB), MSC (EPIDEMIOLOGY), CSCI, CPSYCHOL, AND ANDREW WAKEFIELD, MB, BS, FRCS, FRCPATH

1 Institute of Medicine, *Immunization Safety Review: Multiple Immunizations and Immune Dysfunction* (Inst. of Med., 2002), ix.

2 Institute of Medicine (1991) 'Adverse Effects of Pertussis and Rubella Vaccines', *Institute of Medicine*, 8.

3 Institute of Medicine, *Research Strategies for Assessing Adverse Events Associated with Vaccines*, (Inst. of Med., 1994), 1.

4 Ibid, 16.

5 Ibid, 16–17.

6 Institute of Medicine, *Vaccine Safety Forum: Summaries of Two Workshops* (Inst. of Med., 1997), 1.

7 Ibid, 2.

8 Institute of Medicine, *Immunization Safety Review: Thimerosal-Containing Vaccines and Neurodevelopmental Disorders* (Inst. of Med., 2001), 82.

9 Ibid, 74.

10 Ibid, 65.

11 Ibid, 66.

12 Ibid, 75.

13 Institute of Medicine, *Immunization Safety Review: Multiple Immunizations and Immune Dysfunction* (Inst. of Med., 2002), 18.

14 Ibid, 36.

15 Ibid, 14.

16 Institute of Medicine *Immunization Safety Review: Multiple Immunizations and Immune Dysfunction*, (Inst. of Med., 2002), 2.

17 Advisory Commission on Childhood Vaccines (ACCV) and National Vaccine Advisory Committee (NVAC) Subcommittees on Vaccine Safety, May 31, 1995, Parklawn Building, Conference Room D, Rockville, Maryland, at 75. Transcript available from Division of Vaccine Injury Compensation, Parklawn Building, Room 8A-35, 5600 Fishers Lane, Rockville, Maryland 20857.

18 Louis Z. Cooper et al., "Protecting Public Trust in Immunization," *Pediatrics* 122:1 (2008): 149–153, www.pediatrics.org/cgi/content/full/122/1/149.

19 R. Nordness, *Epidemiology and Biostatistics Secrets* (Philadelphia: Elsevier, 2006), 27.

20 Hennekens, M.D.; Buring, J.E.; Mayrent, S.L., eds., *Epidemiology in Medicine* (Philadelphia: Lippincott Williams & Wilkins, 1987).

21 Institute of Medicine, *Immunization Safety Review: Vaccines and Autism* (Inst. of Med., 2004), www.iom.edu/Reports/2004/Immunization-Safety-Review-Vaccines-and-Autism.aspx.

22 Wikipedia, "Evidence-based medicine," from http://en.wikipedia.org/wiki/Evidence-based_medicine.

23 Wikipedia, "Randomized Controlled Trial," http://en.wikipedia.org/wiki/Randomized_controlled_trial.

24 Passage of H.R. 3069 requires the Secretary of Health and Human Services (HHS), acting through the Director of the National Institutes of Health (NIH), to conduct a comprehensive study to (1) compare the total health outcomes, including the risk of autism, between vaccinated and unvaccinated U.S. populations; and (2) determine whether vaccines or vaccine components play a role in the development of autism spectrum or other neurological conditions. It also requires the Secretary to seek to include in the study U.S. populations that have traditionally remained unvaccinated for religious or other reasons.

25 Mark Blaxill, "From the Roman to the Wakefield Inquisition," *Age of Autism* blog, www.ageofautism.com/2010/01/from-the-roman-to-the-wakefield-inquisition.html.

26 DeLong, G, (2012). Conflicts of interest in vaccine safety research. Accountability in Research 19:2 (2012): 65-88.

27 Hertz-Picciotto, I, Delwiche, L, The Rise in Autism and the Role of Age at Diagnosis. Epidemiology, 20:1 (2009): 84-90.

28 Sharyl Attkisson, "Leading Doctor: Vaccines-Autism Worth Study," *CBS News*, May 12, 2008, www.cbsnews.com/stories/2008/05/12/cbsnews_investigates/main4086809.shtml.

29 Ibid, 9.

30 Ibid, 6.

31 Ibid, 7.

32 Institute of Medicine, *Immunization Safety Review: Vaccines and Autism*, Opening Statement, May 18, 2004, http://vaccinesafety.ecbt.org/ecbt/Website_documents/IOM_Executive_Summary/Opening_Statement.pdf.

33 *Associated Press*, "1 in 4 parents think shots cause autism," March 1, 2010, www.msnbc.msn.com/id/35638229/ns/health-kids_and_parenting/.

34 T. Neale, "Vaccine and Drug Safety Top Parents' Research Priorities," *Medpage Today*, October 12, 2010, www.medpagetoday.com/Pediatrics/GeneralPediatrics/22696.

35 *American Academy of Pediatrics* News Release, "How the Anti-Vaccine Movement Threatens America's Children," October 4, 2010, www.aap.org/pressroom/Offitfinalantivaccine.pdf.

36 Institute of Medicine, *Immunization Safety Review: Measles-Mumps-Rubella Vaccine and Autism* (Inst. of Med., 2001), 1.

36 Institute of Medicine, *Adverse Effects of Pertussis and Rubella Vaccines*, (Inst. of Med., 1991), 7; Stratton et al., "Adverse Events Associated with Childhood Vaccines Other Than Pertussis and Rubella," *JAMA*, 271:20 (1994):1602-1605, http://jama.ama-assn.org/cgi/content/abstract/271/20/1602.

37 National Vaccine Injury Compensation Program Vaccine Injury Table, www.hrsa.gov/vaccinecompensation/table.htm.

38 Last, JM., *A Dictionary of Epidemiology*, 4th ed., (Oxford University Press, 2001).

39 Kennedy, RC, Shearer, MH, Hildebrand, W, "Nonhuman primate models to evaluate vaccine safety and immunogenicity," *Vaccine* 15 (1997): 903–908.

40 Hewitson L, Houser LA, Stott C, Sackett G, Tomko JL, Atwood D, Blue L, White ER, "Delayed acquisition of neonatal reflexes in newborn primates receiving a thimerosal-containing hepatitis B vaccine: influence of gestational age and birth weight," *J. Toxicol. Environ. Health A.*, 73:19 (Jan. 2010): 1298–313.

41 Hewitson L, Lopresti BJ, Stott C, Mason NS, Tomko J, "Influence of pediatric vaccines on amygdala growth and opioid ligand binding in rhesus macaque infants: A pilot study," *Acta Neurobiol. Exp.(Wars)*, 70:2 (2010): 147–64.

42 Gallagher, CM, and Goodman, MS, "Hepatitis B triple series vaccine and developmental disability in U.S. children aged 1–9 years," *Tox. & Envir. Chemistry*, 90:5 (2008): 997 – 1008 www.informaworld.com/smpp/content~content=a905442343~db=all~jumptype=rss.

43 Gallagher CM, Goodman MS, "Hepatitis B Vaccination of Male Neonates and Autism Diagnosis, NHIS 1997-2002," *J Toxicol Environ Health, Part A*, 73 (2010): 1665–1677, www.progressiveconvergence.com/Hepatitis%20B%20Vaccination%20male%20neonates%201997-2002.pdf.

44 McDonald KL, Huq SI, Lix LM, Becker AB, Kozyrskyj AL, "Delay in diphtheria, pertussis, tetanus vaccination is associated with a reduced risk of childhood asthma," *J Allergy Clin Immunol.* 121:3 (2008): 626–31. Epub 2008 Jan 18, www.ncbi.nlm.nih.gov/pubmed/18207561.

45 *CBC News*, "Asthma rates in children have jumped fourfold," Friday, January 27, 2006, www.cbc.ca/canada/story/2006/01/27/asthma-report060127.html.

46 DeSoto, MC and Hitlan, RT, "Sorting out the spinning of autism: heavy metals and the question of incidence," *Acta Neurobiol Exp*, 70 (2010): 165–176 .

47 Palmer R, Wood S, Blanchard R, "Proximity to point sources of environmental mercury release as a predictor of autism," *Health and Place* 15 (2009): 18–24.
DeSoto MC, "Ockham's Razor and autism: The case for developmental neurotoxins contributing to a disease of neurodevelopment," *Neurotoxicology* 30 (2009): 331–337.

48 Eskenazi B, Marks AR, Bradman A, Harley K, Barr DB, Johnson C, Morga N, Jewell NP, "Organophosphate pesticide exposure and neurodevelopment in young Mexican-American children," *Environ Health Perspect*, 115 (2007): 792–798.

49 Tamm C, Duckworth J, Hermanson O, Ceccatelli S, "High susceptibility of neural stem cells to methylmercury toxicity: effects on cell survival and neuronal differentiation," *J Neurochem* 97 (2006): 69–78.

50 Stern AH, Smith AE, "An assessment of the cord blood: maternal blood methylmercury ratio: implications for risk assessment," *Environ Health Perspect* 111 (2003): 1465–1470.; Morissette J, Takser L, St-Amour G, Smargiassi A, Lafond J, Mergier D, "Temporal variation of blood and hair mercury levels in pregnancy in relation to fish consumption history in a population living along the St. Lawrence River," *Environ Res* 95 (2004): 263–274.

51 www.all.org/pdf/McDonaldPaul2010.pdf.

52 Reproduced with permission of *The Autism File*, Issue 37.

53 Dan Olmsted and Mark Blaxill, *The Age of Autism: Mercury, Medicine, and a Man-Made Epidemic* (New York: St. Martin's Press, 2010), 234-43.

54 Falco, M., "10 infants dead in California whooping cough outbreak," *CNN*, October 20, 2010, www.cnn.com/2010/HEALTH/10/20/california.whooping.cough/index.html.

55 Tomljenovic, L, Shaw, CA, Do aluminum vaccine adjuvants contribute to the rising prevalence of autism? *Journal of Inorganic Biochemistry.* 105:11(2011): 1489-99. Epub 2011 Aug 23.

56 For sources regarding aluminium in vaccines, see, e.g., http://www.immunizationinfo.org/issues/vaccine-components/aluminum-adjuvants-vaccines, listing:

- DTP (diphtheria-tetanus-pertussis vaccine)
- DTaP (diphtheria-tetanus-acellular pertussis vaccine)

- Some but not all Hib (*Haemophilus influenzae* type b) conjugate vaccines
- Pneumococcal conjugate vaccine
- Hepatitis B vaccines
- All combination DTaP, Tdap, Hib, or Hepatitis B vaccines
- Hepatitis A vaccines
- Human Papillomavirus vaccine
- Anthrax vaccine
- Rabies vaccine

See also http://www.askdrsears.com/topics/vaccines/vaccine-faqs.

57 Miller, NZ, Goldman, GS, Infant Mortality rates regressed against number of vaccine doses routinely given: Is there a biochemical synergistic toxicity? Human Experimental Toxicology 30:9 (2011): 1420-1428.

58 Hallmayer, J, Cleveland, S, Torres, A, Phillips, J, Cohen, B, Torigoe, T, Miller, J, Fedele, A, Collins, J, Smith, K, Lotspeich, L, Croen, LA, Ozonoff, S., Lajonchere, C., Grether, JK, Risch, N, (2011). Genetic heritability and shared environmental factors among twin pairs with autism. *Arch Gen Psychiatry* 68:11 (2011): 1095-102. doi: 10.1001/archgenpsychiatry.2011.76. Epub 2011 Jul 4.

59 Lathe, R, (2006). *Autism, Brain and Environment* London, UK: Jessica Kingsley Publishers, 2006)

60 Schieve, LA, Boulet, S, Blumberg, S, Kogan, MD, Yeargin-Allsopp, MY, Boyle, CA, Visser, SN, Rice, C, Association between parental nativity and autism spectrum disorder among US-born non-Hispanic white and Hispanic children 2007 National Survey of Children's Health. *Disability and Heath Journal*, 5 (2012): 18-25

61 DeLong, G, A positive association found between autism prevalence and childhood vaccination uptake across the US population. *Journal of Toxicology and Environmental Health* A 74 (2011): 903–916.

62 Hertz-Picciotto, I, Delwiche, L, The Rise in Autism and the Role of Age at Diagnosis. *Epidemiology*, 20:1 (2009): 84-90.

63 Centers for Disease Control and Prevention (CDC) Morbidity and Mortality Weekly Report, (2012). *Prevalence of Autism Spectrum Disorders — Autism and Developmental Disabilities Monitoring Network, 14 Sites, United States, 2008. Surveillance Summaries.* 61(SS03); 1-19. March 30th.

64 See http://www.nj.com/news/index.ssf/2009/04/nj_air_ranked_by_cleanair_grou.html, discussing the American Lung Association's 2009 "State of the Air" report.

65 See the New Jersey Department of Health and Senior Services Minimum Immunization Requirements for School Attendance in New Jersey, at www.nj.gov/health/forms/imm.doc.

CHAPTER 8: A LONG AMERICAN TRADITION —ROBERT JOHNSTON, PHD

1 For a much more detailed treatment of these themes, see Robert D. Johnston, "Contemporary Anti-Vaccination Movements in Historical Perspective," in Johnston, ed., *The Politics of Healing: Histories of Twentieth-Century North American Alternative Medicine* (Routledge, 2004), 259–286.

2 The most recent treatment of this episode is Tony Williams, *The Pox and the Covenant: Mather, Franklin, and the Epidemic that Changed America's Destiny* (Sourcebooks, 2010).

3 Joan Retsinas, "Smallpox Vaccination: A Leap of Faith," *Rhode Island History* 38 (1979): 122.

4 On the Milwaukee riot, see Judith Walzer, Leavitt, *The Healthiest City: Milwaukee and the Politics of Health Reform* (Princeton University Press, 1982).

5 For more information on these issues, and those in the following paragraphs, see Robert D. Johnston, *The Radical Middle Class: Populist Democracy and the Question of Capitalism in Progressive Era Portland, Oregon* (Princeton University Press, 2003), 175–190.

6 On Little, see Johnston, *The Radical Middle Class*, 197-213. For the quotation, see 202.

7 I concentrate on this recent history in "Contemporary Anti-Vaccination Movements."

CHAPTER 9: MEDICAL ETHICS AND CONTEMPORARY U.S. MEDICINE —VERA SHARAV, MLS

1 Leo Alexander, "Medical Science Under Dictatorship," *New England Journal of Medicine*, 241 (1949):39–47, accessible at www.physiciansforlife.org/content/view/1333/33/.

2 Ibid.

3 Sofair AN and Kaldjian LC, "Eugenic sterilization and a qualified Nazi analogy: the United States and Germany, 1930-1945," *Annals of Internal Medicine* 132:4 (2000): 312-319, www.annals.org/content/132/4/312.

4 "Euthanasia in Nazi Germany—The T4 Programme," Website of A New Zealand Resource for Life Related Issues, www.life.org.nz/euthanasia/abouteuthanasia/history-euthanasia6/.

5 "The T-4 Euthanasia Program," website of the Jewish Virtual Library, www.jewishvirtuallibrary.org/jsource/Holocaust/t4.html.

6 Gardella, J., "The Cost-Effectiveness of Killing: An Overview of Nazi 'Euthanasia'" *Medical Sentinel*, July/August 1999, 4(4):132–135.

7 Leon R. Kass, "The Wisdom of Repugnance," *New Republic* 216:22 (June 2, 1997), republished at http://www.catholiceducation.org/articles/medical_ethics/me0006.html. Dr. Leon Kass is an American physician, scientist, educator, and ethicist at the University of Chicago who chaired the Presidents' Council on Bioethics (2001–2005).

8 See, e.g., Rome Statute of the International Criminal Court, Article 7, United Nations website, http://untreaty.un.org/cod/icc/statute/romefra.htm; Forum on Public Policy, "The Racist American Eugenics Program: A Crime Against Humanity," Earnest N. Bracey, Professor of Political Science and African American History, The College of Southern Nevada, www.forumonpublicpolicy.com/archivespring07/bracey.pdf; and Karl Brandt, *Trial of War Criminals before Nuremberg Military Tribunals*, volumes I and II, Case No. 1.

9 Whitney LF, *The Case for Sterilization*, (New York: Frederick A. Stokes, 1934), 193–204. Cited in Sofair AN and Kaldjian LC, note 3 above.

10 Grace Nauman, "Eugenics and Its Ethical Implications Revisited," *The Dartmouth Apologia: A Journal of Christian Thought*, http://dartmouthapologia.org/articles/show/135.

11 Brandon Miller, "Ethicist Urges States to Apologize for Forced Sterilization Policies," *UVA Top News Daily*, March 2, 2004, www.virginia.edu/topnews/03_02_2004/lombardo_paul.html.

12 Sofair, AN and Kaldjian LC, note 3 above.

13 *Jacobson v. Massachusetts*, 197 U.S. 11 (1905).

14 Ibid.

15 See Paul Lombardo, "Eugenic Sterilization Laws, Eugenics Archive," Dolan DNA Learning Center, Cold Spring Harbor Laboratory, www.eugenicsarchive.org/html/eugenics/essay8text.html.

16 Sofair, AN and Kaldjian LC. Eugenic Sterilization and a Qualified Nazi Analogy: The United States and Germany, 1930-1945, *Annals of Internal Medicine*, 2000, 132–4:312–319. www.annals.org/content/132/4/312.

17 Jess Bravin and Louise Radnofsky, "Regrets Only? Native Hawaiians Insist U.S. Apology Has a Price," *The Wall Street Journal*, March 12, 2009, http://online.wsj.com/article/SB123682336964803763.html.

18 Joint Statement by Secretaries Clinton and Sebelius on a 1946–1948 Study, U.S. Department of State, October 1, 2010, www.state.gov/secretary/rm/2010/10/148464.htm.

19 Goldman, D, and Murray, M. "Studies on the Use of Refrigeration Therapy in Mental Disease with Report of Sixteen Cases," *Journal of Nervous and Mental Diseases*, 97 (1947): 152–165.

20 Talbott, JH and Tilotson, KJ, "Effects of Cold on Mental Disorders," April 1941, *Diseases of the Nervous System*; cited by Beam, A. *Gracefully Insane: The Rise and Fall of America's Premier Mental Hospital*, (Perseus Books, 2001), 74–77.

21 Stateville Penitentiary Malaria Study, *Wikipedia*, http://en.wikipedia.org/wiki/Stateville_Penitentiary_Malaria_Study.

22 Philip Pecorino, *Medical Ethics* (Online Textbook), chapter 7, "Human Experimentation, Case: Willowbrook Experiment," www2.sunysuffolk.edu/pecorip/SCCCWEB/ETEXTS/MEDICAL_ETHICS_TEXT/Chapter_7_Human_Experimentation/Case_Study_Willowbrook_Experiments.htm.

23 Shuster, E., "Fifty years later: the significance of the Nuremberg Code," *New England Journal of Medicine* 337 (1997): 1436–1440.

24 "During the Nuremberg War Crime Trials, the Nuremberg Code was drafted as a set of standards for judging physicians and scientists who had conducted biomedical experiments on concentration camp prisoners. This code became the prototype of many later codes intended to assure that research involving human subjects would be carried out in an ethical manner." See The National Commission for the Protection of Human Subjects of Biomedical and Behavioral Research, The Belmont Report, 1979, http://ohsr.od.nih.gov/guidelines/belmont.htm.

25 Glantz, L., "Research with children," *American Journal of Law & Medicine*, 24(1998): 213–244. Ramsey quotation, 235, no. 254.

26 Ibid, 236.

27 Leonard Glantz, "Nontherapeutic Research with Children: *Grimes v Kennedy Krieger Institute*," *Am. Journal of Public Health* 92 (2002): 1070-1073, http://ajph.aphapublications.org/cgi/content/full/92/7/1070; see *Grimes v. Kennedy Krieger Instiute, Inc.*, 782 A.2d 807 (Md. Ct. of Appeals, 2001), www.law.uh.edu/healthlaw/law/StateMaterials/Marylandcases/grimesvkennedykreiger.pdf; see also Alliance for Human Research Protection amicus curiae brief to the Maryland Court of Appeals in support of Grimes, www.ahrp.org/children/AmicusKKI.php.

28 45 Code of Federal Regulations 46 Subpart D: Additional DHHS Protections for Children Involved as Subjects in Research, 48 *Federal Register* 0818, March 8, 1983.

29 Solomon, J., "Researchers Tested AIDS Drugs on Children," *Associated Press*, May 4, 2005. See www.ahrp.org/infomail/05/05/04.php.

30 Vera Sharav, "Children in clinical research: A conflict of moral values," *The American Journal of Bioethics* 3(1) 2003, http:/mitpress.mit.edu/journals/ajob/3/1/sharav.pdf.

31 National Digestive Disease Information Clearinghouse, http://digestive.niddk.nih.gov/ddiseases/pubs/gerdinfant/.

32 Willman, D., "Propulsid: a heartburn drug, now linked to children's deaths," *Los Angeles Times*, Dec. 20, 2000, www.pulitzer.org/archives/6474.

33 The National Childhood Vaccine Injury Act of 1986, 42 U.S.C. 300aa-1 et seq.

34 Routine Vaccinations, *Merck Manual*, www.merck.com/mmpe/sec14/ch169/ch169b.html.

35 Centers for Disease Control and Prevention, "Recent trends in infant mortality in the U.S.," October 2008, www.cdc.gov/nchs/data/databriefs/db09.htm.

36 List of Countries by Infant Mortality Rate, *Wikipedia*, http://en.wikipedia.org/wiki/List_of_countries_by_infant_mortality_rate.

37 Harris, G., "Infant Deaths in U.S.," *New York Times*, Oct. 15, 2008, www.nytimes.com/2008/10/16/health/16infant.html.

38 Pine DS, Coplan JD, Wasserman GA, Miller LS, Fried JE, Davies M, Cooper TB, Greenhill L, Shaffer D, Parsons B., "Neuroendocrine response to fenfluramine challenge in boys," *Archives of General Psychiatry* 54 (1997): 839–846.

39 Halperin JM, "Age-related Changes in the Association between Serotonergic Function and Aggression in Boys with ADHD," *Biological Psychiatry* 41 (1991): 682–689; Halperin JM, "Serotonin Aggression and Parental Psychopathology in Children with Attention-deficit Hyperactivity Disorder," *Journal of the American Academy of Child Adolescent Psychiatry* 36: (1997): 1391–1398.

40 David Shaffer, M.D., Chief, Division of Child and Adolescent Psychiatry, Columbia University Medical Center, http://asp.cumc.columbia.edu/facdb/profile_list.asp?uni=ds18&DepAffil=Psychiatry.

41 McCann UD, Seiden LS, Rubin LJ, Ricaurte G., "Brain serotonin neurotoxicity and primary pulmonary hypertension from fenfluramine and dexfenfluramine," *JAMA*, 278 (1997): 66–672.

42 Washington, HA., *Medical Apartheid: The Dark History of Medical Experimentation on Black Americans from Colonial Times to the Present* (Random House, 2007).

43 Hilts, P., "Experiments on children are reviewed," *New York Times*, April 15, 1998, www.nytimes.com/1998/04/15/nyregion/experiments-on-children-are-reviewed.html. See also ibid.

CHAPTER 10: "THE GREATER GOOD"—ALLEN TATE

1 Recommended Immunization Schedule for Persons Aged 0 Through 6 Years—United States—2010, www.cdc.gov/vaccines/recs/acip.

2 See, e.g., H.R. 3069: Comprehensive Comparative Study of Vaccinated and Unvaccinated Populations Act of 2009, www.govtrack.us/congress/bill.xpd?bill=h111-3069.

3 Michael Sandel, *Justice: What's the Right Thing to Do?* (New York: Farrar, Straus and Giroux, 2009).

CHAPTER 11: JUSTICE DISSERVED: THE HANNAH BRUESEWITZ ODYSSEY—RUSS BRUESEWITZ

1 *Bruesewitz v. Wyeth Inc.*, 561 F.3d 233, 236 (3d Cir. 2009), *aff'd, Bruesewitz v. Wyeth LLC*, 131 Sup. Ct. 1068 (2011) ("Hannah's particular vaccine came from a lot that generated sixty-five reports of adverse reactions with the FDA and Centers for Disease Control and Prevention, including thirty-nine emergency room visits, six hospitalizations, and two deaths. Hannah's physician later indicated, as part of this litigation, that she would not have immunized Hannah had she known of the adverse event reports associated with this lot of the vaccine.")

2 *See O'Connell v. Shalala*, 79 F.3d 170 (1st Cir. 1996); *see also* 60 FR 7678, *et seq.*, February 8, 1995), found at http://www.gpo.gov/fdsys/granule/FR-1995-02-08/95-2945/content-detail.html

3 *Bruesewitz v. Sec'y of Health & Human Servs.*, No. 95– 0266V, 2002 WL 31965744, *1 (Ct. Cl., Dec. 20, 2002).

4 From the U.S. Court of Federal Claims website: Laura D. Millman was appointed Special Master on April 25, 1991. She graduated from the City College of New York, receiving a BA degree cum laude in English in 1966, from Herbert H. Lehman College receiving a MA degree in English in 1971, and from the Fordham School of Law, receiving a JD degree cum laude in 1976. http://www.uscfc.uscourts.gov/laura-d-millman

CHAPTER 13: PEDIATRICS: SICK IS THE NEW HEALTHY —JUDY CONVERSE, MPH, RD, LD

1 American Gastroenterological Society, "Acid Reducing Medicines May Lead To Dependency," July 1, 2009, Bethesda, MD, www.gastro.org/news/articles/2009/07/01/acid-reducing-medicines-may-lead-to-dependency; Cooper WO et al, "Trends in Prescribing of Antipsychotic Medications for US Children," *Academic Pediatrics* 2006; 6:2 (2006): 79–83; Forgacs I and Loganayagam A., "Overprescribing proton pump inhibitors," *BMJ* , 336:7634 (2008): 2; Kuehn BM, Studies Shed Light on Risks and Trends in Pediatric

Antipsychotic Prescribing, *JAMA*, 303:19 (2010): 1901–1903; and Nyquist AC et al, "Antibiotic Prescribing for Children With Colds, Upper Respiratory Tract Infections, and Bronchitis," *JAMA*, 279 (1998): 875–877.

2 Gibney MJ, Margetts BM, Kearney JM, and Lenore A, eds. *Public Health Nutrition.* (Oxford, U.K., Blackwell Science, 2004).

3 Caulfield L et al, "Undernutrition as an underlying cause of child deaths associated with diarrhea, pneumonia, malaria, and measles," *Am J Clin Nutr*, 80:1 (July 2004): 193–198; World Food Programme and United Nations Children's Fund, "Global framework for action: Ending child hunger and undernutrition initiative," (Rome, Italy and New York, NY, USA, 2006); and Department for International Development, *The Neglected Crisis of Undernutrition: Evidence For Action*, (Glasgow, U.K., 2009).

4 Pelletier DL et al., "The effects of malnutrition on child mortality in developing countries," *Bulletin of the World Health Organization* 73:4 (1995): 443-448; Bryce JE et al., "Reducing child mortality: Can public health deliver?" *Lancet*, 362:9387 (July 12, 2003): 159–164; Kinnell HG, "Pica as a feature of autism," *British Journal of Psychiatry* 147 (1985): 80–82; Cubała-Kucharska, M., "The review of most frequently occurring medical disorders related to aetiology of autism and the methods of treatment," *Acta Neurobiol Exp* 70 (2010): 141–146; Cannell, JJ, "On the aetiology of autism," *Acta Paediatr*, 99:8 (August 2010): 1128–1130; Stewart C, Latif A., "Symptomatic nutritional rickets in a teenager with autistic spectrum disorder," *Child Care Health Dev*, 34:2 (March 2008): 276–8; and Conyers R, Efron D., "Agitation and weight loss in an autistic boy," *J Paediatr Child Health*, 43:3 (March 2007): 186–7.

5 Wagener DK et al., "Summary Measures of Population Health: Addressing the First Goal of Healthy People 2010, Improving Life Expectancy," *Healthy People 2010 Statistical Notes*, Centers For Disease Control, Number 22, September 2001.

6 Olshansky SJ et al, "A Potential Decline in Life Expectancy in the United States in the 21st Century," *N Engl J Med*, 352 (2005): 1138-114; *Child Trends (2010). Life Expectancy*, www.childtrendsdatabank.org/?q=node/326;
Meunnig P and Glied S., "What Changes In Survival Rates Tell Us About US Healthcare," *Health Affairs 2010*, http://content.healthaffairs.org/cgi/content/abstract/hlthaff.2010.0073; and Van Cleave J et al., "Dynamics of Obesity and Chronic Health Conditions Among Children and Youth," *JAMA*, 303 (2010): 623–630, 665–666.

7 Hooper E., *The River: A Journey to the Source of HIV and AIDS*, (Boston: Little, Brown and Company, 1999).

8 Converse JM, *When Your Doctor Is Wrong: Hepatitis B Vaccine and Autism*, (Philadelphia: Xlibris, 2002).

CHAPTER 14: MY DAUGHTER IS "ONE LESS"—AMY PINGEL

1 Charlotte J. Haug, "Human Papillomavirus—Reasons for Caution," *New England Journal of Medicine* 359 (August 21, 2008): 8.

2 *Judicial Watch Investigates Side Effects of HPV Vaccine*, www.judicialwatch.org/gardasil.

3 Government denied causation of petitioner's multiple sclerosis but settled the case for $3.5 million. *Jane Doe 89 v. Sec'y of HHS*, Laura D. Millman, Special Master, No. [REDACTED] V, originally filed October 5, 2010, File redacted: October 12, 2010, http://www.uscfc.uscourts.gov/sites/default/files/Millman.Doe%2089%20damages.pdf.

CHAPTER 16: WHO WILL DEFEND THE DEFENDERS? —CAPTAIN RICHARD ROVET, RN, BSN, B-C, (USAF, RET.), WITH AFTERWORD BY COLONEL FELIX M. GRIEDER, (USAF, RET.)

1 "Is Military Research Hazardous to a Veteran's Health? Lessons Learned Spanning Half a Century," *A Staff Report Prepared for by the Committee on Veterans' Affairs*, United States Senate, December 8, 1994, www.gulfweb.org/bigdoc/rockrep.cfm.

2 Lea Steele, "Prevalence and Patterns of Gulf War Illness in Kansas Veterans: Association of Symptoms with Characteristics of Person, Place, and Time of Military Service," *American Journal Epidemiology*, 152:10 (2000): 992–1002 (doi: 10.1093/aje/152.10.992), http://aje.oxfordjournals.org/content/152/10/ 992.full.; Veterans Administration Research Advisory Committee on Gulf War Veterans' Illnesses, *Scientific Progress in Understanding Gulf War Veterans' Illnesses: Report and Recommendations*, September 2004, www1.va.gov/RAC-GWVI/docs/Committee_Documents/ReportandRecommendations_ScientificProgressinUnderstandingGWVI_2004.pdf.; Goss-Gilroy, "Study of Canadian Gulf War Veterans: NR-98.050," Study contracted by the Canadian Department of National Defense, released June 29, 1998 and published on its website, accessed between 1999 and 2001, and archived at http://web.archive.org/web/20000925233153/www.dnd.ca/menu/press/Reports/Health/health_study_eng_1.htm.; Unwin C, Blatchley N, Coker W, Ferry S, Hotopf M, Hull L, et al., "Health of UK servicemen who served in Persian Gulf War," *Lancet*, 353:9148 (Jan. 16, 1999): 169–78.

3 Veterans' Affairs Committee, "Is Military Research," note 1 above, section I para. 2.

4 Pieta Wooley, "Vaccines Show Sinister Side," *Straight.com*, March 23, 2006, www.straight.com/article/vaccines-show-sinister-side; CDC, *Vaccine Excipient & Media Summary*, Part 1 and Part 2, March 2010, www.cdc.gov/vaccines/pubs/pinkbook/downloads/appendices/B/excipient-table-1.pdf and www.cdc.gov/vaccines/pubs/pinkbook/downloads/appendices/B/excipient-table-2.pdf; "VA Research Advisory Committee, *Scientific Progress*, note 2 above.

5 United States General Accounting Office, Report GAO-01-13 to the Chairman, Subcommittee on National Security, Veterans' Affairs, and International Relations, House Committee on Governmental Reform, *COALITION WARFARE: Gulf War Allies Differed in Chemical and Biological Threats Identified and in Use of Defensive Measures* (April 2001).

6 Hearing before the Committee on Veterans' Affairs, United States Senate, "Military Exposures: the continuing challenges of care and compensation," One Hundred Seventh Congress, Second Session, July, 10, 2002.

7 Executive Order 13139, http://frwebgate.access.gpo.gov/cgi-bin/getdoc.cgi?dbname=1999_register&docid=fr05oc99-135.pdf.

8 This doctrine arises from a United States Supreme Court decision, *Feres v. United States*, 340 U.S. 135 (1950), and bars, with limited exception, military personnel or their families from collecting damages from the U.S. government as a result of a soldier's injury or death in the line of duty.

9 "Product Insert for Anthrax Vaccine," archived at http://web.archive.org/web/19991009123133/www.dallasnw.quik.com/cyberella/Anthrax/Product.html.

10 Lee Williams and Hiran Ratanake, "Ex-DAFB Commander says Troops used as Guinea Pigs," *Delaware Online*, October 10, 2004.

11 "Novartis Investigational Adjuvanted (MF59) Pre-pandemic Avian Influenza Vaccine Aflunov® Shows Long Lasting, cross-protective Immune Response," *Medical News Today*, May 2, 2009, www.medicalnewstoday.com/articles/148563.php.

12 http://www.vaccine-a.com.

13 I witnessed this young girl receiving co-administered anthrax and flu vaccines on the "flu shot" line at the Global Activities Center, Dover Air Force base (AFB), Delaware in November 1999.

14 TSGT Clarence Glover died of a mysterious heart condition following anthrax vaccination at Dover AFB. To the best of my knowledge, the military did not release autopsy results to his family.

15 Garth L Nicholson, Meryl Nass, Nancy L Nicholson, "Anthrax Vaccine: Controversy Over Safety and Efficacy," www.gulfwarvets.com/anthrax5.htm.

16 Pratap Catterjee, "Military Contractors' Epic Overcharging," *The Guardian*, October 20, 2010, www.guardian.co.uk/commentisfree/cifamerica/2010/oct/17/pharmaceuticals-industry-anthrax.

17 For information about squalene-containing lots: FAV 008, FAV 020, FAV 030, FAV 038, FAV 043, FAV 047, see http://www.fda.gov/ohrms/dockets/dockets/80n0208/80n-0208-c000037-15-01-vol151.pdf

18 For information about anti-squalene antibodies induced from the following vaccine lots: FAV 041, FAV 070 and FAV 071, see Pamela B. Asa, 1 Russell B. Wilson, 2 and Robert F. Garry, 3 (2000) Antibodies to Squalene in Recipients of Anthrax Vaccine Department of Microbiology, Tulane University Medical School.

CHAPTER 18: THE VACCINE BUBBLE AND THE PHARMACEUTICAL INDUSTRY —MICHAEL BELKIN

1 *The Merck Manual, Online Medical Library*, "Encephalitis," www.merck.com/mmhe/sec06/ch089/ch089f.html.

2 www.immunizationinfo.org/issues/vaccine-safety/cause-or-coincidence.

3 Melody Petersen, *Our Daily Meds* (New York: Picador, 2008), 306.

4 Merriam-Webster Dictionary, www.merriam-webster.com/medical/iatrogenesis.

5 Vaccine Adverse Events Reporting System, December 4, 2010 www.medalerts.org/vaersdb/findfield.php?TABLE=ON&GROUP1=APPY.

6 David Kessler et al, "Introducing MEDWatch: A New Approach to Reporting Medication and Device Adverse Events and Product Problems," *JAMA* 269(21): (1993) 2765–2768. "Only about 1% of serious events are reported to the FDA, according to one study," http://jama.ama-assn.org/cgi/content/summary/269/21/2765?maxtoshow=&hits=10&RESULTFORMAT=&fulltext=David+Kessler+vaccine&searchid=1&FIRSTINDEX=0&resourcetype=HWCIT.

7 These reports can be found at www.medalerts.org/vaersdb/index.php.

8 http://emedicine.medscape.com/article/1176205-overview.

9 www.merckmanuals.com/professional/sec19/ch283/ch283c.html.

10 "Frequently Asked Questions about Febrile Seizures Following Childhood Vaccinations," Centers for Disease Control and Prevention, www.cdc.gov/vaccinesafety/Vaccines/MMRV/qa_FebrileSeizures.html.

11 Michael Belkin, "Hepatitis B Vaccine: Helping or Hurting Public Health?" Hearing before the Subcommittee on Criminal Justice, Drug Policy and Human Resources of the Committee on Government Reform, House of Representatives, One Hundred Sixth Congress, First Session, May 18, 1999, http://frwebgate.access.gpo.gov/cgi-bin/getdoc.cgi?dbname=106_house_hearings&docid=f:63308.wais.

12 Julie Steenhuysen, "CDC backs away from decades-old flu death estimate," *Reuters*, August 26, 2010, www.reuters.com/article/idUSTRE67P3NA20100826?type=domesticNews.

13 Natasha Singer, "Report urges more curbs on medical ghostwriting," *New York Times*, June 24, 2010, www.nytimes.com/2010/06/25/health/25ghost.html?ref=health, www.plosmedicine.org/static/ghost-writing.action.

14 Rob Stein, "Reports accuse WHO of exaggerating H1N1 threat, possible ties to drug makers," *Washington Post*, June 4, 2010, www.washingtonpost.com/wp-dyn/content/article/2010/06/04/AR2010060403034.html.

15 "How Many New Drugs Are Lemons? Ask Donald Light," at website of Howard Brody, MD, PhD, Director of the Institute for the Medical Humanities, University of Texas Medical Branch, and author of *Hooked: Ethics, the Medical Profession, and the Pharmaceutical Industry*, http://brodyhooked.blogspot.com/2010/08/how-many-new-drugs-are-lemons-ask.html.

16 www.merck.com/newsroom/news-release-archive/corporate/2009_1221.html.

17 http://sphweb02.umdnj.edu/sphweb/files/bio/cv/Bresnitz_Aug-05-2008_CV.doc.

18 Kreesten Madsen et al, "A Population-Based Study of Measles, Mumps, and Rubella Vaccination and Autism," *New Engl. J. of Med.* 347 (2002): 1477-1482, www.nejm.org/doi/full/10.1056/NEJMoa021134; Kreesten M. Madsen et al, "Thimerosal and the Occurrence of Autism: Negative

Ecological Evidence From Danish Population-Based Data," *Pediatrics* 112 (2003): 604-606, http://pediatrics.aappublications.org/cgi/content/full/112/3/604.

19 Statement regarding Dr. Poul Thorsen's involvement in Aarhus University projects, Aarhus Universitet, January 22, 2010, www.rescuepost.com/files/thorsen-aarhus-1.pdf.

20 M. Asif Ismail, "Drug Lobby Second to None: how the pharmaceutical industry gets its way in Washington," The Center for Public Integrity, July 7, 2005, http://projects.publicintegrity.org/rx/report.aspx?aid=723.

21 Facts About Childhood Vaccine Ingredients, Pharmaceutical Research and Manufacturers of America (PhRMA), www.immunize.org/concerns/vaccine_components.pdf.

22 Maggie Fox, "Glaxo's herpes vaccine fails in trial, NIH says," *Reuters*, September 30, 2010, www.reuters.com/article/idUSTRE68T33Z20100930 (in testing a herpes vaccine, the "placebo" used was a hepatitis A vaccine).

23 Facts About Childhood Vaccine Ingredients, Pharmaceutical Research and Manufacturers of America (PhRMA), www.immunize.org/concerns/vaccine_components.pdf.

24 Beatrice A. Golomb et al., "What's in Placebos: Who Knows? Analysis of Randomized, Controlled Trials," *Annals of Internal Medicine*, October 19, 2010, www.annals.org/content/153/8/532.abstract.

25 "FACA: Conflicts of Interest and Vaccine Development—Preserving the Integrity of the Process," Hearing before the Committee on Government Reform, House of Representatives, One Hundred Sixth Congress, Second Session, June 15, 2000, http://frwebgate.access.gpo.gov/cgibin/getdoc.cgi?dbname=106_house_hearings&docid=f:73042.wais.

26 "Drugmakers face $140 bln patent 'cliff'—report," *Reuters*, May 2, 2007, www.reuters.com/article/idUSL0112153120070502.

27 Café Pharma, www.cafepharma.com/.

28 www.bloomberg.com/news/2010-11-17/u-s-companies-profit-fret-in-the-chinese-market-survey-says.html.

29 Global Immunization Vision and Strategy, World Health Organization, www.who.int/immunization/givs/en/index.html; and Global Immunization Vision and Strategy, www.gavialliance.org/vision/strategy/gi_vision_strategy/index.php.

30 Bill Hendrick, "Prescription drug use on the rise in U.S.," *Medscape Today*, September 3, 2010, www.medscape.com/viewarticle/728064.

31 "Cardinal Health helps healthcare providers prepare for 2010-2011 flu season," *Medical News Today*, October 3, 2010, www.medicalnewstoday.com/articles/203235.php.

32 "Nearly 400 medicines and vaccines in development to fight infectious diseases," *Medical News Today*, September 12, 2010, www.medicalnewstoday.com/articles/200689.php.

33 "Vaccine sales surge 16% as industry eyes $35B mark," *Fierce Vaccines*, August 13, 2010, www.fiercevaccines.com/story/vax-sales-16-2009/2010-08-13.

34 Kate Kelland, "Pneumococcal shots give space for new strains: study," *Reuters*, September 8, 2010, www.reuters.com/article/idUSTRE6873U420100908.

35 John Treanor, MD, "Influenza Vaccine—Outmaneuvering Antigenic Shift and Drift," *New England Journal of Medicine*, January 15, 2004, 350:3, www.nejm.org/doi/pdf/10.1056/NEJMp038238.

36 Erwin Haas MD et al, "Atypical measles 14 years after immunization," JAMA, 236(9):1050 (1976), http://jama.ama-assn.org/cgi/content/summary/236/9/1050.

37 Donald Light, ed., *The Risks of Prescription Drugs* (New York: Columbia Univ. Press 2010).

38 Kogan, MD, et al., "Prevalence of Parent-Reported Diagnosis of Autism Spectrum Disorder Among Children in the US 2007," *Pediatrics* (published online Oct. 5, 2009).

39 Lorna Wing, "The Definition and Prevalence of Autism: A Review," *European Child and Adolescent Psychiatry*, 2:2 (1993):61–74, Hogrete & Huber Publishers, www.springerlink.com/content/m522g204l111445k/.

40 "US survey: 1 in 10 kids has ADHD, awareness cited," *Associated Press, Nov. 10, 2010, http://abcnews.go.com/Health/wireStory?id=12109129&tqkw=&tqshow=.*

41 "Autism has high costs to U.S. society," Harvard School of Public Health, Press Release, April 25, 2006, www.hsph.harvard.edu/news/press-releases/2006-releases/press04252006.html; Walecia Konrad, "Dealing with the financial burden of autism," *New York Times*, January 22, 2010, www.nytimes.com/2010/01/23/health/23patient.html?ref=health&pagewanted=all.

42 Ray Moynihan and Alan Cassels, *Selling Sickness: How Drug Companies are Turning Us All Into Patients*, (Allen and Unwin 2005).

43 Strategies for Increasing Adult Vaccination Rates, Centers for Disease Control and Prevention, www.cdc.gov/vaccines/recs/rate-strategies/adultstrat.htm.

44 Harry Schwartz, "Swine Flu Fiasco," *New York Times*, December 21, 1976, http://select.nytimes.com/gst/abstract.html?res=F00E1FFA3F5E1B7493C3AB1789D95F428785F9&scp=1&sq=Schwartz%20Swine%20flu%20fiasco%201976&st=cse.

45 Feifei Wei et al, "Identification and characteristics of vaccine refusers," *BMC Pediatrics*, March 5, 2009, 9:18 doi:10.1186/1471-2431-9-18, www.biomedcentral.com/1471-2431/9/18.

46 See www.therefusers.com.

47 "Most U.S. Children Receiving Recommended Vaccinations," *Pediatric Supersite*, September 16, 2010, www.pediatricsupersite.com/view.aspx?rid=70382.

CHAPTER 19: A LICENSE TO KILL? —MARK BLAXILL AND DAN OLMSTED

1 "From Lab to Market: The HPV Vaccine," *The NIH Record*, 59:4 (2007): 5.

2 Ibid.

3 D.R, Lowy and J.T. Schiller. "Prophylactic human papillomavirus vaccines." *Journal of Clinical Investigation*, 116:5(2006):1167-73.

4 http://ppp.od.nih.gov/regulations/coi.asp accessed April 28, 2012.

5 For example, US5437951.

6 Lowy v. Frazer, Patent Interference 104,775, United State Patent and Trademark Office Bureau of Patent Appeals and Interferences, September 20, 2005..

7 "NCI Scientists Recognized as Federal Employees of the Year," *The NIH Record*, 59:21 (2007): 5.

8 http://servicetoamericamedals.org/SAM/recipients/profiles/fym07_lowy_schiller.shtml accessed April 28, 2012.

9 http://www.ott.nih.gov/about_nih/statistics.aspx accessed April 28, 2012.

10 http://www.fda.gov/BiologicsBloodVaccines/Vaccines/ApprovedProducts/ucm111283.htm accessed April 28, 2012.

11 See Nancy B. Miller, "Clinical Review of Biologics License Application for Human Papillomavirus 6, 11, 16, 18 L1 Virus Like Particle Vaccine (*S. cerevisiae*) (STN 125126 GARDASIL), manufactured by Merck, Inc." Center for Biologics Evaluation and Research, Food and Drug Administration, June 6, 2006.

12 See http://dictionary.reference.com/reverseresults?q=efficacy&db=reverse accessed June 27, 2012.

13 L.L. Villa et al. "Prophylactic quadrivalent human papillomavirus (types 6, 11, 16, and 18) L1 virus-like particle vaccine in young women: a randomised double-blind placebo-controlled multicentre phase II efficacy trial." *Lancet Oncology*. 6 (2005):271-8.

14 K.S. Reisinger et al. "Safety and persistent immunogenicity of a quadrivalent human papillomavirus types 6, 11, 16, 18 L1 virus-like particle vaccine in preadolescents and adolescents: a randomized controlled trial." *Pediatric Infectious Disease Journal*. 26 (2007):201-9.

15 Miller, STN 125126.
16 See http://www.merck.com/product/usa/pi_circulars/g/gardasil/gardasil_pi.pdf accessed April 28, 2012.
17 Miller, STN 125126, p.379.
18 http://www.deathriskrankings.com/default.aspx?AspxAutoDetectCookieSupport=1 accessed April 28, 2012.
19 L.E. Markowitz et al. Centers for Disease Control and Prevention (CDC); Advisory Committee on Immunization Practices (ACIP). "Quadrivalent Human Papillomavirus Vaccine: Recommendations of the Advisory Committee on Immunization Practices (ACIP)." *MMWR Recommendations and Reports.* 56 (2007):1-24.
20 Personal communication, April 2010.
21 In April of 2007, the Texas Legislature passed HB 1098, which repealed the mandate put in place by Governor Perry. Perry's executive order subsequently became a liability in the Texas governor's unsuccessful campaign for the 2012 Republican presidential nomination.
22 Sheila M. Rothman and David J. Rothman. "Marketing HPV vaccine: implications for adolescent health and medical professionalism." *JAMA.* 302 (2009):781-6.
23 Cindy Bevington, "Researcher blasts HPV marketing" FW Daily News, March 14, 2007.
24 Ibid.
25 http://www.nvic.org/vaccines-and-diseases/HPV/HPV_Vaccine_Safety_ReptPart_III_081507_rev.aspx accessed April 28, 2012.
26 http://www.judicialwatch.org/documents/2008/JWReportFDAhpvVaccineRecords.pdf accessed April 28, 2012.
27 Charlotte J. Haug. "Human papillomavirus vaccination--reasons for caution." *New England Journal of Medicine.* 359 (2008):861-2.
28 Elizabeth Rosenthal, "Evidence Gap: Drug Makers' Push Leads to Cancer Vaccines' Rise." The New York Times, August 20, 2008 (see http://www.nytimes.com/2008/08/20/health/policy/20vaccine.html?_r=1&ref=health&pagewanted=print)
29 Bevington, 2007.
30 Susan Edelman, "My Girl Died As 'Guinea Pig' For Gardasil." *The New York Post,* July 20, 2008 (see http://www.nypost.com/p/news/regional/item_i4nmIcAdDoQtE4ubmxqGVM)
31 http://www.jenjensfamily.blogspot.com/ accessed April 28, 2012.
32 http://truthaboutgardasil.org/memorial/ accessed April 28, 2012.
33 http://leavittpartners.com/ accessed April 28, 2012.
34 http://www.pharmanet-i3.com/index.htm accessed April 28, 2012.
35 http://www.bcg-usa.com/about/aboutus.php accessed April 28, 2012.
36 For an overview of the problems with "vaccine court," see the chapter in this collection by Mary Holland and Robert Krakow, "The Right to Legal Redress."
37 http://www.iom.edu/~/media/Files/Report%20Files/2011/Adverse-Effects-of-Vaccines-Evidence-and-Causality/summary2.pdf accessed April 28, 2012.
38 http://www.anapolschwartz.com/practices/vaccines/tetlock-petition.asp accessed April 28, 2012.

CHAPTER 20: THE ROLE OF GOVERNMENT AND MEDIA —GINGER TAYLOR, MS

1 Joint Statement of the American Academy of Pediatrics (AAP) and the United States Public Health Service, July 7, 1999, www.hhs.gov/nvpo/vacc_safe/bthi1.htm.
2 National Research Council, *Toxicological Effects of Methylmercury,* (Wash., D.C.: National Academic Press, 2000), 11, www.nap.edu/openbook.php?record_id=9899&page=R1;

Centers for Disease Control and Prevention, "Frequently Asked Questions about Mercury and Thimerosal," www.cdc.gov/vaccinesafety/updates/thimerosal_faqs_mercury.htm; Agency for Toxic Substances and Disease Registry, Public Health Service, U.S. Department of Health and Human Services. *Toxicological Profile for Mercury*, (Atlanta, GA, 1999), www.atsdr.cdc.gov/ToxProfiles/TP.asp?id=115&tid=24, www.atsdr.cdc.gov/ToxProfiles/tp46-c7.pdf.

3 Steven G. Gilbert and Kimberly S. Grant-Webster, "Neurobehavioral Effects of Developmental Methylmercury Exposure," *Environmental Health Perspectives* 135:6 (1995): 103.

4 Arthur Allen, "The Not So-Crackpot Autism Theory," *New York Times*, November 10, 2002, www.nytimes.com/2002/11/10/magazine/10AUTISM.html?ei=1&en=99d1b535fa33bba3&ex=1037894857&pagewanted=print&position=top.

5 Joint Statement, note 1.

6 Verstraeten was an employee of vaccine maker GlaxoSmithKline at the time his study was published. Thomas Verstraeten, "Thimerosal, the Centers for Disease Control and Prevention, and GlaxoSmithKline," *Pediatrics* 113 (2004): 932.

7 Verstraeten T, Davis RL, DeStefano F, Lieu TA, Rhodes PH, Black SB, Shinefield H, Chen RT, Vaccine Safety Datalink Team, "Safety of thimerosal-containing vaccines: a two-phased study of computerized health maintenance organization databases," *Pediatrics* 112 (2003): 1039–48.

8 Thomas Verstraeten, "Thimerosal, the Centers for Disease Control and Prevention, and GlaxoSmithKline," *Pediatrics* 113 (2004): 932, http://pediatrics.aappublications.org/cgi/reprint/113/4/932.pdf.

9 Joint Statement, note 1.

10 Ibid.

11 David Kirby, *Evidence of Harm: Mercury in Vaccines and the Autism Epidemic: A Medical Controversy* (New York: St. Martin's Press, 2005).

12 "Managing Excess Vaccines," PUB-WA 841 (revised 02/2010). Waste and Materials Management Program, Wisconsin Department of Natural Resources, http://dnr.wi.gov/org/aw/wm/publications/anewpub/WA841.pdf.

13 Michael F. Wagnitz, "Reputation Of Vaccines And The People Who Defend Them," *Pediatrics*, 121 (2008): 621–622, eLetters published: (13 March 2008), http://pediatrics.aappublications.org/cgi/eletters/121/3/621#36839.

14 Ibid.

15 "Service Will Incinerate Unused H1N1 Vaccine," *Occupational Health & Safety*, July 23, 2010, http://ohsonline.com/articles/2010/07/23/service-will-incinerate-unused-h1n1-vaccine.aspx?.

16 HRSA is an agency of the U.S. Department of Health and Human Services. Among other responsibilities, it oversees the Vaccine Injury Compensation Program, www.hrsa.gov/about/index.html.

17 Vaccine Injury Table, National Vaccine Injury Compensation Program, Health Resources and Services Administration, U.S. Department of Health and Human Services, www.hrsa.gov/vaccinecompensation/table.htm.

18 Paola Giovanardi Rossi, Antonia Parmeggiania, Véronique Bachb, Margherita Santuccia and Paola Viscontia, "EEG features and epilepsy in patients with autism," *Brain and Development* 17:3 (May–June 1995): 169-174.

19 Jon S. Poling, MD, PhD, Richard E. Frye, MD, PhD, John Shoffner, MD, Andrew W. Zimmerman, MD, "Developmental Regression and Mitochondrial Dysfunction in a Child With Autism," *Journal of Child Neurology* 21:2 (February 2006).

20 David Kirby, "The Vaccine-Autism Court Document Every American Should Read," *The Huffington Post*, February 26, 2008, www.huffingtonpost.com/david-kirby/the-vaccineautism-court-d_b_88558.html.

21 David Kirby, "Government Concedes Vaccine-Autism Case in Federal Court - Now What?" *The Huffington Post*, February 25, 2008, www.huffingtonpost.com/david-kirby/government-concedes-vacci_b_88323.html.

22 Poling, "Developmental Regression," note 19 above.

23 Jim Giles, "Autism payout reignites vaccine controversy," *New Scientist*, March 8, 2008, www.newscientist.com/article/mg19726464.100-autism-payout-reignites-vaccine-controversy.html

24 Sharyl Attkisson, "Family to Receive $1.5M+ in First-Ever Vaccine-Autism *Court Award*," *CBS Evening News/CBS News Investigates Blog*, September 9, 2010, http://www.cbsnews.com/8301-31727_162-20015982-10391695.html

25 Ibid.

26 Ginger Taylor, "CDC Chief Admits that Vaccines Trigger Autism," April 2, 2008. Online video clip, YouTube, www.youtube.com/watch?v=Dh-nkD5LSIg.

27 "Housecall: Unraveling the Mystery of Autism, with Sanjay Gupta," *Cable News Network*, March 29, 2008.

28 David Kirby, "The Next Big Autism Bomb: Are 1 in 50 Kids Potentially At Risk?" *The Huffington Post*, March 26, 2008, www.huffingtonpost.com/david-kirby/the-next-big-autism-bomb-_b_93627.html.

29 These children were part of the same autism treatment program and were not randomly selected.

30 Ibid.

31 Shankar Vedantam, "Fathering Autism: A Scientist Wrestles With the Realities of His Daughter's Illness," *The Washington Post*, July 1, 2008. www.washingtonpost.com/wp-dyn/content/article/2008/06/27/AR2008062703023.html?sid=ST2008070101784&pos=.

32 Anne Schuchat, "Vaccines' benefits outweigh dangers: Autism claim shouldn't deter parents," *The Atlanta Journal Constitution*, March 26, 2008, www.ajc.com/opinion/content/opinion/stories/2008/03/25/vaccinesed_0326.html.

33 David A. Geier, Mark R. Geier, "Autism Spectrum Disorder-associated Biomarkers for Case Evaluation: Mitochondrial Dysfunction Biomarkers," *Medscape Today*, www.medscape.com/viewarticle/584398_6.

34 *NBC*, "The Today Show," March 11, 2008, discussed in *Adventures in Autism* blog, http://adventuresinautism.blogspot.com/2008/03/aap-president-tells-giant-easily.html.

35 Vaccine Injury Table, National Vaccine Injury Compensation Program, www.hrsa.gov/vaccinecompensation/table.htm.

36 *Forehand v. Tayloe*, E. District of North Carolina, verdict May 1, 1985.

37 www.karneylaw.com/Verdicts-Settlements/Top-5-Verdicts-Settlements.shtml#2.

38 "National and state AAP leader, Dr. Tayloe Sr., was medical home advocate," *AAP News*, 31:8 (August 2010), 36, http://aapnews.aappublications.org/cgi/content/full/31/8/36-a.

39 Robert F. Kennedy, Jr. and David Kirby, "The Vaccine Court: Autism Debate Continues," *The Huffington Post*, February 24, 2009, www.huffingtonpost.com/robert-f-kennedy-jr-and-david-kirby/vaccine-court-autism- deba_b_169673.html.

40 Disclosure: Dr. Lisa Masterson of "The Doctors" was my obstetrician and gynecologist and administered thimerosal containing Rhogam to me during pregnancy. Her office's failure to call me with test results may have been a contributing factor in the premature birth of my older son Webster.

41 Mary Holland, Louis Conte, Robert Krakow, and Lisa Colin, Unanswered Questions from the Vaccine Injury Compensation Program: A Review of Compensated Cases of Vaccine-Induced Brain Injury, 28 Pace Envtl. L. Rev. 480 (2011), http://digitalcommons.pace.edu/pelr/vol28/iss2/6

42 Jon Baio, Prevalence of Autism Spectrum Disorders — Autism and Developmental Disabilities Monitoring Network, 14 Sites, United States, 2008, March 30, 2012 / 61(SS03);1-19 http://www.cdc.gov/mmwr/preview/mmwrhtml/ss6103a1.htm?s_cid=ss6103a1_w

43 The Canary Party, The Canary Party and Grassroots Autism Organizations Call for Firings of Health Officials in the Wake of New Autism Numbers, April 2, 2012, http://canaryparty.org/index.php/the-news/100-the-canary-party-and-grassroots-autism-organizations-call-for-firings-of-health-officials-in-the-wake-of-new-autism-numbers

44 Dan Burton, It Is Time to Re-Engage on the Autism Epidemic, The Hill's Congressional Blog, April 24, 2012, http://thehill.com/blogs/congress-blog/healthcare/223265-it-is-time-to-re-engage-on-the-autism-epidemic

45 "Everything You Need To Know About The Flu," *The Doctors*, October 26, 2009, Stage 29, LLC.

46 Vaccines and Thimerosal, Vaccine Safety, CDC, www.cdc.gov/vaccinesafety/concerns/thimerosal/index.html, last modified: February 11, 2010.

47 Ginger Taylor "No Evidence of Any Link," AdventuresInAutism.com, June 14th, 2007, accessed May 1st, 2012, http://adventuresinautism.blogspot.com/2007/06/no-evidence-of-any-link.html.

48 Sharyl Attkisson, "Family to Receive $1.5M+ in First-Ever Vaccine-Autism Court Award," *CBS Evening News/CBS News Investigates Blog*, September 9, 2010, www.cbsnews.com/8301-31727_162-20015982-10391695.html.

49 Arthur Allen, "H1N1: The Report Card," *Readers Digest*, March 2010, www.rd.com/health-slideshows/h1n1-the-report-card/article174741-1.html.

50 Notable Name Database, "Ignaz Semmelweis," www.nndb.com/people/601/000091328/.

51 P Bálint, G Bálint, "The Semmelweis-reflex," *Orv Hetil.* 150:30 (July 26, 2009):1430.

52 "Parents want research on vaccine safety," *UPI.com*, October 12, 2010, www.upi.com/Health_News/2010/10/12/Parents-want-research-on-vaccine-safety/UPI-39721286941024/.

CHAPTER 21: MERCURY TOXICITY AND VACCINE INJURY —BOYD HALEY, PHD

1 DG Fagan, et al, "Organ mercury levels in infants with omphaloceles treated with organic mercurial antiseptic," *Arch. Dis. Child., Br. Med. J.*, 52(12) (1977):962–4.

2 The Food and Drug Administration (FDA) removed mercurochrome from the "generally regarded as safe" and placed it into the "untested" classification in 1998, following a general review of over-the-counter drugs containing mercury that began in 1978.

3 Boyd Haley. "Mercury Toxicity: Genetic Susceptibility and Synergistic Effects." *Medical Veritas.* 2 (2005) 1–8.

4 Branch DR, "Gender-selective toxicity of thimerosal," *Exp. Toxicol. Pathol.*, 61(2) (Mar 20009):133–6; Epub 2008 Sep 3.

5 C. Migdal, et al, "Sensitization effect of thimerosal is mediated in vitro via reactive oxygen species and calcium signaling," *Toxicology*, 2010 July–Aug.; 274(1–3):1–9. Epub 2010 May 10. See also www.generationrescue.org/science for additional studies.

6 SJ James, et al, "Metabolic biomarkers of increased oxidative stress and impaired methylation capacity in children with autism," *Am. J. Clin. Nutr.*, 2004 Dec; 80(6):1611–17. See also www.generationrescue.org/science for additional studies.

7 R. Nataf, et al, "Porphyrinuria in Childhood Autistic Disorder: Implications for Environmental Toxicity," *Toxicol. Appl. Pharmacol.*, 2006 Jul 15; 214(2):99–108, Epub 2006 Jun 16.

8 JS Woods, et al, "The association between genetic polymorphisms of coproporphyrinogen oxidase and an atypical porphyrinogenic resonse to mercury exposure in humans," *Toxicol. Appl. Pharmacol.*, 2005 Aug 7; 206(2):113–20.

9 GC Rampersad, et al, "Chemical compounds that target thiol-disulfide groups on mononuclear phagocytes inhibit immune mediated phagocytosis of red blood cells," *Transfusion*, 2005 Mar; 45(3):384–93.

10 M Waly, et al, "Activation of methionine synthase by insulin-like growth factor-1 and dopamine: a target for neurodevelopmental toxins and thimerosal," *Mol. Psychiatry.*, 2004 Apr; 9(4):358–70.

11 PV Usatyuk, et al, "Redox regulation of 4-hydroxy-2-nonenal-mediated endothelial barrier dysfunction by focal adhesion, adherens, and tight junction proteins," *J. Biol. Chem.*, 2006 Nov 17;281(46):35554-66. Epub 2006 Sep 17.

12 RH Waring and L.V. Klovrza, "Sulphur metabolism in autism," *J. Nutr. Environ. Med.* 2000. 10(1):25–32.

13 Ibid.

14 “The Vaccine War,” *Frontline*, PBS, April 27, 2010, www.pbs.org/wgbh/pages/frontline/vaccines.

15 M Heron, et al, “Annual Summary of Vital Statistics: 2007,” *Pediatrics*, 2010 Jan. 125(1):4–15. Epub 2009 Dec 21; and The World Factbook, Country Comparison—Infant Mortality Rate, Central Intelligence Agency, https://www.cia.gov/library/publications/the-world-factbook/rankorder/2091rank.html.

16 CM Gallagher and MS Goodman, “Hepatitis B triple series vaccine and developmental disability in U.S. children aged 1-9 years,” *Toxicological & Environmental Chemistry*, 2008 Sept–Oct. 90(5):997–1008; and CM Gallagher, MS Goodman, “Hepatitis B Vaccination of Male Neonates and Autism,” *Annals of Epidemiology*, 2009 Sept. 19(9):659.

CHAPTER 22: FLU VACCINE MANDATES FOR U.S. HEALTH CARE WORKERS: POLICY WITHOUT REASON —TONI BARK, MD, MHEM, LEED AP

1 FDA approves vaccines for the 2011-2012 influenza season. (2011, June 11). FDA.gov. http://www.fda.gov/NewsEvents/Newsroom/PressAnnouncements/ucm263319.htm.

2 Key Facts About Seasonal Flu Vaccine, http:www.cdc.gov/flu/protect/keyfacts.htm

3 Healthy People Immunization and Infectious Disease Objective IID-12.9, see benchmark in details, section, http://www.healthypeople.gov/2020/topicsobjectives2020/objectiveslist.aspx?topicId=23.

4 National Vaccine Advisory Committee, Recommendations on Strategies to Achieve the Healthy People 2020 Annual Influenza Vaccine Coverage Goal for Health Care Personnel, Feb. 8, 2012, http://www.hhs.gov/nvpo/nvac/influenza_subgroup_final_report.pdf (NVAC Recommendations), 9.

5 Ibid.

6 Healthy People Immunization and Infectious Disease Objective IID-12.9, see benchmark in details, section, http://www.healthypeople.gov/2020/topicsobjectives2020/objectiveslist.aspx?topicId=23.

7 http://www.hhs.gov/nvpo/nvac/; DeLong, G. (2012). Conflicts of Interest in Vaccine Safety Research. Accountability in Research, 19(2), 65-88.

8 NVAC Recommendations, 1.

9 Ibid.

10 Ibid, 2.

11 Ibid, 3.

12 Ibid, 3, 36, 37.

13 Ibid, 3, footnote 17.

14 Ibid, 23.

15 Influenza Vaccination Information for Health Care Workers. (n.d.). seasonal influenza. http://www.cdc.gov/flu/healthcareworkers.htm.

16 Ibid.

17 Reports, Recommendations and Resolutions. (2012, February 2). U.S. Department of Health and Human Services, from http://www.hhs.gov/nvpo/nvac/reports/index.html.

18 “NVAC Says: Mandate Flu Shots for Health Care Workers,” National Vaccine Information Center, February 21, 2012, http://www.nvic.org/NVIC-Vaccine-News/February-2012/Feds-Recommend-Flu-Shot-Mandates-for-Health-Care-W.aspx.

19 Carmen, W. F. (2000). Effects of influenza vaccination of health-care workers on mortality of elderly people in long-term care: a randomised controlled trial. The Lancet, 355 (9198), 93-7; Bueving, H. (2003). Influenza Vaccination in Children with Asthma. The American Journal of Respiratory and Critical Care, 169(4), 488-493.

20 Ibid.

21 Saxen, H., & Virtanen, M. (1999). Randomized, placebo-controlled double blind study on the efficacy of influenza immunization on absenteeism of health care workers. Pediatric Infectious Disease Journal, (9), 779-83.

22 Brownlee, S., & Lenzer, J. (2009, November). Does the Vaccine Matter? The Atlantic Monthly. 1-3. http://www.theatlantic.com/magazine/archive/2009/11/does-the-vaccine-matter/7723/.

23 Ibid.

24 Ibid.

25 Roos, R. (2011, October 25). Strict meta-analysis raises questions about flu vaccine efficacy. CIDRAP. http://www.cidrap.umn.edu/cidrap/content/influenza/general/news/oct2511lancet.html.

26 Ibid.

27 Majumdar, S. (2008). Mortality Reduction with Influenza Vaccine in Patients with Pneumonia Outside "Flu" Season Pleiotropic Benefits or Residual Confounding? American Journal of Respiratory and Critical Care Medicine, 178(5), 527-533.

28 Ibid.

29 Viboud, C., & Simonsen, L. (2010, April 6). Does Seasonal Influenza Vaccination Increase the Risk of Illness with the 2009 A/H1N1 Pandemic Virus? PLoS Medicine, http://www.plosmedicine.org/article/info:doi/10.1371/journal.pmed.1000259.

30 Djulbegovic, B. (2000). Randomized trials that changed medical practice. RCT-practice change. http://personal.health.usf.edu/bdjulbeg/oncology/RCT-practice-change.htm.

31 Jefferson, T., & Di Pietrantonj, C. (2010, July 7). Vaccines To Prevent Influenza in Health Adults. Cochrane Summaries. http://summaries.cochrane.org/CD001269/vaccines-to-prevent-influenza-in-healthy-adults.

32 Ibid.

33 Cell Press (2012, March 1). Antibodies are not required for immunity against some viruses. ScienceDaily. http://www.sciencedaily.com/releases/2012/03/120301143426.htm.

34 Fedson, D. (1998). Measuring protection: efficacy versus effectiveness. Developments in biological standardization, 95, 195-201. http://www.curehunter.com/public/pubmed9855432.do.

35 Hayes, D. P. (2010). Influenza Pandemics, solar activity cycles, and vitamin D. Medical Hypothesis, 74(5), 831-4; Cannell JJ, Vieth R, Umhau JC, et al. Epidemic influenza and vitamin D. Epidemiol Infect 2006;134:1129–40; Laaksi I, Ruoholo JP, Tuohimaa P, et al. An association of serum vitamin D concentrations <40 nmol/L with acute respiratory tract infection in young Finnish men. Am J Clin Nutr 2007;86:714–7; Karatekin G, Kaya A, Salihoglu O, Balci H, Nuhoglu A. Association of subclinical vitamin D deficiency in newborns with acute lower respiratory infection and their mothers. Eur J Clin Nutr 2009;63:473–7; and Ginde AA, Mansbach JM, Camargo Jr CA. Association between serum 25-hydroxyvitamin D level and upper respiratory tract infection in the Third National Health and Nutrition Examination Survey. Arch Intern Med 2009;169:384–90.

36 *Bruesewitz v. Wyeth*, 562 U.S. __ (2011), www.supremecourt.gov/opinions/10pdf/09-152.pdf.

37 Baxter admits flu product contained live bird flu virus. (2009, February 27). CTV News (Toronto), http://www.ctv.ca/CTVNews/Health/20090227/Bird_Flu_090227/

38 Bigongiari, J. (2011, October 28). Baxter recalls Preflucel flu vaccine. Vaccine News Daily.

39 Goddard, E. (1997). Campylobacter 0:41 isolation in Guillain-Barré syndrome. Archives of Disease In Childhood;,and Nachamkin, L. (1997, December). Microbiologic approaches for studying Campylobacter species in patients with Guillain-Barré syndrome. PubMed. http://www.ncbi.nlm.nih.gov/pubmed/9396692.

40 Haber, P. (2004). Guillain-Barré Syndrome Following Influenza Vaccination. JAMA, 292(20), 2478-2481.

41 Rappuoll, R. (2006). Cell-Culture-Based Vaccine Production: Technological Options. National Academy of Engineering-The Bridge. 3(3). http://www.nae.edu/Publications/Bridge/EngineeringandVaccineProductionforanInfluenzaPandemic/Cell-Culture-BasedVaccineProductionTechnologicalOptions.aspx.

42 Ibid.

43 Patriarca, P. A. (2007, April 10). Use of Cell Lines for the Production of Influenza Virus Vaccines. :WHO. int. http://www.who.int/vaccine_research/diseases/influenza/WHO_Flu_Cell_Substrate_Version3.pdf.

44 Barrett, P. N. (2009, May 28). Vero cell platform in vaccine production: moving towards cell culture-based viral vaccines. PubMed. http://www.ncbi.nlm.nih.gov/pubmed/19397417.

45 FLUAD influenza vaccine for seniors. (n.d.). VRAN. http://vran.org/about-vaccines/specific-vaccines/influenza-vaccine-flu-shot/fluad-influenza-vaccine-for-seniors/.

46 Squalene Induces Autoimmune Disease in Animals. (2009, June 10). Vaclib.org. https://docs.google.com/viewer?a=v&q=cache:5Z0GWbipKnEJ:www.vaclib.org/basic/flu/web-swine/Squalene%2520references%2520for%2520Edda.doc+injected+squalene+and+autoimmunity&hl=en&gl=us&pid=bl&srcid=ADGEESiWLIzNacOeI-21zXMkQnLbLy-ECs16eDAsP352vA1NZ8YkT_B2L_1dPERRpm3b_yvdhmTM-m3OJcNDKUc5DXfyq9Cd6haJZp4YKVKM67nKeshJaQmlKxEoNtc9WAuDLJS-DBf5Ag&sig=AHIEtbRXSaRjd_tzlszx0AajqlIIrhNiiQ&pli=1.

47 Guidice, D. (2006). Vaccines with the MF59 adjuvant do not stimulate antibody responses against squalene. Clinical Vaccine Immunology, 13(9), 1010-1013.

48 Wilson, P. B. (2002). Antibodies to squalene in recipients of anthrax vaccine. Experimental and Molecular Pathology, 73(1), 19-27.

49 Cohen, D. (2010). WHO and the pandemic flu "conspiracies." The British Medical Journal, 340. doi: 10.1136/bmj.c2912; and Flynn, P. (2010, June 7). The handling of the H1N1 pandemic: more transparency needed. assembly.coe. int. http://assembly.coe.int.

50 Matsumoto, G. (2004). VACCINE A: The Covert Government Experiment That's Killing Our Soldiers—And Why GI's Are Only the First Victims. USA: The Perseus Books Group.

51 Ibid.

52 Rubin, R. (2009, May 5). Lessons learned from the 1976 swine flu fiasco. USA Today.

53 Toppo, G. (2009, May 1). USA Today, http://www.usatoday.com/news/health/2009-04-30-swine-flu-us-thursday_N.htm.

54 Cohen, D. (2010). WHO and the pandemic flu "conspiracies." The British Medical Journal, 340. doi: 10.1136/bmj.c2912

55 Cohen, D. (2010). WHO and the pandemic flu "conspiracies." The British Medical Journal, 340. doi: 10.1136/bmj.c2912; and Flynn, P. (2010, June 7). The handling of the H1N1 pandemic: more transparency needed. assembly.coe. int. http://assembly.coe.int.

56 Swine Flu vaccine Narcolepsy Link Confirmed. (2011, September 6). icenews.is; Obrien, C. (2012, April 19). Narcolepsy linked to flu vaccine in 25 children. Irishtimes.com. http://www.irishtimes.com/newspaper/ireland/2012/0419/1224314925694.html

57 Flynn, P. (2010, June 7). The handling of the H1N1 pandemic: more transparency needed. assembly.coe. int. http://assembly.coe.int.

58 Smith, J. (2010). WHO Launches H1N1 Pandemic Review; Critics Watch Closely. The Lancet, http://www.thelancet.com/H1N1-flu/egmn/0c03f4ed.

59 Cohen, D. (2010). WHO and the pandemic flu "conspiracies." The British Medical Journal, 340. doi: 10.1136/bmj.c2912; and Flynn, P. (2010, June 7). The handling of the H1N1 pandemic: more transparency needed. assembly.coe. int. http://assembly.coe.int..

60 NYCRR Subpart 66-3, et seq. (NYS Register, Sept. 2, 2009).

61 Open Letter, Mandatory Flu Vaccine for Health Care Workers, Sept. 24, 2009, http://www.health.ny.gov/press/releases/2009/2009-09-24_health_care_worker_vaccine_daines_oped.htm.

62 ACLU, Maintaining Civil Liberties Protections in Response to the H1N1 Flu, Nov. 2009, 9, www.aclu.org/files/assets/H1N1_Report_FINAL.pdf.

63 Ibid.

64 Andrew T. Pavia, Mandate to Protect Patients from Health Care-Associated Influenza, Clin Infect Dis. (2010) 50 (4): 465-467, 466.

65 Immunization Action Coalition. Honor roll for patient safety, www.immunize.org/laws/influenzahcw.asp.

66 NVAC Recommendations, 18-19.

CHAPTER 23: FORCED CHILD REMOVAL —KIM MACK ROSENBERG, JD

1 New York Social Services Law § 411.

2 Some have raised concerns about potential financial incentives to states from the federal government to remove children from their family homes.

3 An example of this abuse of power is the use of children in foster care in drug trials, in some cases without appropriate informed consent. Such abuse implicates both state agencies and the federal government. For more on this topic, see www.ahrp.org. Another example of misuse of state power and perhaps the good intention to protect children run amok is the 1980s prosecutions of Violet Amirault and her adult children, convicted of crimes against children in their care. Their day care center in Massachusetts had been well-regarded. Journalist Dorothy Rabinowitz has written extensively on this case and the "evidence" against the Amiraults in the *Wall Street Journal* and in *No Crueler Tyrannies: Accusations, False Witness And Other Terrors of Our Times* (Free Press, 2003).

4 Many chapters of this book have cogently addressed issues concerning the role of government in establishing and enforcing vaccine mandates and the doctrines underpinning parental rights. While all states mandate significant vaccination for day care, preschool, or school attendance, they have at least some form of medical exemption to vaccine mandates. Forty-eight recognize some form of religious exemption, and twenty states have philosophical exemptions. While parental rights are strong and there generally is a presumption that a parent acts in the best interest of her child, parental rights are not absolute. In particular, in emergencies, parental rights may cede to governmental interests.

5 It is not just the children and families who are being investigated; doctors who treat these children also are being investigated.

6 www.merckmanuals.com/professional/sec19/ch301/ch301a.html.

7 N.Y. Family Court Act §1012(f)(i)(A).

8 www.nyc.gov/html/acs/downloads/pdf/stateguide_english.pdf.

9 Elissa J. Brown, Ph.D. and Shamir A. Khan, "Child Abuse and Neglect - Definitions, consequences, and treatment," www.aboutourkids.org/files/articles/mar_apr_1.pdf.

10 "Keeping Your Children and Getting Them Back," presented by Edie Mannion, MFT, Mental Health Assoc. of Southeastern Pennsylvania, through the U.Penn. Collaborative on Community Integration. See www.upennrrtc.org/resources/view.php?tool_id=177 as the source for the PowerPoint of this presentation.

11 State law determines mandatory reporters of suspected abuse or neglect.

12 Diekema, Douglas S. and the Committee on Bioethics, "Responding to Parental Refusals of Immunization of Children," *Pediatrics* 115:5 (2005): 1428-31, http://aappolicy.aappublications.org/cgi/reprint/pediatrics;115/5/1428.pdf.

13 Committee on Bioethics, American Academy of Pediatrics, "Religious objections to medical care," *Pediatrics* 99:2 (1997): 279–81, http://aappolicy.aappublications.org/cgi/reprint/pediatrics;99/2/279.pdf. See also Committee on Bioethics, American Academy of Pediatrics, Religious Exemptions From Child

Abuse Statutes. *Pediatrics* 1988; 81(1):169–71, http://aappolicy.aappublications.org/cgi/reprint/pediatrics;81/1/169.pdf.

14 Committee on Bioethics, American Academy of Pediatrics, "Religious objections to medical care," *Pediatrics* 99:2 (1997): 279–81, http://aappolicy.aappublications.org/cgi/reprint/pediatrics;99/2/279.pdf.

15 Ibid.

16 Ibid.

17 For example, *The Merck Manual* states that "[i]t is not clear whether this refusal of vaccination is true medical neglect; it may be considered similar to refusal of non life-saving treatments for religious reasons. In such cases, as long as the children are healthy, there is usually no need to ascertain whether the refusal constitutes medical neglect. However, in the face of illness, refusal of scientifically and medically accepted treatment often requires further investigation and sometimes legal intervention." www.merckmanuals.com/professional/sec19/ch301/ch301a.html.

18 In an article addressing challenges faced by primary care physicians with non-vaccinating patients, the authors suggest that physicians consider "families who refuse vaccine not as annoyances or threats but as complex and unique challenges." Lyren, Anne and Leonard, Ethan, "Vaccine Refusal: Issues for the Primary Care Physician," *Clinical Pediatrics*, June 2006 Issue, 403, http://cpj.sagepub.com/content/45/5/399.full.pdf+html?rss=1. The authors also posit that "[v]accinating a child despite a parent's objection is weakly defensible except in the case of an outbreak when others are at substantial risk of serious harm by the parent's decision." This might form the basis for a neglect investigation although the authors do not state that the child must be at risk of harm. Rather, they suggest that the child should be inoculated for the public good.

19 www.aap.org/immunization/pediatricians/pdf/RefusaltoVaccinate.pdf.

20 Ibid.

21 Ibid.

22 It is more likely that CPS will investigate families who do not follow the mandated vaccination schedule and who also make alternative healthcare choices.

23 *In re Christine M.*, 157 Misc. 2d 4, 595 N.Y.S.2d 606 (Fam. Ct., Kings Co. 1992).

24 Ibid; this case was filed on January 24, 1991, and the decision was issued on December 21, 1992—nearly two years later.

25 United States Government Accountability Office, "Foster Care: State Practices for Assessing Health Needs, Facilitating Service Delivery, and Monitoring Children's Care," (GAO-09-26) (Feb. 2009), 11–12, www.gao.gov/new.items/d0926.pdf; Sheryl Dicker and Elysa Gordon, "Ensuring the Healthy Development of Infants in Foster Care: A Guide for Judges, Advocates and Child Welfare Professionals" (Jan. 2004), 7, 10. See http://main.zerotothree.org/site/DocServer/Infant_Booklet.pdf?docID=1847.

26 New York State Department of Health, Office of Medicaid Management, "EPSDT/CHTP Provider Manual for Child Health Plus A (Medicaid)" (Version 2005-1) ("EPSDT Provider Manual"), www.emedny.org/ProviderManuals/EPSDTCTHP/PDFS/EPSDT-CTHP.pdf.

27 "Working Together: Health Services for Children in Foster Care" (March 1, 2009, 1–10 ("Working Together"), www.ocfs.state.ny.us/main/sppd/health_services/manual/Chapter%201%20Initial%20Evaluation.pdf.

28 Ibid, 2–4, www.ocfs.state.ny.us/main/sppd/health_services/manual/Chapter% 202%20Preventive%20and%20Ongoing.pdf. See also NYS Foster Parent Manual (Jan. 2007), 30 (Medical examinations must be conducted at specified intervals and must include "[a]ssessment of immunization status and provision of immunizations as necessary."), www.ocfs.state.ny.us/main/publications/Pub5011%20NYS%20Foster%20Parent%20Manual.pdf.

29 EPSDT/CTHP Provider Manual, 40, see note 27 above.

30 Ibid.

31 For example, under New York law, when a child is placed in foster care, parental consent is required for routine treatment. If parental consent cannot be obtained, this requirement may be circumvented, including through judicial intervention. "Working Together," note 27, 6–5.

32 *In re: A.Y. v. R.B.*, 2008 N.Y. Misc. LEXIS 7311 (Fam. Ct., Nassau Co., Dec. 18, 2008) (court denied motion by Nassau County Department of Social Services to immunize previously unvaccinated children in foster care pending determination of neglect petition, including medical neglect, where children had previously attended school unvaccinated based on written information supplied by mother and her religious leader); see also the unpublished decision in *In re Shmuel and Ester G*, 6 Misc. 3d 1018A, 800 N.Y.S.2d 357, 2005 N.Y. LEXIS 199 (Fam. Ct., Kings Co. Feb. 7, 2005).

33 While outside the scope of this article, there is a significant body of law on immunization and school attendence. In some cases, the courts look at a parent's refusal to immunize as educational neglect rather than medical neglect if the child is not attending school because he or she is not vaccinated according to state mandates.

34 *In re Shmuel and Ester G*, note 32.

35 Ibid, *2-3.

36 Ibid, *3.

37 Ibid, *11.

38 Ibid, *6.

39 Ibid.

40 N.Y.S. Foster Parent Manual, 31, note 28 above.

41 *In re Isaac J.*, 2010 N.Y. Slip. Op. 5997, 75 A.D.3d 506, 904 N.Y.S.2d 755, 2010 N.Y. App. Div. LEXIS 5880 (2d Dep't July 6, 2010).

42 Ibid.

43 *Diana H. v. Hon. Stephen M. Rubin*, 171 P.3d 200 (Az. App. 2007).

44 Ibid, 208.

45 Ibid, 208-09.

46 *In re Stratton*, 571 S.E.2d 234 (N.C. App. 2002).

47 Ibid, 238.

48 Ibid.

49 *In re C.R.*, 257 Ga. App. 159, 570 S.E.2d 609, 2002 Ga. App. LEXIS 1088 (Ga. Ct. App. Aug. 23, 2002).

50 Ibid, *5-6.

51 Dan Olmsted, "Olmsted on Autism," *Age of Autism* blog, www.ageofautism.com/2008/04/olmsted-on-au-3.html (April 30, 2008).

CHAPTER 24: THE GREATEST THREAT TO OUR COUNTRY —JULIAN WHITAKER, MD

1 Redberg R, et al. Diagnostic tests: another frontier for less is more. *Arch Intern Med.* 2011 April 11;171(7):619.

2 Lawson EH, et al. Appropriateness criteria to assess variations in surgical procedure use in the United States. *Arch Surg.* 2011 Dec;146(12):1433-1440.

3 Welch HG, et al. *Overdiagnosed: Making People Sick in the Pursuit of Health.* Boston, MA: Beacon Press: 2011.

4 Boden WE, et al. Optimal medical therapy with or without PCI for stable coronary disease. *N Engl J Med.* 2007 Apr 12;356(15):1503–1516; and Trikalinos TA, et al. Percutaneous coronary interventions for non-acute coronary artery disease: a quantitative 20-years synopsis and a network meta-analysis. *Lancet.* 2009 March 14;373(9667): 911–918.

5 Gagnon MA, et al. The cost of pushing pills: a new estimate of pharmaceutical promotion expenditures in the United States. *PLoS Med.* 2008 Jan 3;5(1):e1.doi:10.1371/journal.pmed.0050001.

6 Influence and lobbying: pharmaceuticals/health products, industry profile 2011. *Opensecrets.org.* http://www.opensecrets.org/lobby/indusclient.php?id=H04&year=2011

7 Abramson J. *Overdosed America.* 2004, HarperCollins, New York, NY.

8 Lundy P. Prescription drug trends. *Kaiser Family Foundation.* May 2010. www.kff.org/rxdrugs/upload/3057-08.pdf.

9 Starfield B. Is US health really the best in the world? *JAMA* 2000;284(4): 483-485.

10 Sarkar U, et al. Adverse drug events in U.S. adult ambulatory medical care. *Health Services Research,* 2011; 46: 1517–1533. doi: 10.1111/j.1475-6773.2011.01269.x

11 Lucado J, et al. Medication-related adverse outcomes in U.S. hospitals and emergency departments, 2008. Statistical Brief #109. Healthcare Cost and Utilization Project (HCUP). April 2011. Agency for Healthcare Research and Quality, Rockville, MD. www.hcup-us.ahrq.gov/reports/statbriefs/sb109.jsp.

12 Berkrot B. Prescription drug use by U.S. children on the rise. *Reuters.* 2010 May 19. http://www.reuters.com/article/2010/05/19/us-medco-children-idUSTRE64I5N420100519.

13 Centers for Disease Control. Autism Spectrum Disorders (ASDs), data and statistics. http://www.cdc.gov/ncbddd/autism/data.html. Accessed 2012 April 17.

14 Data from Centers for Disease Control, compiled by Shawn Siegel, 2011 July 31.

15 Generation Rescue, Inc. Autism and vaccines around the world: vaccine schedules, autism rates, and under 5 mortality. 2009 April.

16 Doshi P, et al. Japanese childhood vaccination policy. *Cambridge Quarterly of Healthcare, Ethics Special Section: International Voices.* 2010; 19: 283–289. Cambridge University Press.

17 Mary Ann Block, D.O., conversion with physician, 2005 July.

18 Burton, D. Committee on Government Reform. *The Status of Research into Vaccine Safety and Autism.* 2002 June 19. FDA. *Thimerosal in vaccines.* 2005 March 21 *www.fda.gov/cber/vaccine/thimerosal.htm#t1*

19 Olmsted D, and Blaxill, M. *The Age of Autism: Mercury, Medicine, and a Man-Made Epidemic.* Thomas Dunne Books. New York, NY. 2010.

20 Unlike the U.S., many developed countries give the birth dose of hepatitis B vaccine to at-risk infants only, including Denmark, England, France, Germany, Holland, Ireland, Italy, Japan, Norway, and Sweden. http://ecdc.europa.eu/en/Pages/home.aspx, http://venice.cineca.org/Report_Hepatitis_B_Vaccination.pdf, http://www.who.int/immunization_delivery/new_vaccines/4.Coreinformation_Hepatitis%20B.pdf.

21 Centers for Disease Control and Prevention. Polio disease—questions and answers. http://www.cdc.gov/vaccines/vpd-vac/polio/dis-faqs.htm. Accessed 2012 April 17.

22 Data compiled by Roman Bystrianyk based on data from Vital Statistics of the United States, 1937, 1938, 1943, 1944, 1949, 1960, 1967, 1976, 1987, 1992; Historical Statistics of the United States – Colonial Times to 1970 Part 1; Health, United States, 2004, U.S. Department of Health and Human Services; Vital Records & Health Data Development Section, Michigan Department of Community Health; U.S Census Bureau, Statistical Abstract of the United States: 2003; Reported Cases and Deaths f rom Vaccine Preventable Diseases, United States, 1950-2008. www.tinyurl.com/y59zz17.

23 National Cancer Institute. Human Papillomavirus (HPV) vaccines. http://www.cancer.gov/cancertopics/factsheet/prevention/HPV-vaccine

24 Habakus LK, and Holland M. *Vaccine Epidemic.* Skyhorse Publishing. New York, NY. 2011.

25 Sawaya GF, et al. HPV vaccination—more answers, more questions. *N Engl J Med.* 2007 May 10; 356:1991-1993.

26 National Cancer Institute. NCI health information tip sheet for writers:human papillomavirus (HPV) vaccine. 2012 Jan 5. http://www.cancer.gov/newscenter/entertainment/tipsheet/human-papillomavirus-vaccine/print. Accessed 2012 April 18.

27 USPSTF. Screening for Cervical Cancer: U.S. Preventive Services Task Force Recommendation Statement (draft). 2011 Oct 18. http://www.uspreventiveservicestaskforce.org/draftrec4.htm

28 Gostin LO. Mandatory HPV vaccination and political debate. *JAMA*. 2011 Oct 19;306(15):1699-1700.

29 Jason L. Schwartz. HPV Vaccination's Second Act: Promotion, Competition, and Compulsion. American Journal of Public Health: October 2010, Vol. 100, No. 10, pp. 1841[en dash]1844. doi: 10.2105/AJPH.2010.193060. http://ajph.aphapublications.org/doi/abs/10.2105/AJPH.2010.193060.

30 U.S. Cancer Statistics Working Group. *United States Cancer Statistics: 1999–2007 Incidence and Mortality Web-based Report.* Atlanta (GA): Department of Health and Human Services, Centers for Disease Control and Prevention, and National Cancer Institute; 2010. http://www.cdc.gov/uscs. (full site)

31 Gostin LO. Mandatory HPV vaccination and political debate. *JAMA*. 2011 Oct 19;306(15):1699-1700.

32 Centers for Disease Control and Prevention. Genital HPV infection—fact sheet. http://www.cdc.gov/std/hpv/stdfact-hpv.htm.Accessed 2012 April 17.

33 Based on population data and average prices according to the National Cancer Institute's Human papilloma virus (HPV), cancer, and HPV vaccines—frequently asked questions. http://www.cancer.org/Cancer/CancerCauses/OtherCarcinogens/InfectiousAgents/HPV/HumanPapillomaVirusandHPVVaccinesFAQ/hpv-faq-vaccine-cost. Accessed 2012 April 17.

34 Calculation based on number needed to treat, Moore MA. What is an NNT? Bandolier. http://www.medicine.ox.ac.uk/bandolier/painres/download/whatis/NNT.pdf. Accessed 2012 April 18.

35 Moore MA. What is an NNT? Bandolier. http://www.medicine.ox.ac.uk/bandolier/painres/download/whatis/NNT.pdf. Accessed 2012 April 18.

36 National Cancer Institute. Age-adjusted SEER Incidence rates by cancer site all ages, all races, female 1975-2009 (SEER 9). http://seer.cancer.gov/faststats/selections.php?#Output. Accessed 2012 April 18.

37 HPV vaccines—frequently asked questions. http://www.cancer.org/Cancer/CancerCauses/OtherCarcinogens/InfectiousAgents/HPV/HumanPapillomaVirusandHPVVaccinesFAQ/hpv-faq-vaccine-cost. Accessed 2012 April 17.

38 Advisory Committee on Immunization Practices. Recommended adult immunization schedule: United States, 2012.*Ann Intern Med.* 2012 Feb 7;156(3):211-217.

39 Public Law 99-660—Nov. 14, 1986. 100 Stat. 3773. *Wikisource.* http://en.wikisource.org/wiki/Page:United_States_Statutes_at_Large_Volume_100_Part_5.djvu/299

40 National Vaccine Injury Compensation Program. National Law Library. http://www.jurisearch.com/NLLXML/getcode.asp?datatype=S&statecd=US&sessionyr=2008&TOCId=41074&userid=PRODSG&cvfilename=&Interface=&noheader=0

41 Hernandez N. Get kids vaccinated or else, parents told. *Washington Post.* 2007 Nov 14.

42 Serkes K. Doctors oppose Maryland vaccine roundup: expect dangerous reactions when children are treated like cattle. *Association of American Physicians and Surgeons.* 2007 Nov 16. http://www.aapsonline.org/press/nr-11-16-07.php

43 Neale T. Calif. 12-year-olds can get HPV vaccine without parental OK. 2011 Oct. 11.http://www.medpagetoday.com/Pediatrics/Vaccines/28987.

CHAPTER 25: A DOCTOR'S VIEW OF VACCINES AND THE PUBLIC HEALTH —SHERRI TENPENNY, DO

1 James Colgrove, *State of Immunity, the Politics of Vaccination in Twentieth-Century America* (Berkeley: University of California Press 2006), 8.

2 Patient of Dr. Sherri Tenpenny. Her name and specific details were not included to protect privacy.

3 Creation of Vaccine Information Statements (VIS) became a legal mandate under the 1986 National Childhood Vaccine Injury Act, 42 U.S. C. 300aa-27. The CDC makes individual VIS statements available at www.cdc.gov/vaccines/pubs/vis/default.htm#hib.

4 DTaP VIS statement, www.cdc.gov/vaccines/pubs/vis/downloads/vis-dtap.pdf.

5 Statement of David Satcher, MD, PhD., Assistant Secretary for Health and Surgeon General, Department of Human Services before The U.S. House of Representatives Committee on Government Reform. August 3, 1999.

6 Colgrove, *State of Immunity*, 98.

7 "Gail Russell Chaddock, "One Maryland county takes tough tack on vaccinations," *Christian Science Monitor*, Nov. 19, 2007.

8 A Neglect Proceeding under Section 1012, 1031 of the Family Court Act, *Nassau County Dep't of Social Services, Petitioner, on behalf of A.Y., Y.Y., Y.Y. v. R.B., Respondent*, December 18, 2008.

9 "Measles—United States." *Morbidity and Mortality Weekly Report.* 26 (1977): 109–111.

10 Three doses of each of these vaccines: Hib (for *H. influenzae* b), Prevnar (for streptococcus), hepatitis B, rotavirus. Two doses of chickenpox and MMR; one dose of hepatitis A and meningitis. The vaccine for human papillomavirus is mandated in the state of Virginia and there is pending legislation to mandate it in over twenty additional states. The federal government recommends it for sixth grade children.

11 Table "Exemptions from School Immunization Requirements," The National Conference of State Legislatures, www.ncsl.org/IssuesResearch/Health/SchoolImmunizationExemptionLaws/tabid/14376/Default.aspx.

12 Vaccines that contain gelatin include TriPedia, TriHiBit, pediatric Fluzone, adult Fluzone, FluMist, ProQuad, Varivax (chickenpox), Zostavax (shingles), MMR, yellow fever, rabies, Japanese encephalitis.

13 How gelatin is made, www.madehow.com/Volume-5/Gelatin.html.

14 Sakaguchi M, Toda M, et al, "IgE antibody to fish gelatin (type I collagen) in patients with fish allergy," *J Allergy Clin Immunol.*, 2000 Sep;106(3):579–84; Kelso JM, Jones RT, Yunginger JW, "Anaphylaxis to measles, mumps, and rubella vaccine mediated by IgE to gelatin," *J Allergy Clin Immunol.* 1993 Apr; 91(4):867–72; Coop CA, Balanon SK, et al, "Anaphylaxis from the influenza virus vaccine," *Int Arch Allergy Immunol.* 2008; 146(1):85–8. Epub 2007 Dec 14; and Ozaki T, Nishimura N, et al, "Safety and immunogenicity of gelatin-free varicella vaccine in epidemiological and serological studies in Japan," *Vaccine.* 2005 Jan 26; 23(10):1205–8.

15 Sakaguchi M, et al, "Food allergy to gelatin in children with systemic immediate-type reactions, including anaphylaxis, to vaccines," *Journal of Allergy and Clinical Immunology*, 1996 Dec; 98(6 Pt 1):1058–61.

16 Singer, Sanford, et. al, "Urticaria following varicella vaccine associated with gelatin allergy," *Vaccine* 17 (1999) 327–329.

17 Chief Editors, John M. Kelso, MD and James T. Li, MD, PhD, "Adverse reactions to vaccines," *Annals of Allergy, Asthma and Immunology*, 103 (October 2009):83.

18 Insect stings. www.insectstings.co.uk/waspsting.shtml.

19 Thimerosal is nearly 50 percent ethylmercury by weight and continues to be used in the manufacturing process of many vaccines. After filtration, vaccines still have "trace" amounts of thimerosal equivalent to one-half of one microgram or 2,000 parts per billion (ppb). Two hundred ppb meets the EPA classification for hazardous waste and 2 ppb is the EPA limit for safe drinking water. See Michael F. Wagnitz, eLetter, "Reputation of Vaccines: Lessons and the People Who Defend Them," March 13, 2008, in response to Rahul K. Parikh, "Fighting for the Reputation of Vaccines: Lessons from American Politics,"

Pediatrics 2008, 121:621–622, http://pediatrics.aappublications.org/cgi/eletters/121/3/621#36839, and see Taylor, chapter eighteen.

20 What You Need to Know About Mercury in Fish and Shellfish, EPA-823-R-04-005, www.fda.gov/food/foodsafety/product-specificinformation/seafood/foodbornepathogenscontaminants/methylmercury/ucm115662.htm.

21 Zatta PF, Alfrey AC. (Eds), *Aluminium Toxicity in Infants' Health and Disease*, 1997. World Scientific Publishing. See chapter by Simmer, K, "Aluminium in Infancy."

22 Aluminum Toxicity in Infants and Children (RE9607), Pediatrics Volume 97, Number 3 March, 1996, pp. 413–416, http://aappolicy.aappublications.org/cgi/reprint/pediatrics;97/3/413.pdf.

23 Merck package insert information, as of September 28, 2010.

24 Wyeth package insert.

25 FDA Code of Federal Regulations Title 21. Vol. 7. PART 610 -- GENERAL BIOLOGICAL PRODUCTS STANDARDS. Subpart B--General Provisions Sec. 610.15 Constituent materials.

26 Letter to the Editor. "All Star Pediatrics' Vaccine Policy Statement." *AAP News*. May, 2008. Vol. 28. No. 9.

27 Ibid.

28 Kelso and Li, note 17 above.

29 Cannell JJ, Hollis BW, "Use of vitamin D in clinical practice," *Altern Med Rev.*13(1) (March 2008):6–20.

CHAPTER 26: A HOLISTIC HEALTH PERSPECTIVE —ANNEMARIE COLBIN, PHD

1 H. Lindlahr, *Philosophy of Natural Therapeutics*, 1918, quoted in Leon Chaitow, *Vaccination and Immunization: Dangers, Delusions, and Alternatives (What every parent should know)*, (Essex, Great Britain: The C.W. Daniel Company Limited, 1987), 4–5.

2 Wilson, Sir Graham S., MD, Honorary Lecturer in the Department of Bacteriology at the London School of Hygiene and Tropical Medicine, *The Hazards of Immunization* (London: The Athlone Press,1967), 3.

3 J. John and R. Samuel, "Herd immunity and herd effect: new insights and definitions," *European Journal of Epidemiology*, 16:7:601-606, DOI: 10.1023/A:1007626510002.

4 Two recent books to consider are Dr. Mayer Eisenstein, *Make An Informed Vaccine Decision For the Health of Your Child (Santa Fe, New Mexico: New Atlantean Press, 2010)* and Dr. Robert Sears, *The Vaccine Book: Making the Right Decision For Your Child.* (New York: Little, Brown and Company, 2007).

5 Vaccine Excipient and Media Summary, Centers for Disease Control and Prevention, Part 1 and Part 2, March 2010, www.cdc.gov/vaccines/pubs/pinkbook/downloads/appendices/B/excipient-table-1.pdf and www.cdc.gov/vaccines/pubs/pinkbook/downloads/appendices/B/excipient-table-2.pdf.

6 Maurice R. Hilleman, "Yeast Recombinant Hepatitis B Vaccine," *Infection*, Volume 15, Number 1, 3-7, DOI: 10.1007/BF01646107, 1987, www.springerlink.com/content/x3404r658780146n/fulltext.pdf.

7 Thomas McKeown, *The Modern Rise of Population* (New York: Academic Press, 1976).

8 Leonard A. Sagan, MD, *The Health of Nations: True causes of sickness and well-being*, (New York: Basic Books, 1987), 68.

9 Walene James, *Immunization: The reality behind the myth* (Westport, Conn.: Bergin and Garvey, 1995).

10 Sotille, Robert, D.C., "Mandatory Immunization and You?" Life Foundation, Inc., Marietta, GA.

11 Poland, GA; Jacobson, RM, "Failure to reach the goal of measles elimination. Apparent paradox of measles infections in immunized persons," Department of Internal Medicine, Mayo Vaccine Research Group, Mayo Clinic and Foundation, Rochester, MN, *Archives of Internal Medicine*, August 22, 1994; 154(16):1815–20.

12 Celia Christie et al, "The 1993 Epidemic of Pertussis in Cincinnati—Resurgence of Disease in a Highly Immunized Population of Children," *N. Engl. J. Med.* 331:16-21 (July 7, 1994), http://www.nejm.org/doi/full/10.1056/NEJM199407073310104.

13 Christopher Dela Cruz, "Confirmed cases of whooping cough in Hunterdon County increase to 27," *The Star Ledger*, February 21, 2009, www.nj.com/news/index.ssf/2009/02/whooping_cough_outbreak_in_hun.html.

14 Anemona Hartocollis, "Jewish Youths Are At Center of Outbreak of Mumps," *New York Times*, February 11, 2010, www.nytimes.com/2010/02/12/nyregion/12mumps.html.

15 Leon Chaitow, *Vaccination and Immunization*, note 1 above, 4–5.

16 Viera Scheibner, *Vaccination: The Medical Assault on the Immune System* (Victoria, Australia: McPherson's Printing Group, 1993), 205-15.

17 Robert Mendelsohn, MD, *How to Raise a Healthy Child . . . In Spite of Your Doctor*, (New York: Ballantine Books, 1984), 210.

18 Leonard Sagan, MD, *The Health of Nations: True causes of sickness and well*-being, (New York: Basic Books, 1987), 68.

19 McKeown, Thomas, PhD, "Determinants of Health," *Human Nature*, April 1978, 60–67.

20 Farago O. Hanninen et al, "Ignaz Philipp Semmelweis, the Prophet of Bacteriology," *Infect. Control*, 1983 Sept-Oct; 4(5):367:70, http://web.archive.org/web/20080404214853/www.general-anaesthesia.com/semmelweis.htm. The Semmelweis story is told in William Broad and Nicholas Wade, *Betrayers of the Truth: Fraud and Deceit in the Halls of Science*, (New York: Simon and Schuster, 1982); see Taylor, chapter eighteen.

21 Defined as "the automatic rejection of the obvious, without thought, inspection, or experiment." The expression is attributed to author Robert Anton Wilson.

22 Levy, DL, "The Future of Measles in Highly Immunized Populations," *American Journal of Epidemiology*, 1984;120:39–48.

23 "Outbreak of paralytic poliomyelitis in Oman; evidence for widespread transmission among fully vaccinated children," *Lancet*, Sept. 21, 1991;(338);715-720.

24 James Gleick, *Chaos: Making a New Science*, (New York: Penguin Books, 1987), 79.

25 Wiersbitzky, S, Schroder, C, Griefahn, B, et al., "Encephalitis after simultaneous DPT and oral trivalent poliomyelitis (Sabin vaccine) and HiB preventive vaccination," Ernst Moritz Arndt Universität Zentrum für Kinder und Jugendmedizin Klinik und Poliklinik für Kindermedizin, Greifswald, Germany, *Kinderarztl Praxis*, June 1993;61(4-5):172–3; and Chen, Y, "Neural complications of cell culture rabies vaccine prepared from hamster kidney. Clinicopathological report of a case," Nanjung Neuropsychiatric Hospital, Chung Hua Shen Ching Ching Shen Ko Tsa Chih, Aug. 1991;24(4):242–3, 254.

26 The illness depicted in the movie *Lorenzo's Oil*.

27 Harris L. Coulter, *Vaccination, Social Violence, and Criminality: The Medical Assault on the American Brain* (Berkeley, CA: North Atlantic Books, 1990), 157-58.

28 Goldstein, Gary W. and Betz, A Lorris, "The Blood Brain Barrier," *Scientific American*, September 1986.

29 A popular movie, *Awakenings*, deals with post-encephalitic patients of the 1920s.

30 Annell, Anna-Lisa, "Pertussis in infancy — a cause of behavioral disorders in children," *Acta Societatis Medicorum Upsaliensis* LVIII, Supp. 1 (1953), quoted in Coulter, *Vaccination, Social Violence*, note 27, 160.

31 The Emmy-award winning documentary "DPT: Vaccine Roulette" was produced by Lea Thompson at WRC-TV in Washington D.C. in April 1982. Lea Thompson is now a Dateline NBC reporter www.msnbc.msn.com/id/3949442/.

32 Mendelsohn, *How to Raise a Healthy Child*, note 17 above, 209–230.

33 Buttram, Harold E., MD and Hoffman, John Criss, *Vaccinations and Immune Malfunction*, (Quakertown, PA: The Humanitarian Publishing Co.,1982), 42.

34 Live or attenuated viral vaccines include the oral (Sabin) polio, measles, mumps, rubella, varicella (chickenpox), influenza, and hepatitis A.

35 DeLong, Richard, *Live Viral Vaccines: Biological Pollution*, (New York: Carlton Press, 1996), 9.

36 DeLong, Richard, (letter), *Science News*, July 31, 1976.

37 Moriarty, TJ, "The Polio Vaccine and Simian Virus 40: After Thirty Years, Prominent Polio Vaccine Researcher Confirms Suspicions about Monkey Virus Contamination," OLNews@aol.com, December 12, 1996.

38 Kalokerinos, Archie, MD, *Every Second Child*, (New Canaan, CT: Keats Publishing, 1981), 120–150.

39 James Gleick, *Chaos*, note 24 above, 8.

40 Schumacher, W., "Legal/Ethical Aspects of Vaccinations," in *Immunization: Benefit Versus Risk Factors, Proceedings of the 36th Symposium, Organized by the International Association of Biological Standardization at the College of Medicine*, Université Catholique de Louvain, Woluwe St. Lambert, Belgium. November 15–17, 1978, published in *Developments in Biological Standardization*, vol. 43.

41 John Lantos, MD, Director, Children's Mercy Bioethics Center, Mercy Hospital, Kansas City, MO, www.seattlechildrens.org/research/initiatives/bioethics/events/pediatric-bioethics-conference/john-lantos-md/.

42 Health Letter, Public Citizen's Health Research Group, August 1995, 3. Excerpted from *Physician's Weekly*, July 10, 1995. See also, *BMJ* 310 (January 7,1995), www.ncbi.nlm.nih.gov/pmc/articles/PMC2548429/pdf/bmj00574-0006.pdf.

43 Bernard Guyer, Mary Anne Freedman, Donna M. Strobino and Edward J. Sondik, "Annual Summary of Vital Statistics: Trends in the Health of Americans During the 20th Century," DOI: 10.1542/peds.106.6.1307, *Pediatrics* 2000;106;1307–1317, www.ncbi.nlm.nih.gov/pubmed/11099582.

44 Gorman, Christine, "When the Vaccine Causes the Polio." *Time Magazine*, October 30, 1995.

45 Tetanus: Questions and Answers, Immunization Action Coalition, www.immunize.org/catg.d/p4220.pdf.

46 "How is Tetanus Treated?" *MedicineNet.com*, www.medicinenet.com/tetanus/page2.htm#6howis.

47 Seventy-seven percent of World Health Organization member countries with low (<2%) chronic prevalence of hepatitis B virus infection administer the birth dose of hepatitis B vaccine. See "Implementation of Newborn Hepatitis B Vaccination—Worldwide, 2006," *Morbidity and Mortality Weekly Report*, Centers for Disease Control and Prevention, November 21, 2008, www.cdc.gov/mmwr/preview/mmwrhtml/mm5746a1.htm.

48 Tove Ronne, "Measles Virus Infection Without Rash in Childhood is Related to Disease in Adult Life," *Lancet*, 325:8419 (Jan. 5, 1985): 1-5, http://www.thelancet.com/journals/lancet/article/PIIS0140-6736(85)90961-4/abstract.

49 Rudolf Steiner, *Autobiography: Chapters in the Course of My Life: 1861-1907*, (Lantern Books, 2006 ISBN 088010600X), xvi.

40 Personal communication to the author, June 3, 1992.

51 Payer, Lynn, *Medicine and Culture: Varieties of Treatment in the United States, England, West Germany, and France*, (New York: Henry Holt and Company, 1988), 145.

52 Dr. Incao now practices in Crestone, Colorado.

53 Transcript of interview on file with author.

54 www.whale.to/vaccine/olmsted.html.

55 Lynne McTaggart, *The Vaccination Bible* (London, U.K.: What Doctors Don't Tell You Ltd, 1998), 15.

56 Personal communication to the author, August 23, 1995.

57 Christian Weber et al., "Polymorphism of Bordetella pertussis Isolates Circulating for the Last 10 Years in France, Where a Single Effective Whole-Cell Vaccine Has Been Used for More than 30 Years," *Journal of Clinical Microbiology*, December 2001, 4396-4403, vol. 39, no. 12 0095-1137/01/$04.00+0 DOI: 10.1128/JCM.39.12.4396-4403.2001, http://jcm.asm.org/cgi/reprint/39/12/4396; and "Whooping

Cough Returns to Hunterdon County," *The Star Ledger*, February 21, 2009, "The infected children had been vaccinated," www.nj.com/news/index.ssf/2009/02/whooping_cough_returns_to_hunt.html.

58 "Diphtheria, Tetanus, and Pertussis: What You Need to Know, Measles, Mumps, and Rubella: What You Need to Know," and "Polio: What You Need to Know," U.S. Department of Health and Human Services, Public Health Service, Centers for Disease Control, Atlanta, Georgia 30333: October 15, 1991.

59 In 2008, the American Academy of Pediatrics published a sample letter written by All Star Pediatrics in Pennsylvania in *AAPNews*, http://aapnews.aappublications.org/cgi/content/full/29/5/26-a, in which families choosing to modify the recommended vaccine schedule were asked to leave their practice. Member pediatric practices were encouraged to adapt this timesaving letter for their own use. www.immunize.org/catg.d/p2067.doc.

CHAPTER 27: WHAT SHOULD PARENTS DO? —LOUISE KUO HABAKUS, MA

1 Alexander Pope, *An Essay on Man, Moral Essays and Satires*, transcribed from the 1891 Cassell & Company edition, www.gutenberg.org/dirs/2/4/2/2428/2428.txt.

2 Letter to the Editor. "All Star Pediatrics' Vaccine Policy Statement." *AAP News*, 28:9 (May 2008).

3 The CDC's Section 317 Immunization Program is a discretionary federal grant program to all fifty states, six city/urban areas, and eight U.S. territories and protectorates whose purpose is to "support efforts to plan, develop, and maintain a public health infrastructure that helps assure high immunization coverage levels and low incidence of vaccine-preventable diseases. . . . The Section 317 Program remains a significant source of federal funding for most jurisdictional vaccine program operations. . . . CDC supports the immunization efforts of states by providing extramural support and funding through grants and contracts for vaccine purchase and operations/infrastructure activities. More than 90 percent of Section 317 Program funds are provided to states through grants." *Justification of Estimates for Appropriations Committees, Department of Health and Human Services Fiscal Year 2010*, Centers for Disease Control and Prevention, 35.

4 Ibid, 35.

5 Ibid, 38–39.

6 "How the Anti-Vaccine Movement Threatens America's Children," *American Academy of Pediatrics* Press Release, October 4, 2010, www.aap.org/pressroom/Offitfinalantivaccine.pdf.

7 Robert S. Mendelsohn, *How to Raise a Healthy Child...In Spite of Your Doctor* (New York: Ballantine Books, 1984), 233.

8 Kristine Severyn, "*Jacobson v. Massachusetts:* Impact on Informed Consent and Vaccine Policy", 5 *J. Pharmacy & Law*, 249, 270–72.

9 Paul A. Offit and Louis M. Bell, *Vaccines: What You Should Know* (John Wiley & Sons, Inc., 2003). The book included the following vaccine safety statements:

"With the new DTaP vaccine, [the incidence of side effects] is dramatically lower." 36.

"The diphtheria vaccine doesn't cause serious side effects." 41.

"The tetanus vaccine is safe." 44.

"IPV [is] completely safe." 48.

"Is the Hib vaccine safe? Yes. Side effects are mild." 53.

"The chance of serious side effects or death from the measles vaccine is about zero." 60.

"The mumps vaccine does not cause any serious side effects . . . the chance of having a serious reaction from the mumps vaccine is about zero." 62.

"The most serious side effect from rubella vaccine is the development of short-lived—less than one week, and not chronic—arthritis." 65.

"No one has ever died as a result of hepatitis B vaccination." 74.
"The [varicella] vaccine does not cause serious side effects." 82.
"The pneumococcal vaccine does not have serious side effects." 86.
"More than 1 million doses of the rabies vaccine have been given, and the vaccine is safe." 123.
"Because the [flu] vaccine does not cause serious reactions, the benefits of the vaccine clearly outweigh its risk in *all* young children." 130. (emphasis theirs)
"Is the meningococcal vaccine safe? Yes. There are no serious side effects from the vaccine." 133.
"The tuberculosis vaccine is safe."139.
"Although it is relatively new, hundreds of thousands of people have been given the hepatitis A vaccine without any serious side effects." 149.
"The typhoid vaccines used today are safe and fairly effective." 159.
"Side effects from the yellow fever vaccine are rare." 161.

10 Ibid, 18, 22.
11 Ibid, 22.
12 Ibid, 91.
13 Ibid, 179.
14 Ibid, 88–89.
15 Ibid, 101.
16 Ibid, 9.
17 Stephanie Cave, *What Your Doctor May Not Tell You About Children's Vaccinations* (New York: Wellness Central, 2010), xvi.
18 Ibid, 305.
19 Dr. Cave does not recommend the pneumococcal vaccine *Prevnar* for children. She does, however, administer an alternative pneumococcal vaccine, Pneumovax. Author's personal communication with Dr. Cave on December 1, 2010.
20 Aviva Romm, *Vaccinations: A Thoughtful Parent's Guide,* (Rochester: Healing Arts Press, 2001), 71.
21 Ibid, 131. The excerpted paragraph includes some revisions requested by Dr. Romm in her personal email correspondence with the author on December 1, 2010.
22 Randall Neustaedter, *The Vaccine Guide: Risks and Benefits for Children and Adults, Revised Edition,* (Berkeley: North Atlantic Books, 2002).
23 Ibid, xv.
24 Ibid, xiii.
25 Mayer Eisenstein with Neil Z. Miller, *Make an Informed Vaccine Decision for the Health of Your Child,* (Santa Fe: New Atlantean Press, 2010), 211.
26 Ibid,12–13.
27 Ibid, 13.
28 Ibid, 115.
29 Ibid.
30 Ibid, 9.
31 Ibid, 231.
32 Mendelsohn, *How to Raise a Healthy Child . . . In Spite of Your Doctor*, 230.
33 Ibid, 231–232.
34 Ibid, 247.
35 Ibid, 246.
36 Ibid, 233.
37 Ibid.
38 Ibid, 231.

39 Ibid, 242–243.

40 Country Comparison: Infant Mortality Rate, The World Factbook, Central Intelligence Agency, 2012 est., https://www.cia.gov/library/publications/the-world-factbook/rankorder/2091rank.html.

CHAPTER 28: WHO IS DR. ANDREW WAKEFIELD? —MARY HOLLAND, JD

1 *Journal of Medical Virology* 39 (1993), 345–53.

2 *The Lancet* 345 (1995): 1071–74.

3 William R. Long, "Historical Perspective: On Second Looking into the Case of Dr. Andrew J. Wakefield," *The Autism File* 31 (2009): 66–88, n. 14

4 Ibid, 68–73.

5 Ibid, 73.

6 Allan W. Walker, M.D., "A Tribute to Professor John Walker-Smith," *Journal of Pediatric Gastroenterology & Nutrition* 29:5 (1999), 14A, ("From that time [1973] until the present, John has established himself as one of the premier, if not the premier, pediatric gastroenterologists in the world. . . . His abilities as a clinician, clinical investigator, and educator through lectures, review articles, and textbooks have resulted in a worldwide following by former fellows, colleagues, and general pediatricians."), http://journals.lww.com/jpgn/fulltext/1999/11000/a_tribute_to_professor_john_walker_smith,_espghan.6.aspx.

7 Andrew J. Wakefield and James Moody, "Ethics, Evidence and the Death of Medicine, *The Autism File* 34 (2010), *reprinted in* Andrew J. Wakefield, *Callous Disregard: Autism and Vaccines – The Truth Behind a Tragedy*, (New York: Skyhorse Publishing, 2010), 233–46, 236 ("Diagnostic testing was justified by each child's medical history and clinical presentation. Such diagnostic investigations, indicated by clinical need, did not require Ethics Committee approval because they were undertaken in the patient's interest for the purpose of *establishing a diagnosis* and *directing treatment*."(emphasis in original)).

8 A. J. Wakefield et al., "Ileal-lymphoid-nodular hyperplasia, non specific colitis, and pervasive developmental disorder in children," *The Lancet* 351 (Feb. 28, 1998) 637-41, http://briandeer.com/mmr/lancet-paper.htm.

9 Long, "Historical Perspective," note 3 above, 74-75; Wakefield, *Callous Disregard*, note 7 above, 83–100.

10 Ibid 76, n.31.

11 See, e.g., Paul A. Offit, M.D, *Autism's False Prophets: Bad Science, Risky Medicine, and the Search for a Cure*, (New York: Columbia University Press, 2008) 22, citing March 1, 1998 British newspaper headlines of "Alert over Child Jabs," and "Ban Three-in-One Jab, Urge Doctors." For discussion about blaming Dr. Wakefield, see Long, note 3 above, 75–76.

12 Long, "Historical Perspective," note 3 above, 78–80.

13 General Medical Council, "Dr. Andrew Jeremy Wakefield: Determination on Serious Professional Misconduct (SPM) and sanction," May 24, 2010, www.gmc-uk.org/Wakefield_SPM_and_SANCTION.pdf_32595267.pdf.

14 "Comment: Retraction—Ileal-lymphoid-nodular hyperplasia, non specific colitis, and pervasive developmental disorder in children," *The Lancet* (Feb. 2, 2010), http://download.thelancet.com/flatcontentassets/pdfs/S0140673610601754.pdf.

15 General Medical Council, "Fitness to Practice Panel Hearing on Dr. Wakefield, Professor Walker-Smith, and Professor Murch," January 28, 2010, www.gmcuk.org/static/documents/content/Wakefield__Smith_Murch.pdf.

16 General Medical Council, note 13.

17 Polly Tommey, "Discredited Defamation: The Fallacious Case against Dr. Andrew Wakefield," *The Autism File* 34 (2010): 8–10; Long, note 3 above; Wakefield, note 7 above.

18 John Stone, "Smoke and Mirrors: Dr. Richard Horton and the Wakefield Affair," December 22, 2008, *Age of Autism*, at www.ageofautism.com/2008/12/smoke-and-mirrors-dr-richard-horton-and-the-wakefield-affair.html.

19 Long, "Historical Perspective," note 3 above, 70–73; Wakefield, *Callous Disregard*, note 7 above, 25–46.

20 General Medical Council, "Determination," note 13 above, 7 ("Dr. Wakefield caused blood to be taken from a group of children for research purposes at a birthday party, which the Panel found to be an inappropriate social setting. He behaved unethically in failing to seek Ethics Committee approval; he showed callous disregard for any distress or pain the children might suffer, and he paid the children 5 pounds reward for giving their blood. . . . The Panel concluded that his conduct brought the medical profession into disrepute.")

21 Professor John Walker-Smith v. General Medical Council, Committee and Professional Conduct Committee. The High Court of Justice, Queen's Bench Division, Administrative Court. Case No: CO/7039/2010. March 3, 2012.

22 Email from R. Horton to A. Wakefield, March 2012.

23 Ibid, 245.

24 Long, "Historical Perspective," note 3 above, 77–78.

25 Brian Deer published a series of stories in *The Sunday Times* on February 8, 2009: "MMR doctor Andrew Wakefield fixed data on autism," www.thesundaytimes.co.uk/sto/public/news/article148992.ece; "Hidden records show MMR truth," www.thesundaytimes.co.uk/sto/public/news/article148983.ece; and "How the MMR scare led to the return of measles," www.thesundaytimes.co.uk/sto/public/news/article149001.ece.

26 Although there was no direct link to Dr. Wakefield in the Omnibus Autism Proceeding before the U.S. Court of Federal Claims on whether vaccines cause autism, one of the technicians in Dr. Wakefield's laboratories, Dr. Chadwick, was permitted to testify. In closing argument, the lead attorney for the U.S. Department of Justice stated that "all the strands through these cases come back to him [Andrew Wakefield]. He presented bad science." Transcript of *Cedillo v. Sec'y of HHS*, June 26, 2007, 2898-99.

27 Wakefield, *Callous Disregard*, note 7 above, 181–222.

28 P. D'Eufemia, M. Celli, et al, "Abormal intestinal permeability in children with autism," *Acta Paediatrica*, 85 (1996): 1076–9; R. Furlano, A. Anthony *et al*, Colonic CD8 and γδ Tcell filtration with epithelial damage in children with autism," *J. Pediatrics*, 138 (2001): 366–372; "Immune activation of peripheral blood and mucosal CD3+ lymphocyte cytokine profiles in children with autism and gastrointestinal symptoms," *J. Neuroimmunology* 173:1–2 (2006): 126–34; "The significance of ileo-colonic lymphoid nodular hyperplasia in children with autistic spectrum disorder," *European Journal of Gastroenterological Hepatology*, 17:8 (2005): 827–36; "Autistic enterocolitis: Is it a histopathological entity?-reply," *Histopathology* 50 (2007), 380–84; Lenny Gonzalez, "Gastrointestinal Pathology in Autism Spectrum Disorders: the Venezuelan Experience," *The Autism File* 32 (2009), 74–79.

29 *The Lancet*, "Comment: Retraction," note 14 above, states that "the claims in the original paper that children were 'consecutively referred' and that investigations were 'approved' by the local ethics committee have been proven to be false. Therefore we fully retract this paper from the published record."

30 See Mark Blaxill, "From the Roman to the Wakefield Inquisition," www.ageofautism.com/2010/01/from-the-roman-to-the-wakefield-inquisition.html.

31 James Moody, "Postscript," in Wakefield, *Callous Disregard*, note 7 above, 269.

32 Thomas S. Kuhn, *The Structure of Scientific Revolutions* (1962).

CHAPTER 29: THE EXONERATION OF PROFESSOR JOHN WALKER-SMITH —DAVID L. LEWIS, PHD

1 Vaccine Safety: Evaluating the Science, Jamaica, West Indies, (2011 January 3-8), http://www.vaccinesafetyconference.com/index.html.

2 AJ Wakefield, SH Murch, A Anthony, J Linnell, DM Casson, M Malik, M Berelowitz, AP Dhillon, MA Thomson, P Harvey, A Valentine, SE Davies, JA Walker-Smith. (1998) Ileal lymphoid nodular hyperplasia, non-specific colitis and pervasive developmental disorder in children, *Lancet*, 351(9713):637-641. [Retracted]

3 B Deer, "How the vaccine crisis was meant to make money," *BMJ*, 2011;342:c5258; B Deer, "How the case against the MMR vaccine was fixed," BMJ, 2011;342:c5347; and B Deer, "The *Lancet*'s two days to bury bad news," *BMJ*, 2011;342:c7001.

4 DL Lewis (2011), "Letter to the *BMJ* from David Lewis," Rapid response, http://www.bmj.com/rapid-response/2011/11/09/re-how-case-against-mmr-vaccine-was-fixed; F Godlee, "Institutional research misconduct," 2011;Nov 9;343:d7284, http://www.bmj.com/content/343/bmj.d7284?tab=full; B Deer, "Pathology reports solve 'new bowel disease' riddle," *BMJ* 2011;343:d6823, http://www.bmj.com/content/343/bmj.d6823; K Geboes, "I see no convincing evidence of 'enterocolitis, 'colitis,' or a 'unique disease process,'" *BMJ* 2011;343:d6985, http://www.bmj.com/content/343/bmj.d6985; and I Bjarnason, "We came to an overwhelming and uniform opinion that these reports do not show colitis," http://www.bmj.com/content/343/bmj.d6979.

5 ES Reich, "Fresh dispute about MMR 'fraud,'" Nature 2011;479:157-158.

6 DL Lewis (2012 January 8), "Apparent egregious ethical misconduct by British Medical Journal, Brian Deer," UK Research Integrity Office (UKRIO), Reference No. 2011-060, http://www.hallmanwingate.com/fullpanel/uploads/files/david-lewis-bmj.pdf.

7 R Horton (2004 February 20), "A statement by the editors of The Lancet," http://image.thelancet.com/extras/statement20Feb2004web.pdf; and B Deer (2004 February 22), "Revealed: MMR research scandal," The Sunday Times (London).

8 Transcript of the shorthand notes of TA Reed & Co., Ltd. (2010), Day 17-14; and JP Heptonstall (2004 13 April), "Re: Parliamentary Protection and Open Science, Rapid response to 'MP raises new allegations against Andrew Wakefield,'" *BMJ*, http://www.bmj.com/rapid-response/2011/10/30/re-parliamentary-protection-and-open-science.

9 R Horton, note 7 above.

10 "Father of 'Paediatric Gastroenterology' Appealing to High Court for Justice," Top News, February 12, 2012, http://topnews.us/content/246249-father-paediatric-gastroenterology-appealing-high-court-justice.

11 H Hodgson (2004 February 20), "A statement by the Royal Free and University College Medical School and the Royal Free Hampstead NHS Trust," http://image.thelancet.com/extras/statement20Feb2004web.pdf.

12 U.K. Parliament (2004 March 15) "MMR Vaccinations and Autism," Dr. Evan Harris (Oxford, West and Abingdon) at 128, 10.2 pm, http://www.publications.parliament.uk/pa/cm200304/cmhansrd/vo040315/debtext/40315-34.htm.

13 R Barr to R Horton (1997 April 3), 21; and R Horton to R Barr (1997 April 8).

14 B Deer, note 3 above.

15 CA Tarhan to AR Zuckerman (1997 May 20).

16 K Emmerson, M Lohn (2006 August 7), Field, Fisher, Waterhouse, LLP, General Medical Counsel meeting with A. Zuckerman.

17 S Kraus, London Strategic Health Authority, National Health Services (NHS) to CG Miller, (2007 January 15), A description of the ten enclosed documents by DL Lewis (2012) follows: (1) [Letter] John Walker-Smith to Maureen Carroll, Royal Free Hospital Ethics Committee (1997 February 27) in which J Walker-Smith states: "We currently have formal approval to take research biopsies during colonoscopy (code 162-95) and I am writing to organize formal approval for research biopsies to be taken during upper biopsies," (2) [Letter] AD Phillips to M Pegg, Royal Free Hospital and Medical School Ethics Committee (2000 March 15) in which Phillips requests an "updated approval" for continuing to take research biopsies in studies 162-95 (colonoscopy) and 70-97 (upper endoscopy), (3) Same as No. 2 above with handwritten note from Pegg to Carroll requesting she check on the status of the approvals, (4) [Memo] Carroll to Pegg (2000 April 7) re. "Extension to 162-95 & 70-97," in which Carroll states: "...no requests have been received previously for the continuation of these studies... could you please approve this by way of Chairman's Action and return to me," (5) [Letter] Pegg to Phillips (2000 April 28) in which Pegg acknowledges Phillip's March 15 letter and requests a brief annual report for studies 162-95 and 70-97, (6) [Letter] Phillips to Pegg (2000 May 17), which is a cover letter transmitting "1999 Annual Report on Ethical Submissions 162-95 and 70-97," (7) [Report] Phillips to Pegg (2000 May 17), "1999 Annual Report on Ethical Submissions 162-95 and 70-97," (8) Same as No. 6 above with handwritten note from Pegg to Carroll requesting she "create an updated approval letter" and stating: "I think I should sign it," (9) [Letter] Pegg to Phillips (2000 May 25), in which Pegg acknowledges receiving the annual report on the taking of research biopsies, and conveys the committee's approval for Walker-Smith's group to continue taking research biopsies in studies 162-95 and 70-97, and (10) a blank, undated parental consent form titled "CONSENT FORM FOR RESEARCH BIOPSIES."

18 J Walker-Smith to M Carroll, Ethical Practices Subcommittee (1997 February 27).

19 AD Phillips to MS Pegg, Royal Free Hospital Medical School Ethics Committee (2000 May 17).

20 J Walker-Smith, S Murch, AJ Wakefield (1996 August 6), Ethical Practices Subcommittee Protocol and Pro Forma, "A new pediatric syndrome: Enteritis and degenerative disorder following measles/rubella vaccination."

21 MS Pegg to J Walker-Smith (1997 January 7).

22 General Medical Counsel, Exhibit 86(a) (See TA Reed & Company, note 8 above, Day 8-54, which is a letter from J Walker-Smith to Baroness Gardner, Chair, Ethics Committee, Royal Free Hospital (1995 August 24), which states: "The parents have signed a form as attached granting permission. These biopsies are used for a variety of 'research' investigations such as cytokine production where on occasion information of direct and immediate importance to the child's illness has been obtained as well as of research importance. I would be very grateful if you would grant permission for this to continue after our move to the Royal Free;" General Medical Counsel, Exhibit 86(b) (See TA Reed & Company, note 8 above, Day 8-55, which is a blank parental consent form attached to Exhibit 86(a)); GMC Exhibit 86(c) (See TA Reed & Company, note 8 above, Day 8-54 and 55), which is a letter from M Carroll, Royal Free Hospital Ethics Committee, to J Walker-Smith (1995 September 5) stating: "Re The taking of two extra mucosal biopsies for research purposes during the course of colonoscopy in children. I am pleased to be able to inform you that your recent submission to the Ethical Practices Sub-Committee has now received approval by Chairman's Action. This approval will be formally documented at the next meeting of the full committee and meanwhile you are free to carry out the above procedure at the Royal Free;" and TA Reed & Company, Day 12-24, note 8 above.

23 T.A. Reed & Company, Day 9-25, note 8 above.

24 T.A. Reed & Company, Day 8-54 and Day 73-22, note 8 above.

25 G Null, Progressive Radio Network (2011 January 25), interview with B Deer.

26 *Andrew Wakefield v. Channel Four Television Corporation, Twenty Twenty Productions, Ltd. and Brian Deer*, High Court of Justice, Queen's Bench Division, Case No. HQ05X00900.

27 TA Reed & Company, Day 68-57 thru 59, note 8 above.

28 TA Reed & Company, Day 197-2 and 3, note 8 above; General Medical Council, Committee and Professional Conduct Committee (UK) (2010 May 24), "Dr. Andrew Jeremy Wakefield determination on serious professional misconduct (SPM) and sanction," http://www.gmc-uk.org/Wakefield_SPM_and_SANCTION.pdf_32595267.pdf; and General Medical Council, Committee and Professional Conduct Committee (UK). (2010 May 24), "Professor John Angus Walker-Smith determination on serious professional misconduct (SPM) and sanction," http://www.gmc-uk.org/Professor_Walker_Smith_SPM.pdf_32595970.pdf.

29 TA Reed & Company. Day 147-62, note 8 above.

30 TA Reed & Company. Day 136-33, note 8 above.

31 Minutes of the Joint Committee on Vaccination and Immunisation (JCVI 1991-2000) (1992 1 May), Section 7.4, "Report of North Herts Immunogenicity Study (Dr Elizabeth Miller)," http://www.dh.gov.uk/ab/JCVI/DH_095050.

32 U.K. Parliament (2004 March 15), "MMR Vaccinations and Autism," Dr. Evan Harris at Column 128, 10.2 pm. http://www.publications.parliament.uk/pa/cm200304/cmhansrd/vo040315/debtext/40315-35.htm

33 General Medical Council, note 27 above.

34 briandeer.com, "Brian Deer: Solved - the riddle of MMR"; http://briandeer.com/solved/solved.htm, "Judged against a criminal standard of sureness, he was found guilty on four counts of dishonesty and 12 counts involving the abuse of autistic children;" and F Godlee, note 4 above.

35 CNN. *Anderson Cooper 360°* (2011 January 6), interview with B Deer, http://edition.cnn.com/2011/HEALTH/01/06/autism.vaccines/index.html.

36 F Godlee, J Smith, H Marcovitch, "Wakefield's article linking MMR vaccine and autism was fraudulent." *BMJ* 2011:342:7452.

37 CNN. *Anderson Cooper 360°* (2011 January 5), interview with A Wakefield, http://www.cnn.com/2011/HEALTH/01/05/autism.vaccines/index.html.

38 CNN (2011 February 4), "Bill Gates: Vaccine-autism link 'an absolute lie,'" http://www.cnn.com/2011/HEALTH/02/03/gupta.gates.vaccines.world.health/index.html

39 B Deer, note 3 above; Godlee et al., note 35 above; and DJ Opel, DS Diekema, EK Marcuse, "Assuring research integrity in the wake of Wakefield," *BMJ* 2011; 342:d2.

40 General Medical Council, note 27 above.

41 B Deer, note 3 above.

42 L Striukova (2009), "Value of University Patents as a Determinant of Technology Transfer," International Journal of Technology Transfer and Commercialization, 8, 379.

43 U.K. Intellectual Property Office (1997 June 5), Patent Application 9711663.6, "Pharmaceutical composition for treatment of IBD and RBD," Royal Free Hospital School of Medicine, London, Neuroimmuno Therapeutics Research Foundation, Spartanburg, SC USA [undisclosed inventors].

44 U.K. Intellectual Property Office (1998 September 12). Patent Application GB 2325856 A, "Pharmaceutical composition for treatment of MMR virus mediated disease comprising a transfer factor obtained from the dialysis of virus-specific lymphocytes," Royal Free Hospital School of Medicine, London, Neuroimmuno Therapeutics Research Foundation, Andrew Jeremy Wakefield and Hugh Fundenberg, Inventors (Filed Apr. 6, 1998).

45 U.K. Intellectual Property Office (2002 March 2). "Legal status of GB2325856 (A) 1998-12-09:GB, F 9812056 A (Patent of invention) PRS Date: 2002/03/06. PRS Code: WAP Code ... Application withdrawn, taken to be withdrawn or refused after publication under Section 16(1)."

46 SH Polmar (1973), Transfer-Factor Therapy of Immunodeficiencies. *New Eng. J. Med.*, 289(26):1420-1421.

47 B Deer, "Wakefield's 'Autistic Enterocolitis' under the microscope." *BMJ* 2010;340:c1127.

48 Ibid.

49 AP Dhillon (1998), Biopsy score sheets published by the *BMJ*, Nov. 9, 2011. http://www.bmj.com/highwire/filestream/536428/field_highwire_adjunct_files/0

50 DL Lewis, Photomicrographs, note 6 above.

51 DL Lewis, A. Anthony's *PowerPoint* presentation, note 6 above.

52 Godlee et al., note 35 above; Deer, note 46 above.

53 I Booth (2006 November 8), General Medical Council, Fitness To Practice Panel (Misconduct). Wakefield, Walker-Smith, Murch. Second Addendum to Overview Statement.

54 I. Booth (2011 August 10), email to DL Lewis.

55 Godlee et al., note 35 above.

56 B Deer to S Kohn, June 2, 2011, note 6 above.

57 Ibid.

58 T Delamothe to DL Lewis, cc: F. Godlee. October 27, 2011, note 6 above.

59 DL Lewis, note 4 above.

60 ES Reich, "Fresh dispute about MMR 'fraud.'" *Nature* 2011;479:157-158.

61 F Godlee, note 4 above.

62 Ibid.

63 U.K. Parliament 2011 November 10), "UCL and the work of Dr Andrew Wakefield." http://www.parliament.uk/business/committees/committees-a-z/commons-select/science-and-technology-committee/news/111110-ucl--wakefield/

64 DL Lewis, Attachment 4, note 6 above.

65 Ibid.

66 Ibid.

67 DL Lewis, Attachment 1, p.63, note 6 above.

68 Ibid, p.70

69 Ibid, p.49-50

70 F Godlee, note 4 above.

71 B Deer, note 4 above.

72 DL Lewis (2011), "response to Brian Deer...," note 57 above.

73 C Laine, Editor, *Annals of Internal Medicine*, to DL Lewis. May 15, 2010.

74 *Andrew Wakefield v. Channel Four Television*, et al., note 25 above.

75 L Hewitson, LA Houser, C Stott, G Sackett, JL Tomko, D Atwood, L Blue, ER White, AJ Wakefield. (2009) Delayed acquisition of neonatal reflexes in newborn primates receiving a thimerosal-containing Hepatitis B vaccine: Influence of gestational age and birth weight, *Neurotoxicology* doi:10.1016/j.neuro.2009.09.008, http://www.whale.to/vaccine/primates_hep_b-1.pdf.

76 *Andrew Wakefield v. Channel Four Television, et al.*, Judgment (2005 November 4), http://www.bailii.org/ew/cases/EWHC/QB/2005/2410.html, note 25 above.

77 "Retraction—Ileal-lymphoid-nodular hyperplasia, non-specific colitis, and pervasive developmental disorder in children." *Lancet* 375:445. doi:10.1016/S0140-6736(10)60175-4.

78 L Hewitson, et al. (2009), http://www.ncbi.nlm.nih.gov/pubmed/19800915, note 74 above.

79 *Dr. Andrew J. Wakefield v. The British Medical Journal, Brian Deer and Fiona Godlee*, District Court for the 250th District, Travis, TX, Case No. D-1-GN-12-000003, January 3, 2012.

80 *Professor John Walker-Smith v. General Medical Council, Committee and Professional Conduct Committee.* The High Court of Justice, Queen's Bench Division, Administrative Court. Case No: CO/7039/2010. March 3, 2012.

81 Ibid at 20.
82 Ibid at 24-49.
83 Ibid at 50.
84 Ibid at 85, 186.
85 Ibid at 148,186.
86 Ibid at 70.
87 Ibid at 72, 170.
88 Ibid at 77, 183.
89 Ibid at 85, 185.
90 Ibid at 18.
91 Ibid at 63, 150.
92 Ibid at 15, 23.
93 Ibid at 16.
94 TGD França, LLW Ishikawa, SFGZ Pezavento, FC Minicucci, MLRS Cunha, A Sartori. Impact of malnutrition on immunity and infection, *J Venom Anim Toxins incl Trop Dis.* 2009;15(3):374-90.
95 DL Lewis (2012), Personal communication with AJ Wakefield.
96 CNN. *Anderson Cooper 360º*, note 36 above.
97 F Godlee, note 4 above.
98 F Godlee (2012), Correction to "Wakefield's article linking MMR vaccine and autism was fraudulent," *BMJ* 2011;342:d1678, http://www.bmj.com/content/342/bmj.d1678.
99 F Godlee (2012), http://www.bmj.com/content/342/bmj.d1335?tab=responses.
100 CNN. *Anderson Cooper 360º*, note 36 above.
101 *Professor John Walker-Smith v. General Medical Council*, note 78 above.
102 PR Ziring, Congenital rubella: the teenage years, *Pediatr Ann.* 1997, 6:762–770; S Chess, Follow-up report on autism in congenital rubella, *J Autism Child Schizophr*, 1977,7:69 –81; and S Chess, "Autism in children with congenital rubella," *J Autism Child Schizophr* 1971, 1(1):33-47.
103 JS Brown, BP Kotler, RJ Smith, WO Wirtz II, The effects of owl predation on the foraging behavior of heteromyid rodents, *Oecologia* 1988; 76:408-415.
104 Uncorrected Transcript of Oral Evidence, House of Commons Science and Technology Committee, Peer Review, Wednesday 11 May 2011, http://www.publications.parliament.uk/pa/cm201012/cmselect/cmsctech/uc856-ii/uc85601.htm.
105 briandeer.com. "Brian Deer wins a second British Press Award" and "Award-Winning Journalism at *The Sunday Times.*" http://briandeer.com/brian/press-awards-2011-win.htm
106 Ibid.
107 BL Williams, M Hornig, T Buie, ML Bauman, M Cho Paik, I Wick, A Bennett, O Jabado, DL Hirschberg, WI Lipkin. Impaired carbohydrate digestion and transport and mucosal dysbiosis in the intestines of children with autism and gastrointestinal disturbances. *PLoS One.* 2011 September;6(9): e24585.
108 U.S. CDC. Prevalence of Autism Spectrum Disorders — Autism and Developmental Disabilities Monitoring Network, 14 Sites, United States, Morbid Mortal Wkly Rep, 2012;61(3):1-19; YS Kim, BL Leventhal, YJ Koh, E Fombonne, E Laska, EC Lim, KA Cheon, SJ Kim; and YK Kim, H Lee, DH Song, RR Grinker, Prevalence of autism spectrum disorders in a total population sample, *Am J Psychiatry*, 2011 September, 168(9):904-12.
109 DL Lewis, Institutional Research Misconduct: An Honest Researcher's Worst Nightmare, *Autism Sci Digest*, 2012(4):31-40.

CHAPTER 30: THE SUPPRESSION OF SCIENCE —ANDREW WAKEFIELD, MB, BS, FRCS, FRCPATH

1 Brian Martin, "How to Attack a Scientific Theory and Get Away with It (Usually): The Attempt to Destroy an Origin-of-AIDS Hypothesis," *Science as Culture*, 19:2 (June 2010):215-239, www.bmartin.cc/pubs/10sac.html.

2 Colville A and Pugh S, *Lancet* 340 (1992): 786. Farrington P et al., *Lancet* 345 (1995): 567–569; "Children received vaccine despite meningitis risk," *Independent Newspaper*, Sept. 15, 1992.

3 *MSNBC*, "1 in 4 Parents Thinks Shots Cause Autism," March 1, 2010, www.msnbc.msn.com/id/35638229/ns/health-kids_and_parenting/.

4 Center for Personal Rights, "Majority Support Parental Vaccination Choice," *PRNewswire*, May 24, 2010, www.prnewswire.com/news-releases/majority-support-parental-vaccination-choice-according-to-new-harris-poll-94723629.html.

APPENDIX TO CHAPTER 27: WHAT SHOULD PARENTS DO? —LOUISE KUO HABAKUS, MA

1 "Recommendations for combination vaccines modified," *Pediatric Supersite*, November 20, 2010, www.pediatricsupersite.com/view.aspx?rid=77904.

2 This does not include an extra dose for certain children during the 2010-2011 flu season: "All children 6 months through 8 years of age getting a flu vaccine for the first time need two doses four weeks apart. This includes children who received one or two doses of the 2009 H1N1 flu vaccine, but who have never received a seasonal flu vaccine." "What's new about the flu vaccine for the 2010-11 flu season?" Centers for Disease Control and Prevention, last updated October 25, 2010, www.cdc.gov/flu/protect/vaccine/fluvax_whatsnew.htm.

3 This does not include a fifth dose for certain children. Prevnar now contains thirteen serotypes of pneumococcus instead of seven. Children fully immunized with PCV7 should receive an extra single dose of PCV13. "On February 24, 2010 the FDA licensed a new pneumococcal conjugate vaccine (Prevnar 13™ (PCV13), Wyeth Pharmaceuticals Inc., a subsidiary of Pfizer Inc.) for children 6 weeks through 71 months of age. The same day, the Advisory Committee on Immunization Practices (ACIP) of the Centers for Disease Control and Prevention recommended its use in infants and children. . . . The ACIP voted. . .to recommend that healthy children 14 months through 59 months of age who are fully immunized with PCV7 receive a single dose of PCV13 (referred to as a "supplemental" dose) to elicit antibodies against the 6 additional serotypes." Mimi Glodé, MD, "New Pneumococcal Conjugate Vaccine (PCV13) Licensed, Expands Protection Against Invasive Pneumococcal Disease for Infants," Contagious Comments, Department of Epidemiology, The Children's Hospital, April 2010, www.thechildrenshospital.org/pdf/Contagious-Comments-April2010-NewPneumococcalConjugateVaccineLicensed.pdf.; "Licensure of a 13-Valent Pneumococcal Conjugate Vaccine (PCV13) and Recommendations for Use Among Children," Centers for Disease Control and Prevention, MMWR, March 10, 2010, www.cdc.gov/mmwr/preview/mmwrhtml/mm5909a2.htm.

4 This includes the recent addition of a second dose. On October 24, 2010, the CDC Advisory panel announced the addition of a booster dose of the meningitis vaccines to teens at age 16. Changes will be published in the January 2011 issue of *Morbidity and Mortality Weekly*. "CDC Panel Recommends Changes to 2011 Immunization Schedules," *Medscape Today*, October 30, 2010, www.medscape.com/viewarticle/731667.

5 "Recommendations for combination vaccines modified," *Pediatric Supersite*, November 20, 2010, www.pediatricsupersite.com/view.aspx?rid=77904.

6 David Mitchell, "New ACIP Provisional Recs Cover Multiple Immunization Topics – Panel Expresses 'General Preference' for Combination vs. Component Vaccines," American Academy of Family Physicians, *AAFP News Now*, July 8, 2009, www.aafp.org/online/en/home/publications/news/news-now/clinical-care-research/20090708acip-rndup.html.

7 Ibid.

8 Ronald W. Ellis, *Combination Vaccines: Development, Clinical Research, and Approval*, (Totowa: Humana Press, Inc., 1999), ix.

9 Dennis M. Katkocin and Chia-Lung Hsieh,"Pharmaceutical Aspects of Combination Vaccines," in Ellis, *Combination Vaccines*, ibid, 56, 62.

10 Emmanuel Vidor, Agnes Hoffenbach, and Stanley Plotkin, "Pediatric Combination Vaccines," in Ellis, *Combination Vaccines: Development, Clinical Research, and Approval*, (Totowa: Humana Press, Inc., 1999), 1.

11 Katkocin and Hsieh, 57.

12 Ibid, 56.

13 Col SK Jatana and Brig MNG Nair, "Combination Vaccines," MJAFI 2007; 63 : 167–168.; Boaventura Antonio dos Santos et al., "An evaluation of the adverse reaction potential of three measles-mumps-rubella vaccines," *Rev Panam Salud Publica/Pan Am J Public Health* 12(4), 2002, 240–247. Some different mechanisms or routes by which combination vaccines can cause a greater degree of injury include:

Certain bacteria release toxins. The increased endotoxin load of the formulated bulk can cause health problems.

A toxoid is a substance that is modified to become less toxic but is still able to stimulate an immune response. When a toxoid is used as a carrier protein for a conjugate vaccine, it may contain trace levels of residual toxins that, when combined, result in an excessive total level of toxoid.

Adjuvants are added to vaccines to stimulate a greater immune response. When adjuvanted vaccines are added to nonadjuvanted antigens, the combination can be damaging.

The use of other ingredients in combination with each other, such as preservatives, buffers, and stabilizers, can cause injuries.

14 Kathryn M. Edwards and Michael D. Decker, "Practitioners in the Use of Combination Vaccines," *Combination Vaccines*, note 8, 254.

15 Peter R. Paradiso and Robert Kohberger, "Pediatric Combination Vaccines: Clinical Issues," ibid, 96.

16 Laurie Barclay, "Guidelines Updated for Administration of Combination MMRV Vaccine," *Medscape Medical News*, March 18, 2008, http://cme.medscape.com/viewarticle/571595.

17 Mitchell, *AAFP News Now.*

18 "Recommendations for combination vaccines modified," *Pediatric Supersite.*

19 Katkocin and Hsieh, 58.

20 Ellis, ix.; Jatana and Nair, 167–168.

21 Vidor et al., 1.

22 Paradiso and Kohberger, 96.

23 Vaccine package inserts available for all FDA-licensed vaccines at www.immunize.org/fda/.

APPENDIX TO CHAPTER 29: THE EXONERATION OF PROFESSOR JOHN WALKER-SMITH —DAVID L. LEWIS, PHD

1 DL Lewis, M Arens, S Appleton, Cross-contamination potential with dental equipment. Lancet 1992, 340:1252-4.

2 DL Lewis, M Arens, Resistance of microorganisms to disinfection in dental and medical devices. Nature Med 1995, 1: 956-8.

3 SE Mills, JC Kuehne, DV Bradley, Jr., Bacterial analysis of high-speed handpiece turbines, JADA, 1993, 124:59-62.

4 DL Lewis, DK Gattie, ME Novak, S Sanchez, C Pumphrey, Interactions of pathogens and irritant chemicals in land-applied sewage sludges (biosolids), BMC Public Health 2002, 2:11, www.biomedcentral.com/1471-2458/2/11.

5 U. S. House of Representatives, Committee on Science (2000 March 22), "EPA's sludge rule: Closed minds or open debate Science, 106th Congress, 2nd Session, Serial No. 106-95, U.S. Government Printing Office.

6 U. S. House of Representatives, Committee on Science (2000 October 4), "Intolerance at EPA: Harming People, Harming Science," 106th Congress, 2nd Session, Serial No. 106-103, U.S. Government Printing Office.

7 DL Lewis, "EPA Science: Casualty of election politics." Nature 1996, 381:731-2.

8 DL Lewis (2011) "Why EPA was unprepared for the Gulf oil disaster," National Whistleblowers Center, Washington, DC, www.researchmisconduct.org.

9 National Academy of Sciences, National Research Council (2002), "Biosolids Applied to Land: Advancing Standards and Practices," www.nap.edu/books/0309084865/html.

10 DL Lewis, W Garrison, KE Wommack, A Whittemore, P Steudler, J Melillo, "Influence of environmental changes on degradation of chiral pollutants in soils," Nature 1999, 401, 898-901.

11 DL Lewis v. EPA, U.S. Department of Labor, Case Nos. 2003-CAA-6, 2003-CAA-5, Joint stipulations, March 4, 2003.

12 United States of America, ex rel. David L. Lewis, Ph.D., et al. v. John Walker, Ph.D., et al., US District Court, Middle District of Georgia, Athens Division, Case No. 3:06-CV-16, Deposition of Regina Smith, Ph.D., April 27, 2009, p.73, 81-82.

13 RA McElmurray III et al. v. USDA, U.S. District Court, Southern District of Georgia, Case No. CV105-159, Order issued February 25, 2008, p. 35.

14 J Tollefson (2008), "Raking through sludge exposes a stink," Nature 453, 262; and Editorial (2008), "Stuck in the mud—The Environmental Protection Agency must gather data on the toxicity of spreading sewage sludge," Nature 453, 258.

15 United States of America, ex rel. David L. Lewis, Ph.D., et al. v. John Walker, Ph.D., et al., US District Court, Middle District of Georgia, Athens Division, Case No. 3:06-CV-16, Deposition transcript of J Gaskin, January 20, 2009, p. 269, 293-294.

16 Ibid, Order dated May 14, 2012; and DL Lewis (2012), "Some things never change," www.researchmisconduct.org.

CONTRIBUTORS

Toni Bark MD, MHEM, LEED AP, received her undergraduate degree at the University of Illinois. She was awarded her medical doctorate from Rush Medical College and went on to train in pediatrics and rehab medicine. Bark was director of the pediatric emergency room at Michael Reese hospital for two years before going back to school to study naturopathic medical specialties. She continued to work in the emergency room and started a private practice while attending her studies. With an intense interest in environmental impact on health, Bark went on to study for her LEED accreditation. After working three stints in Haiti post-earthquake, she began her studies in health care emergency medicine and disaster planning, which took her into the study of vaccine policy, ethics, safety and production.

Michael Belkin is a financial market strategist and President of Belkin Limited, a financial and economic forecasting firm. He authors the *Belkin Report*, a weekly global market forecasting service for hedge funds, mutual funds, pension funds, investment banks, sovereign wealth funds, and family offices, which has been published since 1992. Belkin is a graduate of the University of California, Berkeley, Haas School of Business and previously served as a vice president and quantitative strategist in the equity department of Salomon Brothers. He also is a former director of the Hepatitis B Vaccine Project of the National Vaccine Information Center (NVIC). His five-week-old daughter, Lyla Rose Douglas Belkin, died hours after receiving the hepatitis B vaccine. He continues to speak out in her memory and to educate parents and medical authorities about issues of vaccine safety and conflicts of interest.

Mark Blaxill, MBA, is the father of a daughter diagnosed with autism, chairman and cofounder of the Canary Party, editor at large for *Age of Autism*, a director of SafeMinds, and a frequent speaker at autism conferences. He has authored a number of scientific publications on autism. He received a bachelor's degree *summa cum laude* from the Woodrow Wilson School of Public and International Affairs at Princeton University and an MBA with distinction from Harvard Business School. He coauthored a book with Dan Olmsted, *The Age of Autism: Mercury, Medicine and a Man-Made Epidemic* (Thomas Dunne, 2010).

Russ Bruesewitz is a senior sales executive from Pittsburgh, Pennsylvania, with a BA in business. Bruesewitz is the father of three daughters. In 1992, his youngest daughter Hannah, started to seize only two hours after receipt of her third DPT shot and continues to suffer from an intractable seizure disorder that has resulted in long-term cognitive impairment. After their claim was rejected by the Vaccine Injury Compensation Program, the family pursued Hannah's case against the vaccine manufacturer in civil court. Eighteen years later, that case ultimately was heard by the U.S. Supreme Court in the landmark decision, *Bruesewitz v Wyeth.* A 6–2 decision in favor of Wyeth ended the Brusewitzs' efforts for legal restitution, however, both Russ and his wife, Robie, continue to advocate for change to the dysfunctional VICP.

Annemarie Colbin, PhD, is an award-winning leader in the field of natural health, and a highly sought-after lecturer and wellness consultant. Colbin received her doctorate in "Interdisciplinary Studies with a Focus on Wholistic Nutrition" from the Union Institute and University in Cincinnati, Ohio. She is Founder and CEO of the Natural Gourmet Institute for Health and Culinary Arts™ in New York City, the oldest natural foods cooking school in the United States, licensed by the New York State Education Department and accredited to offer a Chef's Training Program in natural foods cooking. The associated Natural Gourmet Institute for Food and Health offers avocational classes on food and health available to the general public. Colbin has been teaching widely since the early 1970s. She has been an adjunct professor of nutrition at Empire State College and at Touro College, a visiting lecturer at the Institute for Integrative Nutrition in New York City, and has led workshops all over the country. She is the author of four books. Her best-known book, *Food and Healing* (Ballantine Books, 1986, 1996), on the relationship between food and health, is translated into many languages. The English version was reissued in 1996. She wrote a book on raising healthy children in the 1990s, which remains unpublished; this chapter was adapted from that book.

Judy Converse, MPH, RD, LD, is a registered and licensed dietitian who, in her clinical practice, treats children with growth and feeding problems, autism, food allergies, asthma, and learning and behavior disorders. Converse holds an MA in public health nutrition from the University of Hawaii and a BA in food science and human nutrition from the University of Vermont. Prior to entering private practice, she worked as a nutrition educator for the Hawaii Medical Service Association and as an outpatient dietitian for a major medical care provider. Converse lectures on child nutrition to parental and professional audiences. She also has testified before state and federal legislators in support of safer vaccines. She has authored three books on these topics: *Special Needs Kids Go Pharm-Free*, *Special Needs Kids Eat Right*, and *When Your Doctor Is Wrong: Hepatitis B Vaccine & Autism*, as well as the first accredited learning module on autism and nutrition.

Louise Kuo Habakus, MA, is a cofounder and executive director of the Center for Personal Rights. Habakus has organized and participated in rallies, and her work has been featured in numerous media sources, including *ABC World News Tonight*, *Fox & Friends*, and *The New York Times*. Habakus received an MA in International Policy Studies and a BA in international relations and French studies from Stanford University, where she was elected to Phi Beta Kappa. She was formerly a managing director for Putnam Investments, a corporate vice president for Prudential Financial, Inc., and a consultant with the management consultancy Bain & Company.

Boyd Haley, PhD, is Professor Emeritus of Biochemistry and a former chairman of the chemistry department at the University of Kentucky. He earned a PhD in chemistry and physics from Washington State University, an MS in chemistry from the University of Idaho, and a BA in chemistry and physics from Franklin College. He was a medic in the U.S. Army and later became an NIH Postdoctoral Scholar in Physiology at Yale University Medical School. Prior to joining the University of Kentucky, Haley taught at the University of Wyoming. He is currently the CEO of CTI Science, Inc., a biotechnology firm. For over twenty-five years, Haley has run a number of NIH-funded biomedical research programs, served on NIH Study Section panels, and testified on the effects of mercury toxicity with regard to neurological diseases before numerous governmental agencies.

Alexander Hintz is a thirteen-year-old boy who is in the eighth grade. He was diagnosed with autism at age two and shed the diagnosis after ten years of targeted educational therapy combined with biomedical treatment to address his underlying medical conditions.

Sonja Hintz, RN, is a registered nurse with extensive clinical experience. She is also the mother of a son, Alexander, who no longer has autism. With Alex's diagnosis of autism spectrum disorder eleven years ago, Hintz was compelled to reevaluate her beliefs and knowledge of traditional medicine. Using therapeutic dietary intervention, biomedical treatments, and other therapies, Alex recovered. She currently works at True Health Medical Center, applying what she learned for her own son to help others. She writes and lectures about biomedical treatment for children with autism. Hintz is a graduate of Marquette University with a BS in nursing. She has worked with children with disabilities in various capacities since she was an adolescent.

Mary Holland, JD, teaches and is Director of the Graduate Legal Skills Program at New York University School of Law. She received a JD and an MA from Columbia University and a BA in Russian and Soviet studies from Harvard University. She has worked in the public and private sectors on international legal issues, and has held positions at law firms, the Lawyers Committee for Human Rights, the Aspen Institute Justice and Society Program, and Columbia Law School. She is a founding board member of the Center for Personal Rights and the Elizabeth Birt Center for Autism Law and Advocacy (EBCALA). She is Legal Editor of *The Autism File*, and she recently coauthored an amicus brief to the U.S. Supreme Court in *Bruesewitz v. Wyeth*, a case about the right to sue in the civil justice system for a vaccine design defect.

Robert Johnston, PhD, is an associate professor of history and Director of the Teaching of History Program at the University of Illinois at Chicago. He received his PhD from Rutgers, the State University of New Jersey, and his BA from Reed College. His scholarly research has focused on twentieth-century political history, with special attention to issues relating to class and medicine. He has written and edited numerous books, including the award-winning *The Radical Middle Class: Populist Democracy and the Question of Capitalism in Progressive Era Portland, Oregon*; *The Politics of Healing: Histories of Twentieth-Century North American Alternative Medicine*; and an upper-elementary and middle school history textbook, *The Making of America: The History of the United States from 1492 to the Present*. Johnston is working on his next book, *Crusaders against Vaccination: An American History of Medical Populism*, to be published by Oxford University Press.

David L. Lewis, PhD, has worked for the U.S. Environmental Protection Agency's Office of Research and Development as a senior-level research microbiologist and served on the graduate faculty at the University of Georgia. His research in dentistry and endoscopy published in *The Lancet* and *Nature Medicine* prompted public health organizations worldwide to issue new infection-control guidelines for dentistry. Equally experienced in environmental health studies, he was awarded EPA's Science Achievement Award for his *Nature* article concerning the effects of climate change on health risks posed by environmental pollutants. Lewis currently directs the Research Misconduct Project of the National Whistleblowers Center in Washington, DC (www.researchmisconduct.org).

Robert Krakow, JD, is an attorney in private practice in New York and represents individuals in a variety of legal matters, including injuries from environmental toxins and vaccines. After receiving a JD from George Washington University School of Law in 1980, he began his career in public service as an attorney with the New York Public Interest Research Group (NYPIRG), a consumer advocacy group. From 1981 to 1989, Bob served as an assistant district attorney with the New York County District Attorney's Office, and as Bureau Chief in the special narcotics prosecution division. He founded his law firm in 1989, specializing in the litigation of criminal and civil matters. Krakow has a long history of vigorously advocating for the rights of children with autism and other disabilities. He is a cofounder of the Elizabeth Birt Center for Autism Law and Advocacy. He has served as Chairman of the Board of Directors of Lifespire, Inc., a nonprofit organization serving the needs of developmentally disabled adults and children in New York. He was a founding board member of the National Autism Association, and cofounder of A-CHAMP (now the Autism Action Network), a national organization advocating for the rights of children with neurodevelopmental disorders. He also has served on the board of SafeMinds, an organization focused on environmental causes of neurodevelopmental disorders. Krakow has testified on autism-related issues before state and federal legislative committees.

Kim Mack Rosenberg, JD, received her BA in political science from Carleton College and her JD from Case Western Reserve University School of Law. She currently is a lawyer in private practice in New York City. In addition to being a board member of the Center for Personal Rights, she is an officer and board member of the National Autism Association - New York Metro Chapter and the Elizabeth Birt Center for Autism Law and Advocacy (EBCALA). She writes about autism and wellness at www.embracingwellness.blogspot.com.

Dan Olmsted is Editor of *AgeofAutism.com* and coauthor, with Mark Blaxill, of *The Age of Autism: Mercury, Medicine, and a Man-Made Epidemic* (Thomas Dunne Books, 2010). He began writing about autism in 2005 at United Press International, where his articles on the apparent absence of autism among the Amish and the identities of the first cases of autism reported in the medical literature in the 1940s attracted attention. He was an original staff member and Assistant National Editor at USA Today. He graduated from Yale College.

Amy Pingel is a single mother of four children from Lake Station, Indiana. Her thirteen-year-old daughter Zeda sustained two seizures and lost the ability to eat, walk, and communicate three weeks after receiving her first dose of human papillomavirus (HPV) quadrivalent vaccine, known as Gardasil. Caring for her daughter, who remains on life support, Pingel had to quit her job as the manager of a local gas station. Recovery of quality of life for Zeda remains her top priority. Pingel is part of a growing advocacy movement of parents committed to telling the truth about their children's injuries caused by the HPV vaccine.

Richard Rovet, RN, BSN, B-C (USAF, ret), is a decorated veteran who honorably served in the United States Air Force for over twenty-four years. As a registered nurse in the military, he has served in leadership positions for most of his career, has treated patients in both emergency and trauma settings, and was competitively selected as "Best Officer" several times. Rovet served as a fighter aircraft inspector and mechanic, a flight medic, a clinical nurse and healthcare integrator, Charge Nurse, and manager of the largest internal medicine specialty clinic in the U.S. Air Force. Rovet witnessed, recorded, treated, and interviewed hundreds of individuals with Gulf War Illness and anthrax vaccine reactions on three separate air force bases. He is now retired from active duty. Rovet received his nurse training at Excelsior College. Rovet has testified before Congress on illnesses associated with the military's mandatory anthrax vaccination program. His advocacy on behalf of these veterans continues to this day.

Vera Hassner Sharav, MLS, is a public advocate for human rights. She is the founder and President of the Alliance for Human Research Protection (AHRP), an information resource and catalyst for debate, committed to accountability in biomedical research. AHRP Infomails are followed by healthcare professionals, FDA officials, journalists, lawyers, and patient advocates. Sharav's achievements include opening public debate about the ethics of psychosis-inducing experiments. Her testimony and the testimonies of victims of unethical experiments catalyzed a prize-winning Boston Globe series, which prompted the National Institute of Mental Health to suspend twenty-nine clinical trials. AHRP-filed complaints led to the suspension of a pediatric pesticide experiment, federal investigations on the use of foster-care children in experimental AIDS drug and vaccine trials, suspension of a pediatric smallpox vaccine trial, and Congressional hearings about a "violence prediction" experiment conducted on inner city pre-adolescent boys. Sharav has testified before numerous academic and government-sponsored public policy advisory forums, including the World Congress of Science Journalists. Sharav holds a Master of Library Science from the Pratt Institute. She is a child survivor of the Holocaust and witnessed the death of her father from typhus because he did not have access to penicillin. More information is available at www.ahrp.org.

Lisa Marks Smith was born and raised in Cincinnati, Ohio. She is married and the mother of two sons. She is self-employed, working for Icon Beauty. Following her flu shot and near-death experience in 2005, she learned about alternative therapies that have helped ease some of the problems of post-infectious myositis. When not advocating against mass vaccination policies, she spends time with her family, makes jewelry, and enjoys being alive.

Sookyung Song, JD, graduated from New York University School of Law. With a firm belief in a voluntary vaccination system, she conducted legal research on informed consent and human rights during her internship at the Center for Personal Rights. Song earned a BS in architectural engineering from Seoul National University.

Carol Stott, PhD (Cantab), MSc (Epidemiology), CSci, CPsychol, graduated in Psychology from the University of Manchester. She recently completed an MSc in Epidemiology with the University of London School of Hygiene and Tropical Medicine following an earlier PhD in the Epidemiology of Specific Language Disorders with the University of Cambridge, U.K. Stott spent 13 years (1991 – 2004) in the University of Cambridge, Department of Psychiatry and Autism Research Centre (ARC) as a Research Associate engaged in research into genotype-phenotype associations in developmental disorder and a major epidemiological study of the population frequency of Autism Spectrum Disorders (ASDs). Stott was recently awarded Chartered Scientist status with the U.K. Science Council. She currently works as an independent scientific advisor to a number of U.K. and U.S. organisations, providing input on methodological issues and statistical analysis. Her aim is to bring her experience in epidemiology and measurement theory to new developments in research into environmental factors and their role in the childhood developmental disorders. Stott is also lead training and diagnostic consultant with BeginningwithA, a U.K. training and diagnostic consultancy and Scientific Editor of *The Autism File* magazine.

Allen Tate graduated from New York University with a degree in philosophy. He has two younger siblings who were both adversely affected by vaccine-induced injuries and have been diagnosed with regressive autism. Born in 1989, Tate was vaccinated according to the CDC's 1983 schedule and has remained healthy while his siblings, born in 1992 and 1996 respectively, have suffered the severe consequences of the increased vaccination schedules of the 1990s. His family resides in the suburbs of Philadelphia. Tate is committed to vaccination choice activism and is a board member of The Elizabeth Birt Center for Autism Law and Advocacy.

Gay Tate, PhD, LSW, MLSP, is the mother of Allen Tate and his two younger siblings, Kenny and Olivia. Tate received a PhD in microbiology from Boston University and a BA in biology from Wheaton College. She has published original scientific research in the area of tumor immunology. In 1989, she left research for full-time parenting and soon immersed herself in intensive therapies for Kenny and Olivia. She returned to graduate school in 2005 for graduate degrees in social work, and law and social policy. Tate is currently a licensed psychotherapist at the Women's Therapy Center in Philadelphia with a focus on trauma therapy. She is an advocate for vaccination choice and is committed to speaking out on the inherent dangers of one-size-fits-all, aggressive vaccine policies.

Ginger Taylor, MS, is a conservative, Christian writer, speaker, and activist. On her blog *Adventures in Autism*, Taylor writes about autism parenting, the politics of autism, health, vaccination, informed consent, and both corporate and government corruption. She is a former marriage and family therapist, and holds an MS in clinical counseling from Johns Hopkins University. Taylor served on the steering committee of the first Maine Center for Disease Control and Prevention's Autism Conference to educate medical professionals on the current state of research and treatment of autism. She is also the founder of Greater Brunswick Special Families, a support organization for developmental disabilities.

Sherri Tenpenny, DO, AOBNMM, is a doctor of osteopathic medicine. She was the director of emergency medicine at the Blanchard Valley Regional Health Center in Findlay, Ohio, from 1986 to 1998. She was board certified in emergency medicine from 1995 to 2006 (she chose not to recertify) and is currently board certified in neuromusculoskeletal medicine. She is the founder of Tenpenny Integrative Medical Center (formerly Osteomed II), located in Middleburg Heights, Ohio. A graduate of the University of Toledo, Ohio, she received her medical training at Kirksville College of Osteopathic Medicine in Kirksville, Missouri. Tenpenny is an internationally-known expert on the hazards of vaccines. She is the author of two books: *FOWL! Bird Flu—It's Not What You Think* and *Saying No to Vaccines: A Resource Guide for All Ages.* Tenpenny is a frequent radio and television guest and speaks at national seminars on vaccines and many topics in the area of integrative medicine.

James Turner, JD, is a partner at Swankin & Turner, a Washington, DC law firm organized in 1973, which specializes in food, drug, health, and environmental regulation. He received a JD and a BA in history and political science from Ohio State University. He served as a gunnery officer in the U.S. Navy. Turner was the lead attorney on a successful petition to the U.S. Food and Drug Administration to reclassify acupuncture needles, permitting their legal importation and distribution. He also served as a special counsel to the Senate Select Committee on Food, Nutrition, and Health and to the Senate Government Operations Subcommittee on Government Research. Turner wrote *The Chemical Feast: The Nader Report on Food Protection at the FDA,* which *Time* magazine said "may well be the most devastating critique of a U.S. Government agency ever issued." He also wrote *Voice of the People: The Transpartisan Imperative in American Life.* He is actively involved in the Foundation for Health Choice, an organization advocating choice, information, safety, and redress in healthcare decisions. Turner is Chair of the board of Citizens for Health and a founder of Voice for HOPE (Healers of Planet Earth).

William Wagner, JD, is President of Salt and Light Global and serves on the Executive Board of ParentalRights.org, an organization dedicated to protecting the child-parent relationship. He is currently a tenured professor at a law school, where he teaches ethics and constitutional law. A frequent speaker at world conferences, Wagner has a special interest in protecting free expression and the free exercise of religious conscience. As a lead amicus counsel before the U.S. Supreme Court, he authored briefs on behalf of Christian organizations. He also authored a brief for the Swedish Supreme Court, wrote a testimony submitted to the U.S. Congress, and presented evidence before the U.K. Parliament. He has addressed executive, legislative, parliamentary, and judicial audiences throughout the world, and presented at diplomatic forums, including the United Nations Human Rights Council in Geneva. Wagner's public service includes serving as a federal magistrate judge, a legal counsel in the U.S. Senate, a senior assistant U.S. attorney in the Department of Justice, and an American diplomat. Wagner received a post-doctoral Danforth Fellowship in Law after earning his JD in 1986.

Andrew Wakefield, MB, BS, FRCS, FRCPath, is an academic gastroenterologist from the United Kingdom and was the lead author of the 1998 *The Lancet* paper, a small case study series that was one of the first to posit a possible link between vaccines, intestinal inflammation and neurologic injury in children. Wakefield received his medical degree from St. Mary's Hospital Medical School (University of London). He pursued a career in gastrointestinal surgery with a particular interest in inflammatory bowel disease, for which he has received numerous honors and has published over 140 original scientific articles, book chapters and invited scientific commentaries. As a result of his groundbreaking work concerning vaccines and neurologic injury, Wakefield lost his job, his country, his career and his medical license. He is author of the best-selling book *Callous Disregard: Autism and Vaccines—The Truth Behind a Tragedy.*

Julian Whitaker, MD, is a graduate of Dartmouth College and Emory University Medical School. After postgraduate training in surgery at Emory, Dr. Whitaker worked with nutritional medicine pioneer Nathan Pritikin and in 1979 opened the Whitaker Wellness Institute. To date, more than fifty thousand patients have been treated at the Newport Beach, California, clinic with diet, exercise, targeted nutritional supplements, and safe, noninvasive therapies. Dr. Whitaker is the author of thirteen books, *including Reversing Heart Disease, Reversing Diabetes, Reversing Hypertension, Shed 10 Years in 10 Weeks, The Whitaker Diet, and Dr. Whitaker's Guide to Natural Healing.* He is editor of the monthly newsletter *Health & Healing,* which has reached more than 4 million households since 1991. Dr. Whitaker is also a committed spokesman for freedom of choice in the medical arena and founder of the Freedom of Health Foundation.

INDEX

C

D

E

F